AF321951

Soil Hydrology for a Sustainable Land Management

Soil Hydrology for a Sustainable Land Management

Theory and Practice

Special Issue Editors

Artemi Cerdà
Simone Di Prima
Mirko Castellini
Jesús Rodrigo-Comino

MDPI • Basel • Beijing • Wuhan • Barcelona • Belgrade • Manchester • Tokyo • Cluj • Tianjin

Special Issue Editors

Artemi Cerdà
Universitat de València
Spain

Simone Di Prima
University of Sassari
Italy

Mirko Castellini
Council for Agricultural
Research and
Economics—Research Center for
Agriculture and Environment
(CREA-AA)
Italy

Jesús Rodrigo-Comino
University of Valencia
Spain

Editorial Office
MDPI
St. Alban-Anlage 66
4052 Basel, Switzerland

This is a reprint of articles from the Special Issue published online in the open access journal *Water* (ISSN 2073-4441) (available at: https://www.mdpi.com/journal/water/special_issues/Soil_Hydrology).

For citation purposes, cite each article independently as indicated on the article page online and as indicated below:

LastName, A.A.; LastName, B.B.; LastName, C.C. Article Title. *Journal Name* **Year**, *Article Number, Page Range.*

ISBN 978-3-03936-505-0 (Pbk)
ISBN 978-3-03936-506-7 (PDF)

Cover image courtesy of Artemi Cerdà.

Contents

About the Special Issue Editors

Artemi Cerdà (Ph.D., 1993) has been a Full Professor in Physical Geography that, since 2009, has been researching land degradation processes caused by agriculture, forest fires, and road construction. Soil erosion, runoff generation, and soil degradation are studied within the soil erosion and degradation research team where Dr. Cerdà is the research project and teaching leader. Dr. Cerdà's soil erosion research expertise is based on rainfall simulation and soil erosion plots, laboratory measurements, and a holistic approach from societal, perception, and economic perspectives. His research developed in the Netherlands, Iran, Spain, the USA, and Bolivia demonstrate extensive scientific experience with a passion for Mediterranean landscapes, where humans transformed nature for millennia.

Simone Di Prima (Ph.D., 2016) is a researcher at the Department of Agricultural Sciences of the University of Sassari (Italy). He got a postdoctoral research position at the Laboratoire d'Ecologie des Hydrosystèmes Naturels et Anthropisés (LEHNA, ENTPE, Université Lyon 1, France) focused on stormwater management and the use of non-intrusive geophysical techniques to catch nano-tracers and the pathways of water in urban soils. His main scientific interests focus on soil hydrology and water resources management with specific regard to laboratory and field determination of soil hydraulic properties, infiltration processes, simulation of water flow in the vadose zone. He authored more than 40 scientific papers on international peer reviewed journals. He served as reviewer for several international scientific journals.

Mirko Castellini's research activities focus on the study of soil physical and hydraulic properties. Specific research topics are: (i) soil physical quality, (ii) soil management for sustainable agriculture, (iii) land use change impact on soil properties, (iv) use of soil conditioners (e.g., amendments, composts) to improve the soil water retention, (v) water fluxes in saturated and unsaturated soil conditions, (vi) temporal and spatial variability of physical and hydraulic properties of the soil, and (vii) the main factors affecting soil physical degradation processes (soil surface crusting, soil compaction, etc.).

Jesús Rodrigo-Comino received his Ph.D. at the University of Málaga (Spain). He received his M.S. in land planning and GIS in 2013 in Granada and Málaga Universities. Since 2015, he has written two books related to soil geography, provided several oral presentations at conferences and posters in international meetings, and published his investigations about soil erosion, soil geography, and land degradation processes. He is editor-in-chief of *Air, Soil and Water Research* (SAGE), and also works as an associate editor for *Hydrological Science Journal* (Taylor and Francis), *Journal of Mountain Science* (Springer), and *Hydrology* (mdpi). He is a reviewer for more than 110 international indexed journals. At the moment of the publication of this Special Issue, he was working on an Interreg project about light pollution (Smart Light-HUB) at the Trier University (Germany) and COST-Action Firelinks (CA18135) as a grant holder at the University of Valencia (Spain).

Editorial

Soil Hydrology for a Sustainable Land Management: Theory and Practice

Simone Di Prima [1,2], Mirko Castellini [3,*], Jesús Rodrigo-Comino [4,5] and Artemi Cerdà [4]

[1]   Agricultural Department, University of Sassari, Viale Italia, 39, 07100 Sassari, Italy; sdiprima@uniss.it
[2]   Université de Lyon, UMR5023 Ecologie des Hydrosystèmes Naturels et Anthropisés, CNRS, ENTPE, Université Lyon 1, 3 rue Maurice Audin, 69518 Vaulx-en-Velin, France
[3]   Council for Agricultural Research and Economics–Research Center for Agriculture and Environment (CREA–AA) Via C. Ulpiani, 570125 Bari, Italy
[4]   Soil Erosion and Degradation Research Group, Department of Geography, University of Valencia, 46010 Valencia, Spain; rodrigo-comino@uma.es (J.R.-C.); artemio.cerda@uv.es (A.C.)
[5]   Physical Geography, Trier University, 54286 Trier, Germany
*   Correspondence: mirko.castellini@crea.gov.it

Received: 27 March 2020; Accepted: 9 April 2020; Published: 13 April 2020

Abstract: Soil hydrology determines the water–soil–plant interactions in the Earth's system, because porous medium acts as an interface within the atmosphere and lithosphere, regulates main processes such as runoff discharge, aquifer recharge, movement of water and solutes into the soil and, ultimately, the amount of water retained and available for plants growth. Soil hydrology can be strongly affected by land management. Therefore, investigations aimed at assessing the impact of land management changes on soil hydrology are necessary, especially with a view to optimize water resources. This Special Issue collects 12 original contributions addressing the state of the art of soil hydrology for sustainable land management. These contributions cover a wide range of topics including (i) effects of land-use change, (ii) water use efficiency, (iii) erosion risk, (iv) solute transport, and (v) new methods and devices for improved characterization of soil physical and hydraulic properties. They involve both field and laboratory experiments, as well as modelling studies. Also, different spatial scales, i.e., from the field- to regional-scales, as well as a wide range of geographic regions are also covered. The collection of these manuscripts presented in this Special Issue provides a relevant knowledge contribution for effective saving water resources and sustainable land management.

Keywords: soil hydrology; sustainable land management; soil water content; water fluxes; soil erosion; runoff; spatial variability; BEST-procedure; Hydrus-1D; Arduino

---

1. Introduction

The United Nations defines sustainable land management as "the use of land resources, including soils, water, animals and plants, for the production of goods to meet changing human needs, while simultaneously ensuring the long-term productive potential of these resources and the maintenance of their environmental functions".

Soil hydrology determines the water-soil interactions in the Earth's system. Specifically, soil acts as an interface within the atmosphere, biosphere and lithosphere, and regulates main processes of the hydrosphere as runoff discharge, aquifer recharge and soil water content [1–3], represents a critical part of environmental sciences. However, due to the ongoing climate change, more sensitive agro-environments will have to adapt to the changed thermo-pluviomentrical trends (among others, the increase in crop evapotranspiration, impact of rainfall erosivity) [4–6], and sustainable land management will the main issue, especially for countries of Mediterranean basin. As a consequence, there is a need to develop and test new methods and experimental procedures to assess those changes from a soil hydrology perspective.

The main goal of this Special Issue (SI) was to present advanced researches on soil infiltration methods, grand water recharges, soil water content dynamics, and about the impact of vegetation type on physical and hydraulic properties of the soil (i.e., water repellency, hydraulic conductivity, soil water content). Contributions of this SI were focused on: BEST-procedure [7] application, for a simple and expeditious method for estimating hydraulic properties of the soil [8–11]; Steady version of the Simplified method based on a Beerkan Infiltration run (SSBI method) [12] for field saturated hydraulic conductivity (K_s) estimation [13]; Hydrus-1D and SWAP codes for modelling soil water dynamics [11,14,15]; a new low-cost device for the automation of K_s measurements in the laboratory [16]. Also, investigations on the topic of rainfall erosivity [6,17], or the impact of land-vegetation cover interaction on soil water content [18] and soil water repellency [19], increased our knowledge on the sustainable management of specific agro-environments. A synthesis of main results and/or innovative methods to assess those changes from a soil hydrology point of view were reported in the following section.

2. Overview of This Special Issue

This Special Issue collects 12 original contributions focused on soil hydrology and aimed to address the challenging topic of sustainable land management. From a methodological point of view, the contributions involve both field [8–11,13,19] and laboratory [14,15] experiments, and modelling [6,16–18] studies. The Special Issue includes studies carried out at different spatial scales, from the field- to regional-scales. A wide range of geographic regions are also covered, including Brazil [11,13], Mexico [16], Mediterranean basin [6,8,10,14,18], and Central [15,17,19] and Western [9] Europe. Specifically, contributions focus on five main topics including (i) land-use change [8,12,18,19], (ii) water use efficiency [14], (iii) erosion risk [6,17], (iv) solute transport [15], and (v) new methods and devices for improved characterization of soil physical and hydraulic properties [9–11,16].

Topic (i) comprises four papers. Hewelke et al. [19] assessed the influence of the abandoning arable use and the spontaneous afforestation with a pine stand on soil hydraulic properties. This author showed evidence of the occurrence of soil water repellency on the surface layer.

Lozano-Baez et al. [13] investigated the recovery of top-soil saturated soil hydraulic conductivity (K_s), soil physical and hydraulic properties in five land-use types in the Brazilian Atlantic Forest. The studied land-use types included (i) a secondary old-growth forest; (ii) a forest established through assisted passive restoration 11 years ago; (iii) an actively restored forest, with a more intensive land-use history and 11 years of age; (iv) a pasture with low-intensity use; and (v) a pasture with high-intensity use. They used the Beerkan method to determine K_s values in the field and also measured tree basal area, canopy cover, vegetation height, tree density and species richness in forest covers. These authors reported that K_s estimates decreased when land use was more intense before forest restoration actions.

Castellini et al. [8] assessed the impact of alternative soil management strategies (conventional tillage and no-tillage) on physical and hydraulic properties of fine-textured soils, applying both field and lab procedures.

Lozano-Parra et al. [18] investigated the effect of the interactions between soil moisture and vegetation covers on soil temperature. These authors monitored for two and a half hydrological years of soil water content and soil temperature of open grasslands and below tree canopies.

Topic (ii) comprises one paper. Ventrella et al. [14] presented a method based on a physically-based Hydrus-1D model to increase the water use efficiency of cropping systems. This model was calibrated by optimizing the hydraulic parameters based on the comparison between simulated and measured soil water content values. The model allowed us to simulate the soil water contents measured under a typical cultivation scheme of a drip-irrigated horticultural system.

Topic (iii) comprises two papers. Gericke et al. [17] used the universal soil loss equation (USLE) to identify areas of erosion risk in the federal state of Brandenburg, NE Germany. Using an ensemble of climate scenarios, these authors assessed the impact of climate change on rainfall erosivity and the potential soil erosion risk.

Baiamonte et al. [6] applied the Revised Universal Soil Loss Equation (RUSLE) model to two Sicilian (Italy) vineyards subjected to different management practices. These authors studied the interactions of rainfall erosivity and cover management factors, as well as their time scale effects, for the vineyard crop.

Topic (iv) comprises one paper. Szymkiewicz et al. [15] used the SWAP model to simulate transient water flow and solute transport for ten layered soil profiles composed of materials ranging from gravel to clay. The simulated scenarios were compared with simplified approaches for estimating solute travel time.

Topic (v) comprises four papers. Castellini et al. [10] investigated the relationships between soil physical and hydraulic properties and wheat yield at the field scale and tested the Beerkan estimation of soil transfer parameter (BEST) method for the spatialization of soil hydraulic properties.

Silva Ursulino et al. [11] investigated the dynamics of soil water content in two plots in the Gameleira Experimental River Basin, Northeast Brazil. Specifically, Time Domain Reflectometry (TDR) probes and Hydrus-1D for modelling one-dimensional flow were used in two stages: with hydraulic parameters estimated with the Beerkan Estimation of Soil Transfer Parameters (BEST) method and optimized by inverse modelling. The performance analysis of the simulations provided strong indications of the efficiency of parameters estimated by BEST to predict the soil moisture variability in the studied river basin without the need for calibration or complex numerical approaches.

Bouarafa et al. [9] assessed the hydraulic properties of sustainable urban drainage systems (SuDS) located in the urban zone of Lyon (France). They used the BEST method to analyze infiltration data and for the determination of both shape and scale parameters of the soil water retention curve $h(\theta)$ and the hydraulic conductivity curve $K(\theta)$. This study allowed us to reveal the infiltration inefficiency of some of the structures.

Rodríguez-Juárez et al. [16] presented a new automated laboratory infiltrometer for the determination of the saturated hydraulic conductivity. The device consisted of low-cost components and was realized using the popular Arduino microcontroller board and commercially available sensors.

3. Conclusions

The 12 original manuscripts collected in this SI have reported experimental results, based on both standard and innovative methodologies, for the sustainable land management and from soil hydrology.

The collected contributions were summarized by grouping them into five main topics to show research advances in specific fields as effects of land-use change, water use efficiency, erosion risk, solute transport, and new methods and devices for improvements on the characterization of physical and hydraulic properties of the soil.

Highlights of this SI suggest, or confirm, that the main hydrological processes can be affected, both at the small or medium-large scale, by the changes in soil use by the stakeholders. These changes can worsen the optimal balance between water and air into the soil, can have relevance for the erosion processes of the soil, but also can change the hydrometeorology of specific environments. Investigations on specific environments, i.e., agricultural, forestry or of transition between them, were presented in this SI, and the man-made impacts were quantified to account for possible environmental effects.

Overall, the manuscripts of this Special Issue reported results for poorly investigated environments. However, for some relatively more investigated agro-environments (i.e., extensive or high-income crops), findings highlighted the need to establish further comparisons to select and evaluate eco-sustainable agricultural practices. A direct or indirect common thread among manuscripts was to share viable solutions to optimize the water resource, or to increase the water use efficiency in specific environments.

Funding: This research received no external funding.

Conflicts of Interest: The authors declare no conflicts of interest.

References

1. Bormann, H.; Klaassen, K. Seasonal and land use dependent variability of soil hydraulic and soil hydrological properties of two Northern German soils. *Geoderma* **2008**, *145*, 295–302. [CrossRef]
2. Manici, L.M.; Castellini, M.; Caputo, F. Soil-inhabiting fungi can integrate soil physical indicators in multivariate analysis of Mediterranean agroecosystem dominated by old olive groves. *Ecol. Indic.* **2019**, *106*, 105490. [CrossRef]
3. Skopp, J.; Jawson, M.D.; Doran, J.W. Steady-state aerobic microbial activity as a function of soil water content. *Soil Sci. Soc. Am. J.* **1990**, *54*, 1619–1625. [CrossRef]
4. Niedda, M.; Pirastru, M.; Castellini, M.; Giadrossich, F. Simulating the hydrological response of a closed catchment-lake system to recent climate and land-use changes in semi-arid Mediterranean environment. *J. Hydrol.* **2014**, *517*, 732–745. [CrossRef]
5. Garofalo, P.; Ventrella, D.; Kersebaum, K.C.; Gobin, A.; Trnka, M.; Giglio, L.; Dubrovský, M.; Castellini, M. Water footprint of winter wheat under climate change: Trends and uncertainties associated to the ensemble of crop models. *Sci. Total Environ.* **2019**, *658*, 1186–1208. [CrossRef] [PubMed]
6. Baiamonte, G.; Minacapilli, M.; Novara, A.; Gristina, L. Time Scale Effects and Interactions of Rainfall Erosivity and Cover Management Factors on Vineyard Soil Loss Erosion in the Semi-Arid Area of Southern Sicily. *Water* **2019**, *11*, 978. [CrossRef]
7. Lassabatère, L.; Angulo-Jaramillo, R.; Ugalde, J.M.S.; Cuenca, R.; Braud, I.; Haverkamp, R. Beerkan estimation of soil transfer parameters through infiltration experiments: BEST. *Soil Sci. Soc. Am. J.* **2006**, *70*, 521–532. [CrossRef]
8. Castellini, M.; Fornaro, F.; Garofalo, P.; Giglio, L.; Rinaldi, M.; Ventrella, D.; Vitti, C.; Vonella, A.V. Effects of No-Tillage and Conventional Tillage on Physical and Hydraulic Properties of Fine Textured Soils under Winter Wheat. *Water* **2019**, *11*, 484. [CrossRef]
9. Bouarafa, S.; Lassabatere, L.; Lipeme-Kouyi, G.; Angulo-Jaramillo, R. Hydrodynamic Characterization of Sustainable Urban Drainage Systems (SuDS) by Using Beerkan Infiltration Experiments. *Water* **2019**, *11*, 660. [CrossRef]
10. Castellini, M.; Stellacci, A.M.; Tomaiuolo, M.; Barca, E. Spatial Variability of Soil Physical and Hydraulic Properties in a Durum Wheat Field: An Assessment by the BEST-Procedure. *Water* **2019**, *11*, 1434. [CrossRef]
11. Silva Ursulino, B.; Maria Gico Lima Montenegro, S.; Paiva Coutinho, A.; Hugo Rabelo Coelho, V.; Cezar dos Santos Araújo, D.; Cláudia Villar Gusmão, A.; Martins dos Santos Neto, S.; Lassabatere, L.; Angulo-Jaramillo, R. Modelling Soil Water Dynamics from Soil Hydraulic Parameters Estimated by an Alternative Method in a Tropical Experimental Basin. *Water* **2019**, *11*, 1007. [CrossRef]
12. Bagarello, V.; Di Prima, S.; Iovino, M. Estimating saturated soil hydraulic conductivity by the near steady-state phase of a Beerkan infiltration test. *Geoderma* **2017**, *303*, 70–77. [CrossRef]
13. Lozano-Baez, S.E.; Cooper, M.; Frosini de Barros Ferraz, S.; Ribeiro Rodrigues, R.; Castellini, M.; Di Prima, S. Recovery of Soil Hydraulic Properties for Assisted Passive and Active Restoration: Assessing Historical Land Use and Forest Structure. *Water* **2019**, *11*, 86. [CrossRef]
14. Ventrella, D.; Castellini, M.; Di Prima, S.; Garofalo, P.; Lassabatère, L. Assessment of the Physically-Based Hydrus-1D Model for Simulating the Water Fluxes of a Mediterranean Cropping System. *Water* **2019**, *11*, 1657. [CrossRef]
15. Szymkiewicz, A.; Savard, J.; Jaworska-Szulc, B. Numerical Analysis of Recharge Rates and Contaminant Travel Time in Layered Unsaturated Soils. *Water* **2019**, *11*, 545. [CrossRef]
16. Rodríguez-Juárez, P.; Júnez-Ferreira, H.E.; González Trinidad, J.; Zavala, M.; Burnes-Rudecino, S.; Bautista-Capetillo, C. Automated Laboratory Infiltrometer to Estimate Saturated Hydraulic Conductivity Using an Arduino Microcontroller Board. *Water* **2018**, *10*, 1867. [CrossRef]
17. Gericke, A.; Kiesel, J.; Deumlich, D.; Venohr, M. Recent and Future Changes in Rainfall Erosivity and Implications for the Soil Erosion Risk in Brandenburg, NE Germany. *Water* **2019**, *11*, 904. [CrossRef]

18. Lozano-Parra, J.; Pulido, M.; Lozano-Fondón, C.; Schnabel, S. How do Soil Moisture and Vegetation Covers Influence Soil Temperature in Drylands of Mediterranean Regions? *Water* **2018**, *10*, 1747. [CrossRef]
19. Hewelke, E. Influence of Abandoning Agricultural Land Use on Hydrophysical Properties of Sandy Soil. *Water* **2019**, *11*, 525. [CrossRef]

Article

Hydrodynamic Characterization of Sustainable Urban Drainage Systems (SuDS) by Using Beerkan Infiltration Experiments

Sofia Bouarafa [1,2,*], Laurent Lassabatere [1], Gislain Lipeme-Kouyi [2] and Rafael Angulo-Jaramillo [1]

[1]   UMR5023 Ecologie des Hydrosystèmes Naturels et Anthropisés, Université de Lyon, 3 rue Maurice Audin, 69518 Vaulx-en-Velin, France; laurent.lassabatere@entpe.fr (L.L.); rafael.angulojaramillo@entpe.fr (R.A.-J.)

[2]   Laboratory of Wastes Waters Environment and Pollutions (DEEP), University of Lyon, INSA Lyon, 69621 Villeurbanne, France; gislain.lipeme-kouyi@insa-lyon.fr

*   Correspondence: sofia.bouarafa@entpe.fr

Received: 8 February 2019; Accepted: 27 March 2019; Published: 30 March 2019

Abstract: Stormwater management techniques in urban areas, such as sustainable urban drainage systems (SuDS), are designed to manage rainwater through an infiltration process. In order to determine the infiltration capacities of different SuDS and to identify their unsaturated hydraulic properties, measurements with the Beerkan method (i.e., single ring infiltration tests) were carried out on four types of common infiltration structures in an urban zone of Lyon (France): A drainage ditch with an underlying storage structure, a parking lot with a waterproof pavement that transfers runoff water toward the ditch, a vegetated hollow core slab, and an embankment of a grass-covered garden that was used as a reference for rainwater infiltration capacity. The novelty of this study lies in the use of three Beerkan estimation of soil transfer parameters (BEST) algorithms: BEST-slope, BEST-intercept, and BEST-steady to analyze infiltration data. The BEST methods are based on the analysis of the infiltration rate from transient to steady-state flow. They allow the determination of both shape and scale parameters of the soil water retention curve $h(\theta)$ and the hydraulic conductivity curve $K(\theta)$. The three BEST methods are efficient and simple for hydraulic characterization of SuDS. The study of the hydrodynamic behavior of the four structures revealed the infiltration inefficiency of some of them. Their average infiltration rates are considerably lower than the reference infiltration rain garden. The results confirmed the impact of some physical conditions, such as pore structure modification due to invasive vegetation colonization and the presence of soil organic matter, on soil hydrodynamic behavior degradation.

Keywords: infiltration; SuDS; urban runoff; Beerkan; BEST algorithm

---

1. Introduction

The guidelines for urban stormwater management have undergone several changes over the course of time. They were based first on hydraulic and hygienist premises, which consist of quick evacuation of stormwater toward natural aquatic environments through separate or combined sewer systems. These conventional drainage systems only consider water quantity management issues, and they are economically and ecologically costly. Growth of waterproofed surfaces, as well as climate change, has generated an important rise in flood events in urban areas because of the limited capacity of sewage networks. Diverse activities pursued in cities produce a large variety of pollutants that are disposed in air and on surfaces. They can be organic, such as hydrocarbons, oils, and grease; inorganic, such as metals and dissolved nutrients; or pathogenic microorganisms, such as bacteria and viruses [1–3]. All of these contaminants end up in the receiving water bodies [4–6].

Currently, stormwater management in urban areas embodies a qualitative approach that consists of restoring rainwater into a hydrological cycle that has to be close to the natural process, by limiting runoff and fast water accumulation. New techniques have been innovated to fulfill the double criteria of quantitative and qualitative water management. They are commonly called SuDS: Sustainable urban drainage systems [7].

SuDS are designed to substitute and/or supplement pipe network systems. They are based on decentralization of the stormwater treatment point, i.e., stormwater is managed close to its drop point through an infiltration process. There are several types of SuDS: e.g., infiltration basins, ditches, rain gardens, and porous pavements. Their hydraulic efficiency relies on two main standards that are infiltration and retention capacity [8–10]. Globally, the infiltration capacity should be high enough to prevent flooding while enabling pollution removal processes through settling and adsorption.

The infiltration capacity of a soil is usually estimated by measuring the rate at which water soaks away from test pits or boreholes [11,12]. However, further information about the hydrodynamic characteristics of the soil would be helpful to assimilate water transfer profiles. Depending on their texture, porosity, particle size distribution, and hydric history, some soils will provide better contaminant retention, while others are more likely to develop preferential pathways promoting direct transfer of contaminated water toward the underground water table [13–15].

Attention will be devoted in this paper to the assessment of SuDS infiltration capacity through the determination and analysis of their hydrodynamic characteristics by using the Beerkan estimation of soil transfer parameters (BEST) method, which is an effective well-tried method in the field of hydrodynamic property characterization of different soil textures [16–18]. The BEST method relies on the analysis of in situ infiltration data built by the Beerkan infiltration protocol. The Beerkan infiltration method was introduced by Haverkamp et al. (1996) [19]. It is a simple, inexpensive, and repeatable method that quantifies water infiltration curves in porous media. It consists of infiltrating known water volumes under saturated conditions through a ring, until reaching steady-state infiltration. By exploiting the resulting infiltration curve, bulk density, particle size distribution, and hydric conditions, the BEST method provides an estimation of saturated hydraulic conductivity K_s, sorptivity S, and shape and scale parameters of soil–water relationships.

The BEST method outputs allow the determination of hydraulic conductivity $K(\theta)$ and hydraulic retention curves $h(\theta)$ by Equation (1) of van Genuchten et al. (1980) [20] under the Burdine condition [21], Equation (2), and Equation (3) of Brooks and Corey [22], respectively.

$$\frac{(\theta - \theta_r)}{(\theta_s - \theta_r)} = \left[1 + \left(\frac{h}{h_g} \right)^n \right]^{-m} \tag{1}$$

$$m = 1 - \frac{2}{n} \tag{2}$$

$$K(\theta) = K_s \left(\frac{\theta - \theta_r}{\theta_s - \theta_r} \right)^{\eta} \tag{3}$$

where θ_r and θ_s ($L^3 L^{-3}$) are the residual and saturated volumetric water contents, respectively. θ_r is assumed to be zero. K_s ($L T^{-1}$) is the saturated hydraulic conductivity; n, m, and η are the shape parameters; and h_g is the pressure head scale parameter of $h(\theta)$ calculated from the sorptivity, as follows:

$$S^2(\theta_0, \theta_s) = -c_p \theta_s K_s h_g (1 - \frac{\theta_0}{\theta_s})[1 - (\frac{\theta_0}{\theta_s})^{\eta}] \tag{4}$$

where θ_0 is the initial volumetric water content and c_p is a constant derived by Haverkamp et al. (1999) [23]:

$$c_p = \Gamma \left(1 + \frac{1}{n} \right) \left[\frac{\Gamma(m\eta - 1/n)}{\Gamma(m\eta)} + \frac{\Gamma(m\eta + m - 1/n)}{\Gamma(m\eta + m)} \right] \tag{5}$$

where Γ is the incomplete gamma function.

There are three different BEST methods to estimate parameters K_s and S: BEST-slope [18], BEST-intercept [24] and BEST-steady [25]. They differ according to the fitting method of infiltration equations to experimental data.

By using the parameters K_s and S obtained by the three BEST methods, the macroscopic capillary length scale λ_c (L) and the average characteristic size of hydraulically activated pores λ_m (L) can be calculated by using the equations of White and Sully (1987) [26] and Warrick and Broadbridge (1992) [27]:

$$\lambda_c = \frac{bS^2}{\left(\theta_f - \theta_i\right)K_s} \tag{6}$$

$$\lambda_m = \frac{\sigma}{\rho_w g}\,\frac{1}{\lambda_c} \tag{7}$$

where b is a constant depending on soil water diffusivity function, and it is frequently considered that $b = 0.55$. θ_i and θ_f are the initial and final volumetric water contents, respectively. σ is the surface tension of water ($\sigma = 73$ mN m^{-1}), ρ_w is water density, and g is the acceleration due to gravity.

The BEST methods have been successfully used to characterize different soils with different textures. Lassabatere et al. (2006) [18] applied the method on three different types of soils: An agricultural soil, a sandy soil, and a fluvioglacial deposit. Acceptable estimations of the hydraulic parameters were provided for the three soils. Yilmaz et al. (2010) [24] used the BEST method to define unsaturated hydraulic properties of a Basic Oxygen Furnace (BOF) slag in order to study the impact of spatial heterogeneity and follow its evolution through time. Bagarello et al. (2012) [28] used the BEST method to estimate the soil water retention curve of 199 Sicilian soils and discussed the efficiency of the method by comparing the results to reference soils from a known database. The study concluded that BEST water retention model can be appropriate for most soils. Di Prima et al. (2015) [17] used a Beerkan automatic device combined with the BEST algorithms to study the hydraulic properties of three agriculture soils. The results showed that the BEST methods can be a good substitute to laboratory measurements to define the hydraulic properties of soils.

In this study, the BEST methods were applied to four current urban stormwater management structures located in Lyon (France) in order to evaluate their infiltration capacity and identify possible malfunctioning through the study of their hydrodynamic parameters and curves. The studied structures include: (i) A drainage ditch with an underlying storage structure, (ii) a parking lot with a waterproof pavement that transfers runoff water toward a ditch, (iii) a vegetated hollow core slab, and (iv) an embankment of a grass-covered garden that was used as a reference for rainwater infiltration capacity. The observations and conclusions related to these structures cannot be generalized. The hydrodynamic functioning of each SuDS depends on multiple intrinsic conditions such as PSD, water content, vegetation cover, etc. Similar functioning cannot be expected for two SuDS of the same design. The main objective of this paper is to evaluate the convenience of using the BEST methods to determine the characteristic hydrodynamic parameters of SuDS in order to monitor their efficiency. The impact of erosion and some other physical conditions influencing the infiltration capacity of SuDS are discussed through the obtained results. A comparison of the outcomes of the three BEST methods was performed by a statistical analysis.

2. Materials and Methods

2.1. Studied Sites and Structures

The city of Lyon experiences significant rainfall events. The average annual precipitation height is around 831.9 mm, corresponding to 104.1 rainy days (1981–2010). The studied stormwater management structures were located in the east of Lyon in Lyon university campus, which is a pilot site for urban green city renovation as part of the Lyon city Field Observatory for Urban Water Management (OTHU). The studied structures were a reference grass-covered embankment, an impervious parking lot, a drainage ditch, and a vegetated hollow core slab. They receive the same type of runoff, mainly originating from parking lots, sidewalks, and circulation lanes.

The reference structure was a grass-covered garden embankment that had no specific infiltration or storage design (Figure 1a,b). Urban surfaces of this kind (gardens, parks, etc.) play an important role in ensuring rainwater infiltration. This structure was considered as a reference because it is the closest to natural soil. Infiltration tests were conducted in two different areas of this embankment, located within a dozen meters. They were called "Embk 1" and "Embk 2".

The first stormwater management structure was a parking lot incorporating a waterproof pavement (Figure 1c). This structure is not an infiltration system. The parking lot surface is only expected to grant water runoff to a nearby receiving ditch. After four years of use, some parts of the parking lot surface were subject to alterations and cracks because of invasive plant colonization (Figure 1d). In this study, it was referred to as "PKG-Int" and "PKG-Alt" for the intact and the altered areas, respectively. Both areas were studied separately to acknowledge the impact of invasive vegetation and surface deterioration on impervious surfaces.

The second structure was the drainage ditch, also called a dry swale, that receives the nearby waterproof parking lot runoff (Figure 1e). It is an experimental structure that receives stormwater from a 302.3 m^2 catchment area. The bottom of this structure is waterproofed. Rainwater infiltrates through 300–600 mm of topsoil, then through a filtration geotextile membrane, before reaching a calibrated gravel layer connected to a 160 mm diameter draining pipe. The drain outlet effluents are gathered in an underground storage structure where water quality is monitored. This structure was referred to as "the Ditch."

The last studied structure was the vegetated hollow core slab, also called a concrete grid parking lot (Figure 1f). It is a grass–concrete reinforced structure that is able to support significant loads (the weight of cars, passage of pedestrians, etc.) and infiltrate rainwater and runoff through topsoil-filled cells. The structure underneath the topsoil is permeable and includes a geotextile filtration layer. The depth of topsoil in the cells is well conceived to ensure that the soil will not be compacted through usage. It was referred to as "Grid-PKG".

Figure 1. Soil surface of studied structures: (**a**) Embk 1, (**b**) Embk 2, (**c**) PKG-Int, (**d**) PKG-Alt, (**e**) Ditch, and (**f**) Grid-PKG.

2.2. Infiltration Tests

Before starting a Beerkan test, the ground surface must be prepared. If the studied area is vegetated, plants must be cut while keeping the roots in situ. A ring is then positioned on the surface and embedded carefully, without destroying the soil's structure, to a depth of less than 1 cm in order to avoid lateral water losses. Bentonite can be used to secure the outer edges of the ring, especially in the case of coarse-textured soils. Constant water volumes are infiltrated successively through the ring under a surface-saturated condition, and the time needed for the entire infiltration of each volume is recorded. The test is stopped once the infiltration time is stabilized, i.e., when at least three consecutive infiltration times are identical [29]. In general, a set of 8–15 water volumes will achieve steady-state infiltration [18]. A sample is quickly collected at the end of the test from the ring center to determine the final volumetric water content. An undisturbed core is collected nearby the test spot to determine both soil bulk density and initial volumetric water content. Another soil sample is collected to determine the particle size distribution (PSD).

Beerkan tests were carried out on five different spots of Embk 1, two spots of Embk 2, seven spots of PKG-Int, six spots of PKG-Alt, seven spots along the Ditch, and eight cells of Grid-PKG. Infiltration rings of 15 cm diameter were used for all the tests, except for Grid-PKG for which 7.5 cm diameter rings were used so as to fit the cells' dimensions. 100 mL volumes of water were used for infiltrating.

To define the particle size distribution of each soil, the collected samples were oven dried before being manually sieved to a 1.3 mm particle size. Thereafter, a fine proportion of the soil matrix was analyzed by a laser particle size analyzer (Mastersizer 2000, Malvern, UK).

2.3. BEST Method for Hydraulic Parameter Estimation

The BEST method was developed to estimate hydraulic parameters K_s and S, as well as shape parameters n, m, and η, and to define the water retention curve $h(\theta)$ and the hydraulic conductivity curve $K(\theta)$, determined by Equations (1) and (3), respectively.

The scale parameter D_g and shape parameters M and N are estimated by fitting the cumulative frequency $F(D)$ (Equations (8) and (9) in Table 1) to the particle size distribution of the soil's fine fraction (<2 mm). The estimation of the soil fractal dimension (Equations (10) and (11) in Table 1) then allows the computation of the shape index parameter p_m (Equation (12) in Table 1) and subsequently the shape parameters n, m, and η directly by solving Equations (12)–(16) in Table 1.

Table 1. Shape parameter estimation from particle size distribution analysis.

Shape Parameters from Particle-Size Analysis		
Step-1: Fitting equation		
$F(D) = \left[1 + \left(\frac{D_g}{D}\right)^N\right]^{-M}$ (8) $M = 1 - \frac{2}{N}$ (9)		$F(D)$ is the cumulative frequency D diameter M and N shape parameters Dg scale parameter
M, N and Dg are defined by optimizing the fit to the PSD (fraction < 1.3 mm) by the least square technique		
Step-2: Solving for fractal dimension s		
$(1-\varepsilon)^s + \varepsilon^{2s} = 1$ (10) $\kappa = \frac{2s-1}{2s(1-s)}$ (11)		s fractal dimension of media ε soil porosity
Step-3: Shape parameters		
$m = \frac{1}{p_m}\left(\sqrt{1+p_m^2}-1\right)$ (12) $p_m = \frac{mn}{1+m}$ (13)		p_m shape index m, n pore distribution index
$p_m = \frac{MN}{1+M}(1+\kappa)^{-1}$ (14) $n = \frac{2}{1-m}$ (15)		
$\eta = \frac{2}{n \times m} + 2 + p$ (16)		η hydraulic conductivity shape parameter p tortuosity parameter *

(10) and (11) Fuentes et al. 1998 [30]; (12) and (13) Zatarain et al. 2003 [31]; (16) Haverkamp et al. 1999 [23]; * $p = 1$ (Burdine 1953) [21].

The hydrodynamic parameters K_s and S are estimated by fitting experimental infiltration data to the analytical axisymmetric infiltration model of Haverkamp et al. (1994) [32] (Equations (17)–(21) in Table 2). They can be calculated by the three BEST methods: Slope, intercept, and steady. The three methods define K_s and S by making different use of the parameters i_{exp} and b_{exp}, which are both estimated by a linear regression analysis of the data that describes the steady-state condition of the cumulative infiltration curve $I(t)$. BEST-slope and BEST-intercept are based on the model of Haverkamp et al. (1994) [32] that fits transient-state experimental data to estimate sorptivity S and then the saturated hydraulic conductivity K_s. BEST-slope uses the slope i_{exp} (Equations (22) and (23) in Table 2), and BEST-intercept uses the intercept b_{exp} (Equations (24) and (25) in Table 2), whereas BEST-steady uses both parameters and does not require additional experimental data fitting (Equations (26) and (27) in Table 2).

Table 2. Estimation of hydrodynamic parameters K_s and S by the Beerkan estimation of soil transfer parameters (BEST) methods using the model of Haverkamp et al. (1994) [32].

Axisymetrical infiltration model of Haverkamp et al. (1994)

$$I(t) = S\sqrt{t} + (AS^2 + BK_s)t \qquad (17) \qquad\qquad q(t) = \frac{S}{2\sqrt{t}} + (AS^2 + BK_s) \qquad (18)$$

Constants $A.B$ and C^*

$$A = \frac{\gamma}{r_d(\theta_s - \theta_0)} \quad (19) \qquad B = \frac{(2-\beta)}{3}\left[1 - \left(\frac{\theta_0}{\theta_s}\right)^{\eta}\right] + \left(\frac{\theta_0}{\theta_s}\right)^{\eta} \quad (20) \qquad C = \frac{1}{2\left[1 - \left(\frac{\theta_0}{\theta_s}\right)^{\eta}\right](1-\beta)}\ln\left(\frac{1}{\beta}\right) \quad (21)$$

$\beta \approx 0.6$ and $\gamma \approx 0.75$ **

BEST methods

i_{exp} steady state regression line slope

b_{exp} steady state regression line intercept

BEST *slope*   **BEST *intercept***   **BEST *steady***

$$K_s = i_{exp} - AS^2 \qquad (22) \quad K_s = C\frac{S^2}{b_{exp}} \qquad (24) \quad K_s = \frac{C\, i_{exp}}{Ab_{exp} + C} \quad (26)$$

$$I(t) = S\sqrt{t} + \left[A(1-B)S^2 + Si_{exp}\right]t \qquad (23) \quad I(t) = S\sqrt{t} + \left[AS^2 + BC\frac{S^2}{b_{exp}}\right]t \qquad (25)$$

The Haverkamp equation fitting on experimental data allows to calculate the sorptivity S and subsequently the saturated hydraulic conductivity Ks.

$$S = \sqrt{\frac{i_{exp}}{A + \frac{C}{b_{exp}}}} \qquad (27)$$

* A, B and C defined under the Burdine et al. (1953) [21] condition [2] by the equations of Haverkamp et al. (1994) [32]; for the specific case of Brooks & Corey (1964) [22] equation [3]; ** β and γ values are valid for $\theta_0 < 0.25\,\theta_s$ [33].

2.4. Statistical Analysis

Estimates of K_s and S by the three BEST methods were statistically analyzed. In order to evaluate the influence of the studied structure on the variance of parameters K_s and S, one-way ANOVA tests were performed. These tests can be used to determine any statistical significance between variables by calculating two main parameters: The *F-value*, which is the ratio of the variation between sample means to the variation within the sample, and *Pr(>F)*, which represents the non-significance hypothesis probability. *Pr(>F)* should be compared to a critical significance level, usually 5%, in order to assess the non-significance of the hypothesis.

Normal/log-normal distribution verifications of both parameters K_s and S were elaborated by using the Quantile-Quantile (QQ) plots and the Kolmogorov–Smirnov tests (R core team, 2013). QQ plots represent the quantiles of experimental data against theoretical probability distribution quantiles, in our case, the normal or log-normal distribution. The distribution is considered to be normally or log-normally distributed if the points in the QQ plot are arranged close to the first bisector $y = x$. The Kolmogorov–Smirnov (K-S) test quantifies the distance between the reference distribution and the studied sample distribution. This test calculates the parameter *D-value* that represents the maximal distance between the two distributions and the parameter *p-value* that refers to the probability of observing a higher distance than the *D-value*. The *p-value* should be compared to a critical statistics threshold in order to decide if the sample is normally or log-normally distributed. Usually, 5% is used.

For each test, the estimations of parameters K_s and S were plotted as a function of the average of the BEST-slope, intercept, and steady estimates in order to compare the three methods. Linear regression models were then used to test the correlation between K_s and S and assimilate soil capillary and gravity drainage functioning.

BEST graphs and estimations were elaborated by using the open-source software Scilab, while R was used for the statistical analysis (R Core Team, 2013). Box-and-whisker plots were used to represent the estimates of K_s, S, and λ_m. They are able to depict each structure's statistic values: The median, the upper, and lower quartiles and the highest and lowest values. This representation can be used to compare the studied structures and assimilate variability within the same structure.

3. Results and Discussion

3.1. Hydrodynamic Characteristics of SuDS

All soils have mainly a silty-sandy texture (Figure 2a,b). Particle size distributions of the fine fractions (<1.3 mm) revealed the bi-modality of the waterproof parking lot and the Ditch. The proportion of coarser particles (>1.3 mm) varied between 5% (Embk 2) and 66% (waterproof parking lot, PKG-Int, or PKG-Alt) of the sample's total weight.

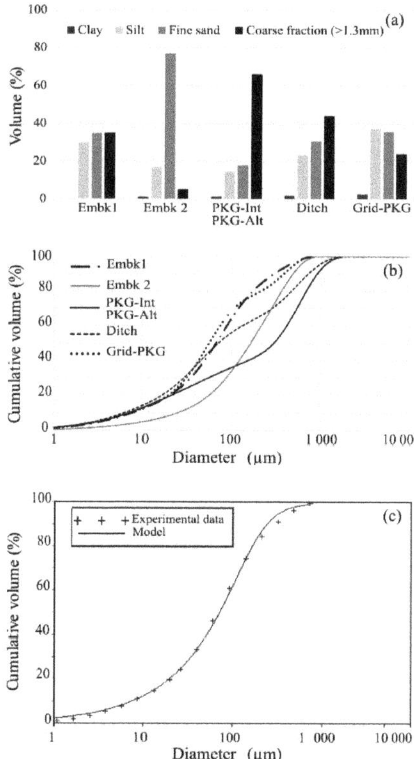

Figure 2. (a) Granulometry fractions of the studied soils, (b) particle size distributions of the fine fractions (<1.3 mm) and (c) example of particle size distribution fitting by the BEST method (Embk 1).

Shape parameters n, m, η, and c_p were defined from fine fraction PSD fitting (Figure 2c). The waterproof parking lot and the Ditch had similar shape parameter values, slightly lower than the other soils (Table 3). Grid-PKG and both the studied embankments had close values.

Table 3. Hydraulic shape parameters (n, m, and η) and shape parameter function c_p.

	n	m	η	c_p
Embk 1	2.36	0.15	8.61	2.06
Embk 2	2.33	0.14	8.97	2.09
PKG-Int	2.26	0.12	10.64	2.20
PKG-Alt	2.26	0.12	10.64	2.20
Ditch	2.28	0.12	10.08	2.17
Grid-PKG	2.40	0.17	7.94	2.00

Examples of experimental Beerkan cumulative infiltration curves of each structure are presented in Figure 3. The conducted tests on both areas of the reference garden embankment showed a significant infiltration flow in Embk 1 (infiltration rate $\approx 10^{-1}$ mm s^{-1}), whereas a low infiltration rate was recorded in Embk 2 (infiltration rate $\approx 10^{-2}$ mm s^{-1}). This finding can be explained by the narrow particle size distribution of Embk 2 and its low proportion of coarse particles (>1.3 mm). Apart from the potential presence of biological pores, such as earthworm channels, macropores were accordingly almost inexistent in this structure. Besides, a comparison of the initial and final water contents indicated that $\theta_{final} \approx 1.5\,\theta_{initial}$, which suggests that the soil was initially very wet. Microporosity, which was the overriding porosity in this structure, must have been nearly saturated from the beginning of infiltration. This observation also explains the absence of an infiltration transient state on the infiltration curves (Figure 3a). These conditions are susceptible to lead later to erroneous estimations of the hydraulic parameters by the BEST methods.

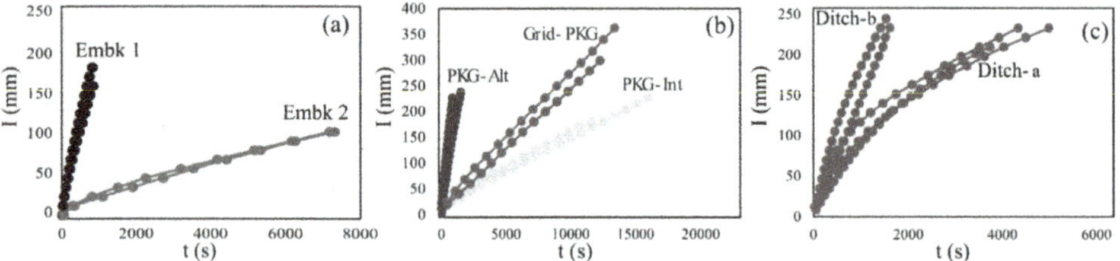

Figure 3. Examples of Beerkan infiltration curves of the studied soils: (**a**) Embk 1 and Embk 2; (**b**) PKG-Int, PKG-Alt, and Grid-PKG; (**c**) Ditch.

The Beerkan infiltration curves revealed a big difference in infiltration capacity between PKG-Int and PKG-Alt. The flow at the level of the altered area was approximately four times greater than that of the intact area (Figure 3b). This was due to the presence of cracks and macropores created by plant roots. It is important to note that the mean infiltration rate of PKG-Int was 2.2×10^{-2} mm s^{-1}. This is greater than 10^{-5} mm s^{-1}, which is the reference value to qualify a soil as pervious [34]. These results show that the surface coating of this parking lot was inefficient in both altered and intact areas because of a design failure and erodible material choice.

Infiltration curves of the drainage ditch revealed infiltration rates in the order of 9×10^{-2} mm s^{-1} (Ditch-a in Figure 3c). The marked concavity at the first moments of infiltration reflected the importance of capillary forces (i.e., significant sorptivity). At some spots for this structure, the infiltration curves showed a sudden increase after a certain infiltration time (Ditch-b in Figure 3c). This was probably related to the presence of swelling organic matter in the soil creating a capillary barrier effect [16,35], which favored evolution of preferential pathways around the organic matter zones. It might also have resulted from pore clogging induced by siltation in some areas. Structures with infiltration rates greater than 10^{-2} mm s^{-1} are potentially exposed to this phenomenon [9]. The risk of pore clogging was more important in the Ditch since the nearby parking lot ensured that the runoff held a high proportion of silt and was, above all, subject to erosion. Another interpretation is possible regarding the consistence of the pollutants the Ditch was susceptible to receive. In fact, the parking lot surface could potentially have been covered by greasy car emissions. These hydrophobic pollutants were carried by rainwater and end up on the Ditch surface, leading in the same way as organic matter to a barrier effect [16].

In Grid-PKG, the mean infiltration rate was 3.4×10^{-2} mm s^{-1}. This low infiltration capacity was probably due to clogging [36–38] or hydrophobicity effects. The extremely low initial volumetric water content in this structure ($\theta_{final} = 13\,\theta_{initial}$) may have led to water repellency at the beginning of the infiltration tests. Water repellency can be expected when soil water content is below a threshold value [39]. This case illustrates the importance of water content history knowledge to predict the hydrodynamic behavior of a soil. Vegetation was also dense in the Grid-PKG cells. Biofilms on plant roots may have been the origin of this hydrophobic behavior [40,41].

The fitting of Haverkamp's equations (Equations (17) and (18) in Table 2) [32] on Beerkan curves could be used to estimate S and K_s by the three BEST methods; this is shown in Figure 4. Embk 1 had significant values of K_s and S. Its saturated hydraulic conductivity was significant due to the wide particle size distribution favoring the presence of macropores and, thus, gravity flow [42–44]. At the same time, the presence of a large proportion of fine materials (silt) led to significant sorptivity values [45]. The lowest values of K_s and S were obtained by Embk 2 as a consequence of its narrow particle size and low proportion of fine particles.

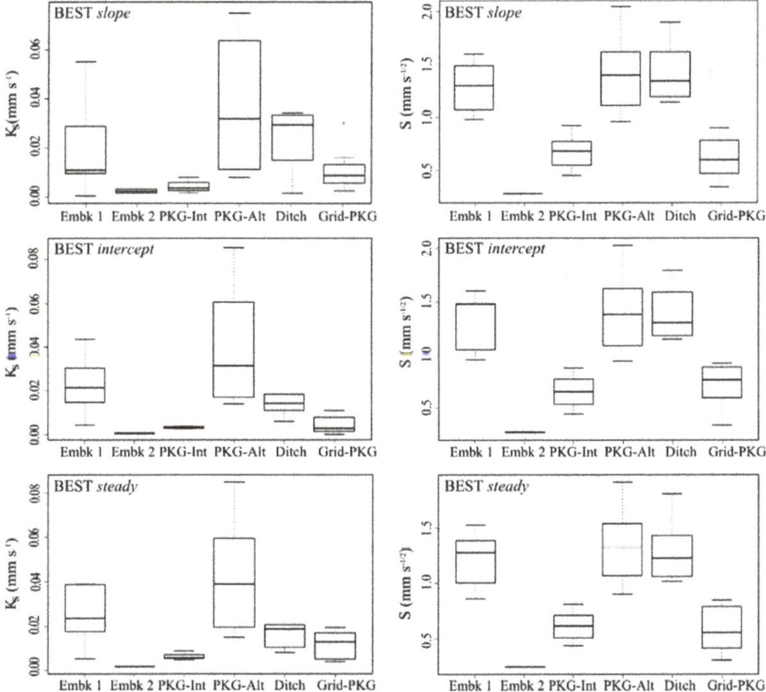

Figure 4. Estimates of K_s and S of the studied soils obtained by the three BEST methods.

PKG-Alt had the highest values of K_s and S (Figure 4). Its saturated hydraulic conductivity was significant due to the presence of cracks created by plants roots favoring gravity flow. This structure was also composed of a high proportion of fine materials (silt and clay), leading to significant sorptivity values. The parking lot area had the largest variation range of both parameters K_s and S because of the soil heterogeneity, regarding vegetation density and erosion extent. Lower values of saturated hydraulic conductivity and sorptivity were observed for PKG-Int.

Estimates of K_s and S for Ditch were in approximately the same order as the reference structure Embk 1, unlike Grid-PKG that had lower values. This result was in agreement with its low infiltration rates.

By considering the average values of saturated hydraulic conductivity and sorptivity for each BEST method, characteristic curves $h(\theta)$ and $K(\theta)$ were established by using Equation (1) of van Genuchten [20] and Equation (3) of Brooks and Corey [22] (Figure 5). The three BEST methods gave similar results. Water retention curves of PKG-Int, PKG-Alt, and the Ditch had a sudden variation with θ, represented by a steeper slope. Their hydraulic conductivity curves also evolved with θ quicker than the other structures. This resulted from their high proportion of coarse particles and their bimodal particle size distribution (Figure 2a,b). This proportion of coarse material presented a wide range of macropores that were the first to be hydraulically activated at the beginning of infiltration at high pressure and accordingly had higher saturated hydraulic conductivities [18] (Figure 5). This explains the rapid response of $h(\theta)$ and $K(\theta)$. Moreover, because of their bimodal PDS, once the water pressure reached the activation value of the average size pores corresponding to the first distribution mode, most of the pores became saturated. Consequently, $h(\theta)$ and $K(\theta)$ varied steeply with θ [18].

For the other structures, the water retention curves had smoother slopes with a plateau because of their fine and extended particle size distribution. Their pores were gradually saturated, which explains the slower variation of $K(\theta)$ and $h(\theta)$.

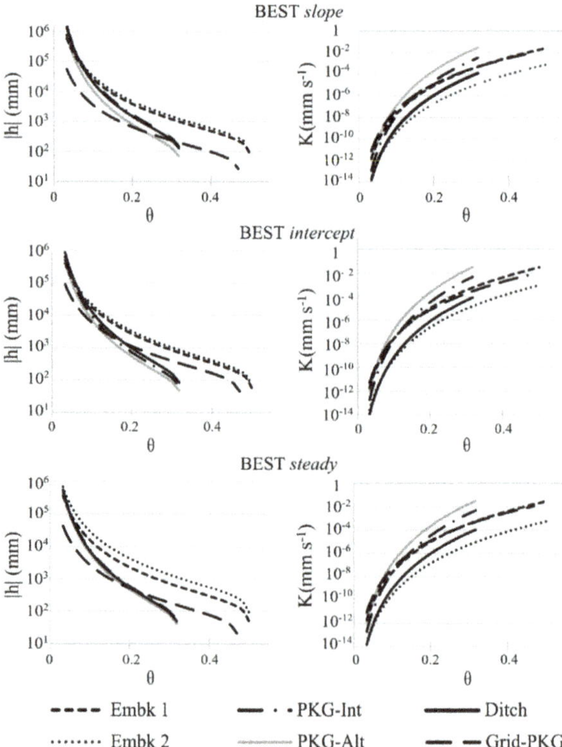

Figure 5. Hydrodynamic characteristic curves $K(\theta)$ and $h(\theta)$ of the studied soils obtained by the three BEST methods.

Parameter λ_m, the mean characteristic radius of hydraulically activated pores, was calculated for each structure after Equation (7) by using hydraulic conductivity, K_s, and sorptivity, S, estimates by the three BEST methods. The mean pore radii of the studied structures varied in the intervals [18–260 µm] by BEST-slope, [39–390 µm] by BEST-intercept, and [0.05–890 µm] by BEST-steady (Figure 6).

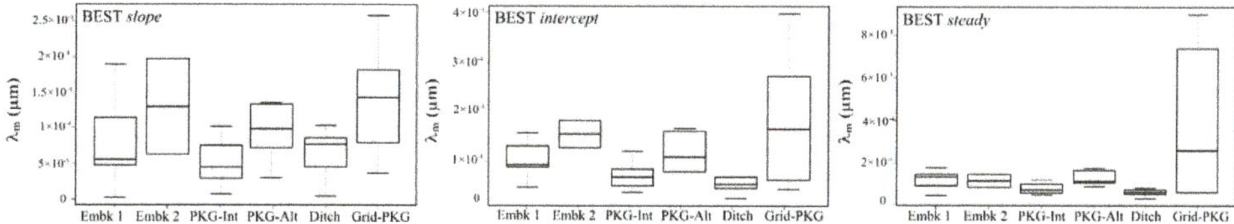

Figure 6. Average characteristic size of hydraulically activated pores, λ_m, estimates by the three BEST methods.

Apart from Grid-PKG and Embk 2, the three BEST methods gave proximate results for the rest of the studied structures. The obtained values also matched their corresponding observed mean infiltration rates (Table 4). Structures with high mean hydraulically activated pore radii were most likely to have a significant infiltration flow and vice versa. The largest differences between the three methods and the least coherent results were noticed in Grid-PKG and Embk 2. The calculated mean radii of these two structures were very significant and did not correspond to their slow infiltration flows. Their special hydric initial conditions explain the erroneous estimates for K_s and S. For Embk 2, the soil was initially very wet, a steady state was quickly reached, and thus, the transient state was poorly described. This implies an incorrect estimation of sorptivity, especially in the cases of BEST-slope and BEST-intercept, which rely on transient-state data fitting. For Grid-PKG, soil water repellency made steady-state infiltration hardly reachable. Hence, too few data points were collected, which confused the BEST-slope and intercept results [16,17]. Parameter K_s was then overestimated, and S was under-estimated [46], which explains the overestimation of λ_m.

Accordingly, it is important to make sure that while conducting a Beerkan infiltration test a steady state is reached, and the soil is initially dry enough to acquire transient-state describing data. In general, it must meet the condition $\theta_{initial} < 0.25\ \theta_{saturated}$ [32], without being overly dry to avoid water repellency problems [39].

Table 4. Comparison of the observed infiltration rates to the obtained values of parameters K_s, S, and λ_m.

| | λ_m (mm) | | | K_s (mm s^{-1}) | | | S (mm s$^{-1/2}$) | | | IR (mm s^{-1}) | Accuracy |
| | BEST | | | BEST | | | BEST | | | | |
	slope	*intercept*	*steady*	*slope*	*intercept*	*steady*	*slope*	*intercept*	*steady*		
Embk 1	0.102	0.118	0.139	0.021	0.025	0.025	1.290	1.315	1.211	0.108	+
Embk 2	0.129 *	0.155 *	0.118 *	0.002	0.003	0.002	0.282	0.274	0.248	0.012 *	−
PKG-Int	0.059	0.076	0.087	0.004	0.006	0.006	0.672	0.656	0.616	0.025	+
PKG-Alt	0.108	0.126	0.136	0.037	0.042	0.043	1.425	1.408	1.346	0.132	+
Ditch	0.078	0.071	0.069	0.036	0.025	0.033	1.412	1.496	1.265	0.088	+
Grid-PKG	0.166 *	0.213 *	0.443 *	0.011 *	0.007	0.012	0.619	0.718	0.585	0.031 *	−

(+) Accurate (−) Inaccurate; * λ_m does not correspond to the observed IR.

3.2. Statistical Analysis of the Three BEST Methods

A comparison between K_s and S values, as well as the characteristic curves $K(\theta)$ and $h(\theta)$ obtained by the three BEST methods, shows globally coherent results. This outcome was confirmed by the correlation coefficients. A considerable correlation between BEST-slope and BEST-steady was observed in terms of both parameters K_s and S (*coeff*(K_s) = 0.93 and *coeff*(S) = 0.99). A slightly weaker correlation was noticed between BEST-slope and BEST-intercept for the parameter K_s (*coeff(K_s)* = 0.60). BEST-slope failed to properly calculate K_s and S in some cases (4 tests out of 43). This was probably due to the lack of precision when choosing steady-state points, especially in low permeability soils where steady-state could hardly be achieved. It could also result from the soil's initial hydric conditions that could distort the description of the transient state [16,17]. The first infiltration moments are the moments when capillary forces are predominant. When initial water content was high, the infiltration curve did not represent the sorptivity impact which influenced the calculations of both S and K_s.

One-way ANOVA tests on parameters K_s, S, and the average infiltration rate (*IR*) allowed *F-values* and *Pr(>F)* to be calculated. For K_s, *F-value* = 4.325 and *Pr(>F)* = 0.00462. For S, *F-value* = 16.4 and *Pr(>F)* = 1.09×10^{-7}. For *IR*, *F-value* = 7.284 and *Pr(>F)* = 1.59×10^{-4}. These results suggest that the variability of K_s, S, and *IR* of the different structures (with a threshold of significance of 5%) was significant compared to the variability within the same structure.

The normality tests showed that sorptivity, S, is normally distributed, while saturated hydraulic conductivity is log-normally distributed, as shown in the QQ plots (Figure 7). This result was confirmed by the Kolmogorov–Smirnov tests, for which we considered a threshold value of 5% (Table 5). Similar observations were made as part of other studies of the BEST methods [17].

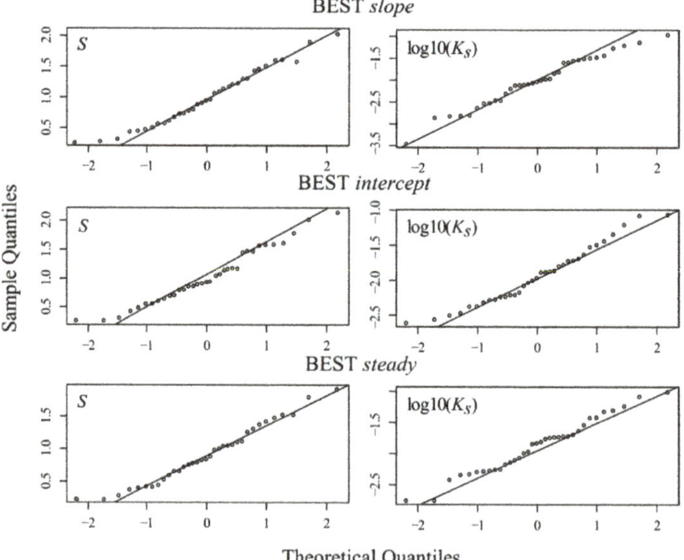

Figure 7. QQ plots of parameters K_s and S by the three BEST methods.

Table 5. Normality test Kolmogorov–Smirnov on the estimates of K_s and S by the three BEST methods.

Parameter	BEST	Normality		Log-Normality	
		D	**p-value**	**D**	**p-value**
K_s	slope	0.241	0.028	0.111	0.736 *
	intercept	0.236	0.034	0.114	0.711 *
	steady	0.257	0.016	0.089	0.922 *
S	slope	0.084	0.946 *	0.098	0.861
	intercept	0.110	0.749 *	0.088	0.926
	steady	0.077	0.976 *	0.107	0.781

** p-value > 0.05 Distribution is normally/Log-normally distributed.*

For each test, K_s was represented as a function of the geometric average of the estimations by the three BEST methods (K_s being log-normally distributed), and S was represented as a function of the arithmetic average (S being normally distributed) (Figure 8). The three methods gave results ranging around the mean with a slight underestimation of the sorptivity by BEST-slope and a minor overestimation of the same parameter by BEST-steady. Therefore, use of the geometric average of K_s and the arithmetic average of S to plot K_s vs. S can be justified.

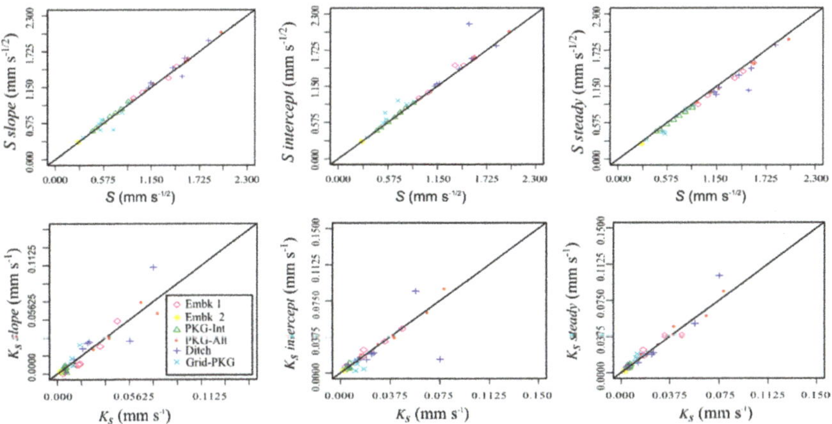

Figure 8. Parameters K_s and S vs. their mean values by the three BEST methods.

The obtained K_s vs. S plot defines a point cloud with a positive slope tendency (dashed line in Figure 9). This plot displays two groups of soils. The first group contained tests with low values and low variance of both parameters K_s and S. They belonged mainly to Embk 2, PKG-Int, and Grid-PKG. These structures had a low infiltration capacity due to their initial hydric condition and/or their homogenous particle size distribution. The second group contained the rest of the structures that had significant and diffuse values of K_s and S due to their heterogeneous textures and structures: Extended PSD, large fraction of fine particles, presence of macropores, cracks, vegetation, etc.

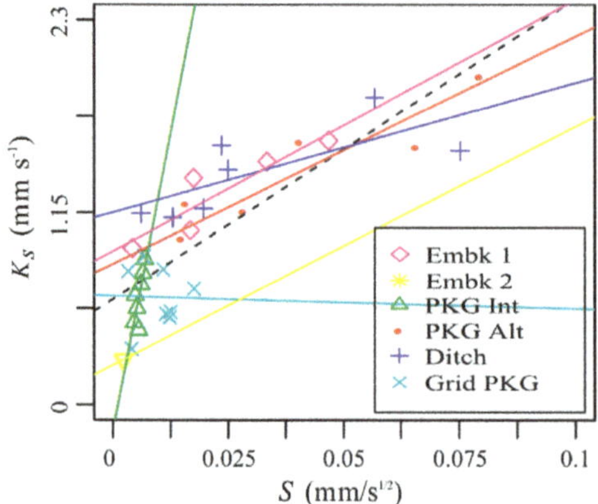

Figure 9. Mean K_s vs. mean S of the studied soils.

This representation of K_s vs. S enables the infiltration flow to be indexed, in a general way, as predominantly sorptive or gravity driven. For SuDS, this information can be crucial to pertinently choosing conception materials that allow both good infiltration and retention capacity. Soils with high values of K_s and S would be the best options to ensure good functioning. Following the evolution of the point cloud of a SuDS through time with this representation could also be interesting, in order to identify the origin of a possible malfunction.

4. Conclusions

The three BEST methods gave globally similar results and enabled the most important hydrodynamic parameters of the studied SuDS to be successfully determined. However, under extreme water content conditions, either initially very dry or very wet soils, the obtained results of the three methods were less accurate.

SuDS are accumulation points of rainwater, which makes of them a source of possible contaminant transfer toward the water table. The application of the BEST method allowed to evaluate the risk of dysfunctioning, related to their infiltration capacity or contaminant retention, through the study of hydrodynamic parameters K_s and S, water retention curve $h(\theta)$, and hydraulic conductivity curve $K(\theta)$. It also allowed the representation of soils in terms of their sorptivity/gravity infiltration capacity, in order to identify malfunctioning SuDS.

This study showed that continuous monitoring of SuDS infiltration capacity is essential, especially when structures are subject to erosion, as was the case of the studied waterproof parking lot that turned out to be permeable because of a design defect. The invasive vegetation in the altered areas led to high infiltration flows. This means that a significant proportion of the parking lot runoff was infiltrated before reaching the ditch, which presents an important hazard of contaminants (e.g., automobile emissions) being transferred to the underlying groundwater. The increase of the infiltration flow in the altered areas by invasive vegetation has already been the subject of other research [47]. This observation shows the interest in regular maintenance of these kinds of structures in order to ensure their proper functioning.

The potential presence of organic matter in the Ditch caused a slope variation in some infiltration curves. Residues of plants and leaves and the presence of hydrophobic substances in the runoff can enhance local hydrophobic behavior in soils [16,48].

Embk 1 displayed an infiltration flow and hydrodynamic behavior almost identical to the Ditch. This observation confirms that urban natural surfaces can be as effective as sustainable drainage structures in terms of infiltration capacity. However, in some of the studied soils, we noted low infiltration flows in both natural surfaces and drainage systems, as was the case in Embk 2 and

Grid-PKG, because of their extremely high/low initial water content conditions, as well as their specific particle size distributions. When local soil is used in such systems, its hydrodynamic characteristics should be checked.

The obtained results on the studied SuDS are not meant to be generalized. This study shows that the hydrodynamic properties of each structure depend on multiple conditions (PSD, hydric content, vegetation cover, etc.). Similar functioning cannot be expected for two SuDS of the same design. This case study only shows that Beerkan tests combined with the BEST method can be effective in following the evolution of the hydrodynamic functioning of SuDS throughout their usage.

Author Contributions: S.B., L.L., and R.A.-J. carried out the infiltration tests. L.L. developed the BEST algorithm. S.B. carried out the data analysis by using the BEST method. S.B., L.L., and R.A.-J. finalized the data interpretations. S.B. prepared the original draft. L.L., R.A.-J., and G.L.-K. supervised and reviewed the drafting process.

Funding: The French National Research Agency (ANR) contributed to the funding of the INFILTRON Project (ANR-17-CE04-0010).

Acknowledgments: The authors wish to thank the Field Observatory in Urban Water Management (OTHU) for technical and scientific support. This work was performed within the INFILTRON project supported by the French National Research Agency (ANR-17-CE04-010). We would also like to thank the trainees Rachid Ridouani and Saint-Louis Saint-Martin who helped acquire the experimental infiltration data.

Conflicts of Interest: The authors declare no conflict of interest.

References

1. Martinelli, I. *Infiltration des eaux de ruissellement pluvial et transfert de polluants associés dans un sol urbain: Vers une approche globale et pluridisciplinaire*; INSA Lyon: Lyon, France, 1999.
2. Mikkelsen, P.S.; Häfliger, M.; Ochs, M.; Tjell, J.C.; Jacobsen, P.; Boleer, M. Experimental assessment of soil and groundwater contamination from two old infiltration systems for road run-off in switzerland. *Sci. Total Environ.* **1996**, *189/190*, 341–347. [CrossRef]
3. Scholz, M. *Wetland Systems to Control Urban Runoff*, 1st ed.; Elsevier: Oxford, UK, 2006.
4. Arora, D.; Jindal, N.; Shukla, R.K.; Bansal, R. Water borne Hepatitis A and Hepatitis E in Malwa Region of Punjab, India. *J. Clin. Diagn. Res.* **2013**, 2163–2166. [CrossRef]
5. Hatt, B.E.; Fletcher, T.D.; Deletic, A. Treatment performance of gravel filter media: Implications for design and application of stormwater infiltration systems. *Water Res.* **2007**, *41*, 2513–2524. [CrossRef] [PubMed]
6. Wang, T.; Zlotnik, V.A.; Wedin, D.; Wally, K.D. Spatial trends in saturated hydraulic conductivity of vegetated dunes in the Nebraska Sand Hills: Effects of depth and topography. *J. Hydrol.* **2008**, *349*, 88–97. [CrossRef]
7. Fletcher, T.D.; Shuster, W.; Hunt, W.F.; Ashley, R.; Butler, D.; Arthur, S.; Trowsdale, S.; Barraud, S.; Semadeni-Davies, A.; Bertrand-Krajewski, J.-L. SUDS, LID, BMPs, WSUD and more—The evolution and application of terminology surrounding urban drainage. *Urban Water J.* **2014**, *12*, 525–542. [CrossRef]
8. Palhegyi, G.E. Designing storm-water controls to promote sustainable ecosystems: Science and application. *J. Hydrol. Eng.* **2010**, *15*, 504–511. [CrossRef]
9. Woods Ballard, B.; Wilson, S.; Udale-Clarke, H.; Illman, S.; Scott, T.; Ashley, R.; Kellagher, R. *The SuDS Manual*; CIRIA C753; CIRIA: London, UK, 2015.
10. Boogaard, F.C.; van de Ven, F.; Langeveld, J.G.; Kluck, J.; van de Giesen, N. Removal efficiency of storm water treatment techniques: Standardized full scale laboratory testing. *Urban Water J.* **2015**, *14*, 255–262. [CrossRef]
11. Bettess, R. *Infiltration Drainage-Manual of Good Practice*; CIRIA R156; CIRIA: London, UK, 1996; ISBN 978-0-86017-457-8.
12. BRE Digest 365. *Soakway Design*; Buildings Research Establishment: Bracknell, UK, 1991; ISBN 0-85125-502-7.
13. Allaire, S.E.; Roulier, S.; Cessna, A.J. Quantifying preferential flow in soils: A review of different techniques. *J. Hydrol.* **2009**, *378*, 179–204. [CrossRef]
14. Lamy, E.; Lassabatere, L.; Bechet, B.; Andrieu, H. Modeling the influence of an artificial macropore in sandy columns on flow and transfer. *J. Hydrol.* **2009**, *376*, 392–402. [CrossRef]
15. Lassabatere, L.; Spadini, L.; Delolme, C.; Fevrier, L.; Cloutier, R.G.; Winiarski, T. Concomitant Zn-Cd and Pb retention in a carbonated fluvio-glacial deposit under both static and dynamic conditions. *Chemosphere* **2007**, *69*, 1499–1508. [CrossRef]

16. Angulo-Jaramillo, R.; Bagarello, V.; Iovino, M.; Lassabatere, L. Soils with specific features infiltration. In *Measurements for Soil Hydraulic Characterization*; Springer: Cham, Switzerland, 2016; pp. 289–346.

17. Di Prima, S.; Lassabatere, L.; Bagarello, V.; Iovino, M.; Angulo-Jaramillo, A. Testing a new automated single ring infiltrometer for Beerkan infiltration experiments. *Geoderma* **2015**, *262*, 20–34. [CrossRef]

18. Lassabatere, L.; Angulo-Jaramillo, R.; Soria Ugalde, J.M.; Cuenca, R.; Braud, I.; Haverkamp, R. Beerkan estimation of soil transfer parameters through infiltration Experiments—BEST. *Soil Sci. Soc. Am. J.* **2006**, *70*, 521–532. [CrossRef]

19. Haverkamp, R.; Arrue, J.L.; Vandervaere, J.-P.; Braud, I.; Boulet, G.; Laurent, J.P.; Taha, A.; Ross, P.J.; Angulo-Jaramillo, R. *Hydrological and Thermal Behaviour of the Vadose Zone in the Area of Barrax and Tomelloso (Spain): Experimental Study, Analysis and Modeling*; Project UE 1996; n8 EV5C-CT 92 00 90; European Union: Brussels, Belgium; p. 1996.

20. van Genuchten, M.T. A closed form equation for predicting the hydraulic conductivity of unsaturated soils. *Soil Sci. Soc. Am. J.* **1980**, *44*, 892–898. [CrossRef]

21. Burdine, N.T. Relative permeability calculation from pore size Distribution data. *Petr. Trans. Am. Inst. Min. Metall. Eng.* **1953**, *198*, 71–77. [CrossRef]

22. Brooks, R.H.; Corey, C.T. *Hydraulics Properties of Porous Media*; Hydrology Paper 3; Colorado State University: Fort Collins, CO, USA, 1964.

23. Haverkamp, R.; Bouraoui, F.; Zammit, C.; Angulo-Jaramillo, R.; Delleur, J.W. Soil properties and moisture movement in the unsaturated zone. In *The Handbook of Groundwater Engineering*; CRC: Boca Raton, FL, USA, 1999; pp. 2931–2935.

24. Yilmaz, D.; Lassabatere, L.; Angulo-Jaramillo, R.; Deneele, D.; Legret, M. Hydrodynamic characterization of basic oxygen furnace slag through an adapted best method. *Vadose Zone J.* **2010**, *9*, 107–116. [CrossRef]

25. Bagarello, V.; Di Prima, S.; Iovino, M.; Provenzano, G. Estimating field-saturated Soil hydraulic conductivity by a simplified beerkan infiltration experiment. *Hydrol. Process.* **2014**, *28*, 1095–1103. [CrossRef]

26. White, I.; Sully, M.J. Macroscopic and microscopic capillary length and time scales from field infiltration. *Water Resour. Res.* **1987**, *23*, 1514–1522. [CrossRef]

27. Warrick, A.W.; Broadbridge, P. Sorptivity and macroscopic capillary length relationships. *Water Resour. Res.* **1992**, *28*, 427–431. [CrossRef]

28. Bagarello, V.; Iovino, M. Testing the BEST procedure to estimate the soil water retention curve. *Geoderma* **2012**, *187–188*, 67–76. [CrossRef]

29. Mubarak, I.; Angulo-Jaramillo, R.; Mailhol, J.C.; Ruelle, P.; Khaledian, M.; Vauclin, M. Spatial analysis of soil surface hydraulic properties: Is infiltration method dependent? *Agric. Water Manag.* **2010**, *97*, 1517–1526. [CrossRef]

30. Fuentes, C.; Vauclin, M.; Parlange, J.-Y.; Haverkamp, R. Soil water conductivity of a fractal soil. In *Fractals in Soil Science*; Baveye, P., Crawford, J.W., Rawls, W.J., Eds.; Lewis Publisher: Boca Raton, FL, USA, 1998; pp. 333–340.

31. Zatarain, F.; Fuentes, C.; Haverkamp, R.; Antonino, A.C.D. Prediccion de la forma de la caracteristica de humedad del suelo a partir de la curva granulometrica, 1-9. In Proceedings of the XII Congreso Nacional De Irrigacion, Zacatecas, Mexico, 13–15 April 2003.

32. Haverkamp, R.; Ross, P.J.; Smetten, K.R.J.; Parlange, J.Y. Three-dimensional analysis of infiltration from the disc infiltrometer: 2. Physically based infiltration equation. *Water Resour. Res.* **1994**, *30*, 2931–2935. [CrossRef]

33. Smetten, K.R.J.; Parlange, J.Y.; Ross, P.J.; Haverkamp, R. Three-dimensional analysis of infiltration from the disc infiltrometer: 1. A capillary-base theory. *Water Resour. Res.* **1994**, *30*, 2925–2929. [CrossRef]

34. Interpave. *Permeable Pavements. Guide to the Design, Construction and Maintenance of Concrete Block Permeable Pavements*, 6th ed.; Interpave: Leicester, UK, 2010.

35. Täumer, K.; Stoffregen, H.; Wessolek, G. Determination of repellency distribution using soil organic matter and water content. *Geoderma* **2005**, *125*, 107–115. [CrossRef]

36. Sañudo-Fontaneda, L.A.; Andrés-Valeri, V.C.; Rodriguez-Hernandez, J.; Castro-Fresno, D. Field Study of Infiltration Capacity Reduction of Porous Mixture Surfaces. *Water* **2014**, *6*, 661–669. [CrossRef]

37. Razzaghmanesh, M.; Beecham, S. A Review of Permeable Pavement Clogging Investigations and Recommended Maintenance Regimes. *Water* **2018**, *10*, 337. [CrossRef]

38. Andrés-Valeri, V.C.; Marchioni, M.; Sañudo-Fontaneda, L.A.; Giustozzi, F.; Becciu, G. Laboratory Assessment of the Infiltration Capacity Reduction in Clogged Porous Mixture Surfaces. *Sustainability* **2016**, *8*, 751. [CrossRef]

39. Ebel, B.A.; Moody, J.A. Rethinking infiltration in wildfire-affected soils. *Hydrol. Process.* **2013**, *27*, 1510–1514. [CrossRef]

40. FAWB. *Advancing the Design of Stormwater Biofiltration*; Facility for Advancing Water Biofiltration, Monash University: Victoria, Australia, 2008.

41. FAWB. *Stormwater Biofiltration Systems, Adoption Guidelines. Planning, Design and Practical Implementation*; Facility for Advancing Water Biofiltration, Monash University: Victoria, Australia, 2009.

42. Lassabatere, L.; Yilmaz, D.; Peyrard, X.; Peyneau, P.E.; Lenoir, T.; Šimůnek, J.; Angulo-Jaramillo, R. New analytical model for cumulative infiltration into dual-permeability soils. *Vadoze Zone J.* **2014**, *13*. [CrossRef]

43. Youngs, E.G.; Leeds-Harrison, P.B. Aspects of transport processes in aggregated soils. *J. Soil Sci.* **1990**, *41*, 665675. [CrossRef]

44. Mangala, O.S.; Toppo, P.; Ghoshal, S. Study of Infiltration Capacity of Different Soils. *Int. J. Trend Res. Dev.* **2016**, *3*, 388–390.

45. Leeds-Harrison, P.B.; Youngs, E.G.; Uddin, B. A device for determining the sorptivity of soil aggregates. *Eur. J. Soil Sci.* **1994**, 269–272.

46. Lassabatere, L.; Angulo-Jaramillo, R.; Yilmaz, D.; Winiarski, T. BEST method: Characterization of soil unsaturated hydraulic properties. In *Advances in Unsaturated Soils*; Caicedo, B., Murillo, C., Hoyos, L., Colmenares, J.E., Berdugo, I.R., Eds.; CRC Press: Boca Raton, FL, USA, 2013.

47. Mitchell, A.R.; Ellsworth, T.R.; Meek, B.D. Effect of root systems on preferential flow in swelling soil. *Commun. Soil Sci. Plant Anal.* **1995**, *26*, 2655–2666. [CrossRef]

48. Tu, M.C.; Traver, R.G. Water table fluctuation from green infrastructure sidewalk planters in Philadelphia. *J. Irrig. Drain E-ASCE* **2019**, *145*, 05018008. [CrossRef]

Article

Assessment of the Physically-Based Hydrus-1D Model for Simulating the Water Fluxes of a Mediterranean Cropping System

Domenico Ventrella [1,*], Mirko Castellini [1], Simone Di Prima [2], Pasquale Garofalo [1] and Laurent Lassabatère [2]

[1] Council for Agricultural Research and Economics-Research, Center for Agriculture and Environment (CREA-AA), Via C. Ulpiani 5, 70125 Bari, Italy

[2] Université de Lyon, UMR5023 Ecologie des Hydrosystèmes Naturels et Anthropisés, CNRS, ENTPE, Université Lyon 1, 3 rue Maurice Audin, 69518 Vaulx-en-Velin, France

* Correspondence: domenico.ventrella@crea.gov.it; Tel.: +39-080-5475023

Received: 7 June 2019; Accepted: 7 August 2019; Published: 10 August 2019

Abstract: In a context characterized by a scarcity of water resources and a need for agriculture to cope the increase of food demand, it is of fundamental importance to increase the water use efficiency of cropping systems. This objective can be meet using several currently available software packages simulating water movements in the "soil–plant–atmosphere" continuum (SPAC). The goal of the paper is to discuss and optimize the strategy for implementing an effective simulation framework in order to describe the main soil water fluxes of a typical horticultural cropping system in Southern Italy based on drip-irrigated watermelon cultivation. The Hydrus-1D model was calibrated by optimizing the hydraulic parameters based on the comparison between simulated and measured soil water content values. Next, a sensitivity analysis of the hydraulic parameters of the Mualem–van Genuchten model was carried out. Hydryus-1D determined simulated soil water contents fairly well, with an average root mean square error below 9%. The main fluxes of the SPAC were confined in a restricted soil volume and were therefore well described by the one-dimensional model Hydrus-1D. Water content at saturation and the fitting parameters α and n were the parameters with the highest impact for describing the soil/plant water balance.

Keywords: Hydrus-1D; TDR probe; saturated hydraulic conductivity; soil water flux; watermelon

---

1. Introduction

In many regions of the world, scarcity of water resources is one of the most important concerns for agriculture, together with increasing food demand due to increasing population and rapid economic growth. These concerns are exacerbated by the impacts of climate change on crop yield [1], fresh water availability, and soil fertility. In this scenario, above all in environments with high degrees of evapotranspiration, it becomes of fundamental importance to increase the water use efficiency of cropping systems by acting on the optimization of irrigation (method, amount, and timing) and the application of suitable irrigation scheduling for crops cultivated in dry environments, where irrigation is one of the essential agronomical practices to obtain sustainable yields. For these reasons, many software packages, with different complexity levels, are available and are used for simulating water movement in the "soil–plant–atmosphere" continuum and estimating indicators useful for increasing water use efficiency.

Among the available models, the physically-based models ones, i.e., based on numerical solutions of the Richards equation, are increasingly being used to simulate water and solute movement in the vadose zone for a variety of common applications in research and soil/water management (see for example [2–6]).

Hydrus software packages are finite element models for simulating the one- and two- or three-dimensional movement of water, heat, and multiple solutes in variably saturated media, and since their implementation they have been extensively applied in various applications. Hydrus-1D, with related manuals and case studies as examples, can be freely downloaded from the Hydrus website (https://www.pc-progress.com/en/Default.aspx?hydrus-1d) [7].

In the Mediterranean environment of Southern Italy, one of the first applications of Hydrus-1D concerned the simulation of water movement and solute transport in a variably saturated water flow of fine-textured soil subjected to a fluctuating saline groundwater table [8]. Since then, the use of Hydrus has involved the estimation of the water fluxes under several cropping systems differentiated from a spatial and agronomical point of view.

Plastic mulch is an agronomic technology widespread in the world because it allows farmers and engineers: (1) to increase soil temperature, (2) to reduce weed pressure and certain insect pests, (3) to increase the soil moisture, and (4) to improve nutrient use efficiency [9–11]. Under cotton cultivation, for example, the effect of plastic mulch on soil water balance under drip irrigation has been examined by Han et al. [9], who revealed a minimal effect on soil water distribution but a real effectiveness on soil water conservation due to a great reduction of soil evaporation.

Hydrus was used also to check out the optimal drip lateral depth under a cropping system of eggplant characterized by large inter-row distance and localized irrigation [12]. Recently, the dynamics of soil water contents and water fluxes in an olive orchard, submitted to two irrigation systems, were assessed by Autovino et al. [2]. The abovementioned studies were based on Hydrus-2D, which simulates two-dimensional water movements, with the assumption of the absence of water pressure head gradient along the plant rows. The use of Hydrus-2D can lead to higher complexity in terms of discretization of the domain and running simulation time when compared to the one-dimensional Hydrus-1D version. However, using physically-based models can be useful for increasing water use efficiency, saving water, and minimizing risks of water percolation and leaching.

In the case of horticultural systems, characterized by wide inter-rows, a one-dimensional domain could be suitable if the infiltration of irrigated water, root absorption, and percolation took place in a confined volume and not on the entire surface of the soil. This condition generally occurs with localized irrigation, which aims to distribute water in a limited volume of soil explored by the roots. However, this condition also depends on soil properties and agronomic management (presence or absence of plastic mulch, tillage, methods of fertilization, etc.), and the optimization of these factors needs to be carefully investigated for specific agro-environments.

Although physically-based models for simulating the one- and two- or three-dimensional movements of water are effective for estimating field-scale water fluxes, they generally require complex model parameterizations and input variables, some of which are not readily available. In order to address the issue of missing soil hydraulic parameters, which is particularly important for regional studies, a common approach is to use pedotransfer functions to convert available soil information, such as texture, bulk density, etc., to soil hydraulic parameters [13]. Hydrus packages make use of the pedotransfer functions (PTFs) based on neural networks [14] to predict van Genuchten's [15] water retention parameters and the saturated hydraulic conductivity (Ks) by means of textural information, bulk density, and soil water content at a pressure head (h) of -300 and $-15,000$ cm (i.e., at field capacity and permanent wilting point, respectively).

The aim of this study is to discuss and optimize the strategy regarding the characterization of soil, the impact of its accurate description and the implementation of an effective simulation framework in order to describe the main soil water fluxes of the "soil–plant–atmosphere" continuum for a typical horticultural cropping system based on drip-irrigated watermelon cultivation in Southern Italy.

2. Materials and Methods

2.1. Experimental Site

In a field of a private farm cultivated with drip-irrigated watermelon (*Citrullus lanatus*, Thunb) located in Castellaneta (Southern Italy, 40°35′29.01″ N, 16°55′26.18″ E), an automated Time Domain Reflectometry (TDR) system, installed in May 2006, was used to continuously measure soil-moisture content variations on hourly basis (Figure 1). The TDR system included a cable tester (TDR100) interfaced to a data logger (CR10) and multiplexers, a solar panel, and a storage battery (Campbell Scientific, Logan, UT, USA). The waveguides probes consisted of three 15-cm stainless-steel rods. Three probes were inserted in the middle between two plant rows, horizontally from field level and perpendicular to the rows. Another 16 probes were horizontally inserted below the drip lines, and plant rows in four profiles each consisted of four probes inserted at −10, −30, −50, and −70 cm depths. The TDR system was removed in July. Analysis of the waveforms curves was automatically performed by adopting the procedure implemented in CR10 and the volumetric soil water content (SWC) was estimated by the general equation of Topp et al. [16].

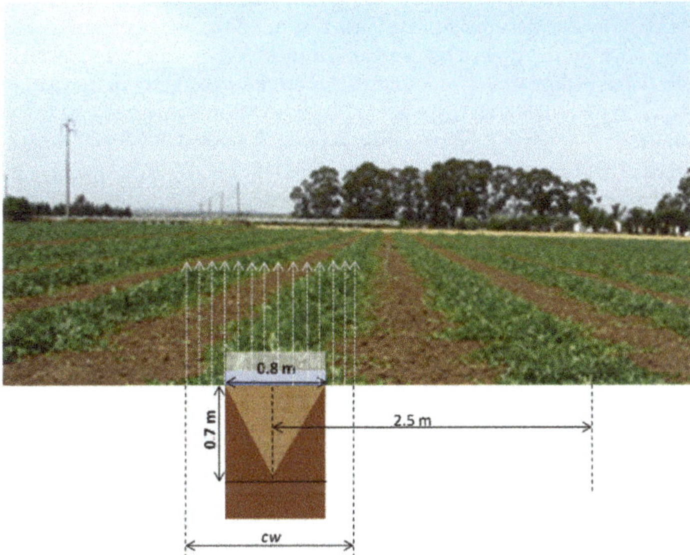

Figure 1. Photo of a watermelon field at Castellaneta (TA, Italy) and a schematic layout of the water/soil/plant domain. The isosceles triangle indicates the maximum root growth, the brown rectangle the soil domain in which the one-dimensional soil water fluxes take place, and the vertical grey arrows the transpirative fluxes of watermelon. *cw* is the canopy width measured during the monitored cropping cycle.

Table 1 reports the main characteristics for two soil layers (0–20 and 20–40 cm depth). The soil is a clay with a total organic carbon close to the USDA lower normal limit (10 g kg^{-1}). Comparable values of field capacity and permanent wilting point were detected in the soil profile, although it was slightly more compact in the lower layer than in the superficial one (Table 1). A visual inspection of soil profile (1 m depth) during the installation of the TDR probes showed that the soil was quite homogeneous and consequently we extended the hydraulic properties of second layer to all the domain profile up to the considered bottom boundary.

Table 1. Physical and chemical of soil characteristics.

Parameter	Unit	First Layer 0–20 cm	Second Layer 20–40 cm
pH		8.23	8.15
Electrical conductivity	dS m^{-1}	0.366	0.4632
Total organic carbon	g kg^{-1}	10.0	9.7
Total nitrogen	g kg^{-1}	0.99	0.99
Sand	%	16.48	16.33
Silt	%	36.10	37.32
Clay	%	47.43	46.35
Dry bulk density	g cm^{-3}	1.27	1.36
Field capacity	cm^3 cm^{-3}	0.37	0.38
Permanent wilting point	cm^3 cm^{-3}	0.19	0.20

The climate of the area is typically Mediterranean, with mean monthly minimum temperatures of 2.7 °C in the winter and maximum temperature of 27 °C in the summer. The 30-year mean annual precipitation of the area is 638 mm, with more than 60% of the rainfall occurring from October through March (www.ilmeteo.it/portale/medie-climatiche/Castellaneta).

In the field of watermelon, the inter-row distance was of 2.50 m with a plant distance in the row of 0.90 m. Before the transplanting time, a black plastic mulch, 35 cm wide, was placed on the rows in order to reduce the water evaporation losses and to control the weed germination and growth. Below the plastic mulch, a soft drip-line with holes every 30 cm and flow rate (q) of 1.2 l h^{-1} was placed on each row. The duration (t_i, hour) of each irrigation event was estimated by analysing the daily trend of soil water content (SWC) measured by the shallowest TDR probe (−10 cm) which, throughout the irrigation, indicated a rapid increase of SWC. The irrigation depth (v_i, mm) was estimated, through the following equation:

$$v_i = \frac{t_i n_e q}{A} \tag{1}$$

where n_e is the number of the emitters of the drip line (130 m long) equipped with volumetric meter and A is the product of 130 m by 0.8 m, with this last parameter being the width of the area served by the drip hole, visually estimated and constant throughout the crop cycle. Each v_i was verified by the water volume measured by the volumetric meter installed on the drip hole of the monitored row. The irrigation volumes supplied by the farmer were of 3239 m^3 ha^{-1}.

During the watermelon cultivation, the following measurements were carried out at a weekly or ten-day scale: gravimetric SWC at two depths using disturbed soil cores; perpendicular variation in shallow SWC through regular measurements between two adjacent rows utilizing the 5-cm "Theta Probes ML2x" transept in the inter-row; and soil bulk density with undisturbed soil cores (5 cm in diameter by 5 cm in height). Regarding the crop growth, every week the canopy width (cw) along the rows was monitored and a ceptometer (LP-80, Decagon Devices, Inc., Pullman, WA, USA) was utilized for measuring the canopy cover index (i).

To estimate reference evapotranspiration, ET_0, close to TDR probes, a standard meteorological station was installed for monitoring hourly air temperature, humidity, global radiation, precipitation and wind speed at a 2-m height. Daily values of ET_0 were determined according to the FAO Penman–Monteith equation [17].

Because of the plastic mulch placed on the row, the soil evaporation was considered equal to 0 and the adjusted potential daily plant transpiration (T'_p) was estimated by:

$$\begin{aligned} T_P &= ET_0 i, \\ T'_P &= c T_P \end{aligned} \tag{2}$$

where c (>1) is the ratio between cw, the measured width of the transpiring canopy, and the reference width (0.8 m) of the considered soil domain in which the watermelon roots were developed and concentrated (Figure 1). The adjustment of the potential transpiration is necessary to be able to use a one-dimensional model for crops with root apparatus concentrated below a soil surface smaller than that of the canopy and in a volume of soil wetted by drip irrigation.

2.2. Modelling Approach

The Hydrus-1D software (version 4.17) [18] was used to simulate one-dimensional vertical isothermal variably-saturated flow at the experimental site and below the drip line and plastic mulch, considering the soil uniformly irrigated and explored by the roots.

Simulations were performed from 30 May to 25 July 2006. The one-dimensional flow domain extended to a depth of 2.3 m, and was divided into two separate soil layers, from 0 to −20 cm and from −20 to −230 cm, comprising a total of 74 finite elements. A finer discretization was used near the soil surface to accommodate relatively steep gradients in the pressure head. Four observational nodes were seated in the soil profile in correspondence with the TDR probe at depths of −10, −30, −50, and −70 cm. As a soil surface boundary condition, we used a system-dependent atmospheric condition in accordance with the approach of Feddes [19] and Šimůnek et al. [18]. The time variable conditions of top boundary were defined at daily scale by irrigation depth (v_i, Equation (1)) and adjusted potential transpiration (T'_p, Equation (2)). At the bottom boundary a "free drainage" condition was adopted. It is determined by an unit vertical hydraulic gradient implemented in Hydrus-1D as a form of a variable flux boundary condition, suitable for cases where the water table is very deep and does not affect the crop/soil water balance. The root water uptake stress response function of Feddes [19] was used optimizing the related parameters in a preliminary calibration phase. The initial conditions were considered as the hydraulic potential for a profile in equilibrium after the heavy irrigations were carried out in the area during the first phase of melon cultivation.

The daily output variables simulated by Hydrus-1D and utilized in this study were: (1) transient SWC at observational points, (2) water potential in the soil explored by the roots, (3) cumulated actual soil water uptake, (4) infiltration, and (5) deep percolation.

The soil hydraulic functions were described according to van Genuchten [15]:

$$S_e = \frac{\theta - \theta_r}{\theta_s - \theta_r} = \left[1 + (ah)^n\right]^{-m} \tag{3}$$

$$K = K_s S_e^{0.5}\left[1 - \left(1 - S_e^{1/m}\right)^m\right]^2 \tag{4}$$

where S_e is the relative saturation, θ_s and θ_r are the saturated and residual water content (m^3 m^{-3}), h is the soil water pressure head (cm), a (cm^{-1}) and n are fitting parameters $m = 1 - (1/n)$, and K_s is the saturated hydraulic conductivity (cm d^{-1}).

In order to calibrate Hydrus-1D, four sets of hydraulic parameters (C1 to C4) obtained by applying the pedotransfer function of Rosetta [14] were considered. In particular, in C1 the parameters were estimated utilizing as independent variables the percentages of clay, silt, and sand of the two soil layers considered (0/–20 cm, −21/–230 cm). C2 also takes into account the measured bulk density, whereas the measured soil water contents at soil water potential of −300 cm and −15,000 cm were included as independent variables for C3 and C4, respectively. In this investigation, we considered as a reference a fifth set of parameters (C5) obtained by the Wind evaporation method [20] and the Unit hydraulic gradient (Uhg) method [21,22]. Thus, the four considered sets (C1–C4) were compared for validation purposes with the reference parameters (C5). In details, the Wind and Uhg methods allowed us to obtain an accurate soil hydraulic characterization, namely to estimate the water retention curve and the hydraulic conductivity function of investigated soil. For this purpose, two soil cores (0.075 m in height by 0.085 m in diameter) were subjected, in sequence, to a 1D unsaturated infiltration process and to a transient stage of evaporation to estimate the hydraulic properties of the soil in the 0–10 and 11–20 cm layers. Practically, during the transient of evaporation, applied methods provides θ-h-K values which need to be interpolated to obtain the water retention curve (θ-h) and hydraulic conductivity (K-h)

functions. As a consequence, hydraulic functions were simultaneously parameterized by applying the Metronia code [23]. Details about applied methodologies, i.e., Unit hydraulic gradient and Wind evaporation methods, can be found in Bagarello et al. [22,24], respectively.

The Table 2 reports the parameters of the 5 approaches. For C5 a single layer was considered because lab experiments provided similar results between soil layers.

Table 2. Van Genuchten parameters of the five approaches.

Approach	Layer (cm)	θr	θs	α	n	Ks
C1	0–20	0.0911	0.4554	0.0204	1.2546	11.36
	21–230	0.0910	0.4577	0.0193	1.2667	10.76
C2	0–20	0.0992	0.5015	0.0185	1.3257	24.74
	21–230	0.0955	0.4738	0.0172	1.325	15.75
C3	0–20	0.0901	0.4986	0.021	1.2525	37.84
	21–230	0.0890	0.4714	0.0177	1.2536	20.10
C4	0–20	0.0722	0.4916	0.0076	1.378	30.75
	21–230	0.0706	0.4694	0.0061	1.392	18.05
C5	0–230	0.1784	0.5049	0.0522	1.324	116.41

An accurate analysis of SWC profiles and moisture temporal fluctuations, following the infiltration and drying–redistribution processes, irrigation/rainfall events, and crop evapotranspiration, allowed us to obtain useful information on the root spatial distribution, both in terms of root density and maximum root depth reached during the crop growth cycle [25]. The first step of this methodology was to identify in the transient of $\theta(z,t)$ a suitable time frame between two irrigations/rain events in which the soil water reservoir has been depleted by roots. Figure 2 shows the SWC soil profiles of the selected time frame from 5 July to 13 July. The root water uptakes (RWU_z, m^3 m^{-3}), for each TDR probe location, were determined by calculating the differences in SWC between the irrigation of 5 July, and 13 July, the day prior to the next irrigation. $RWUz$ values were scaled by total water uptake calculated on a daily basis, $\rho(z,t)$ (m^3 m^{-3} d^{-1}), and the normalized uptake volume $\rho'(z,t)$, under the hypothesis that it is a useful indicator of root density in agreement with Coelho and Or [25], was calculated as:

$$RWUz \equiv \rho'(z,t) = \frac{\rho(z,t)}{\sum_z \rho(z,t)} \tag{5}$$

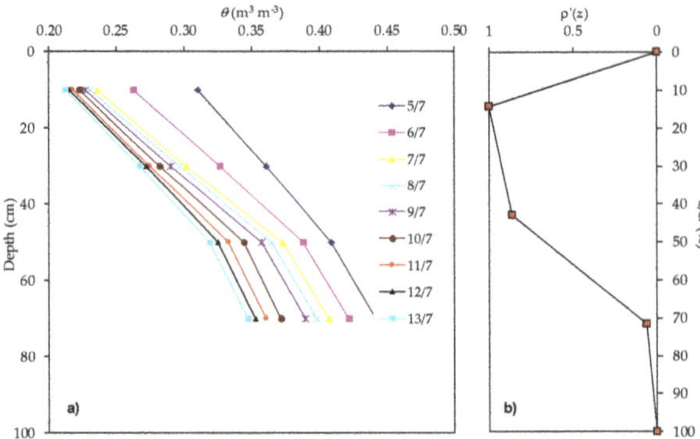

Figure 2. Measured soil water content profiles from 5 to 13 July (**a**) and root density profile (**b**) in a watermelon field.

Figure 2 shows $\rho'(z,t)$ as a function of the relative depth, z/z_r, where z_r was 70 cm. The watermelon root apparatus deepened in the first 70 cm of the soil profile, with most of it (about 70%) developed in the first 50 cm. These two parameters were used to define the root sub-model available in Hydrus-1D.

2.3. Hydraulic Conductivity Measurements

Single ring infiltration measurements of the Beerkan type [26,27] were carried out on a clay soil at the experimental farm of CREA-AA, Foggia (41°27′03″ N, 15°30′06″ E). This fine textured soil was chosen with the aim (1) to apply several models for single ring infiltration data, to estimate the saturated hydraulic conductivity of the soil, Ks, and (2) to use these Ks values as an alternative to synthetically generated data to evaluate the impact of Ks on soil hydraulic parameter optimization.

Specifically, a wide range of Ks values was obtained on undisturbed and tilled plots between December and June (five sampling dates), during the crop season of durum wheat; this allowed us to consider a relatively wide variability of the physical and hydraulic properties of the soil. For a given date or plot, 5–12 infiltration experiments were carried out, and a total of 78 cumulative infiltrations were determined. The single-ring infiltration model of Stewart and Abou Najm [28] and the SSBI method (steady version of the simplified method based on a Beerkan infiltration run) proposed by Bagarello et al. [29] were used to estimate Ks from Beerkan infiltration experiments. More specifically, we considered nine different calculation approaches corresponding to scenarios vi-xiv described in detail by Di Prima et al. [5]. Briefly, those calculation approaches differed by: (1) the way they constrained the macroscopic capillary length, (2) the use of transient or steady-state infiltration data, and (3) the fitting methods applied to transient data. The obtained Ks data set ranged more than two orders of magnitude between min and max Ks values.

2.4. Parametrization Evaluation

Statistical indicators were used to compare the performance of Hydrus-1D regarding the modelling of SWC and TDR observations at different depths. For this purpose, we applied the IRENE software, a tool designed to carry out statistical analysis for use in model evaluation [30]. Specifically, the five approaches for estimating the van Genuchten parameters were compared on the basis of 24 statistical indicators. The five approaches (C1 to C5) were ranked according to the best value reached by each statistical indicator by attributing a weight of 5 and 1 for the best and worst performance, respectively.

The ranking index for each approach was calculated by averaging the 24 relative weights. In this paper only 11 statistical indicator will be reported: absolute and relative root mean squared error [31], modeling efficiency [32], index of agreement [33], coefficient of determination and of residual mass [34], mean absolute error [35], regression coefficient with slope and intercept, and standard error. The aforementioned references are reported in the documentation included in the software package of IRENE [30].

3. Results

3.1. Field Meterological Characterization

The dynamics of meteorological variables (minimum and maximum temperature, reference evapotranspiration (ETref), and rainfall) recorded during the crop cycle season are shown in Figure 3. Daily temperatures reached the highest values in the last ten days of June (about 20 °C and 35 °C for minimum and maximum temperature, respectively), followed by a colder period which lasted until the end of soil monitoring. From the beginning, ETref increased from 4 to more than 10 mm in the last ten days of June, remaining higher than 7.5 mm for the remaining period except for the first ten days of July when it dropped to 4 mm. In the first 12 days the rainfall was 23 mm, followed by a dry period followed interrupted by a 65-mm event recorded on 4 July. After that, significant rainfall for soil water balance did not follow up to the end of the monitoring period, except for one episode of rainfall of about 35 mm on 26 July. Throughout the study period, against an atmospheric evaporative demand of almost 400 mm, 135 mm of rain were recorded with a corresponding water deficit of 265 mm, which made irrigation necessary to meet the transpiration demand of the watermelon.

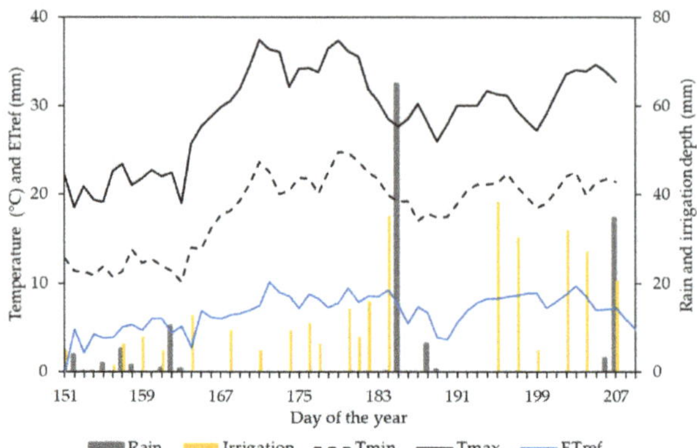

Figure 3. Daily values of minimum and maximum temperature (Tmin and Tmax), reference evapotranspiration (ETref), rain, and irrigation.

During the monitored period, i (Equation (2)), the soil cover measured through ceptometer, increased from 0.60 to 0.85. The canopy width cw ranged from 1 to 2 m, with values of c, the correction fraction of Equation (2), between 1.2 and 2.5. The potential transpiration of Equation (2), T_P, calculated over the all soil surface, was 302 mm, while the adjusted potential transpiration, T'_p, calculated for defining the time variable top boundary condition of monodimensional soil domain, was 631 mm.

Figure 3 shows the 21 irrigation events carried out by the farmer for a total irrigation depth of about 300 mm, with an average height of 14 mm and an irregular temporal distribution. In fact, in the first half of the observed cycle (32 days), compared with the next one, the irrigations were more frequent (one every 2.3 days) and with an average height of 8 mm. Afterward (26 days starting from 1 July), the farmer carried out less frequent irrigations (one every 3.6 days) but with a higher height (on average 27 mm).

With the TDR100 it was possible to characterize the time evolution of soil water content (Figure 4). As a result of infiltration processes, due to rain or irrigation and root uptake, under the dripper lines there were large temporal changes that, in deeper layers, were more damped and with higher values than the shallower ones. In the inter-row space, SWC remained low and almost constant until the rain event of 65 mm of 4 July; after that SWC increased and then gradually decreased. Because of this rainfall, the farmer did not irrigate for 10 days, causing the soil water reserve to be emptied for the entire monitored profile, as detected by TDR probes installed horizontally under the plant rows at 10- and 70-cm depths. These observations strongly support the hypothesis of one-dimensional fluxes along the soil profile below the plant rows. In fact, the water fluxes of the soil–plant–atmosphere's continuum occurred mainly in a confined soil volume, well explored by the roots and placed below the plastic mulch.

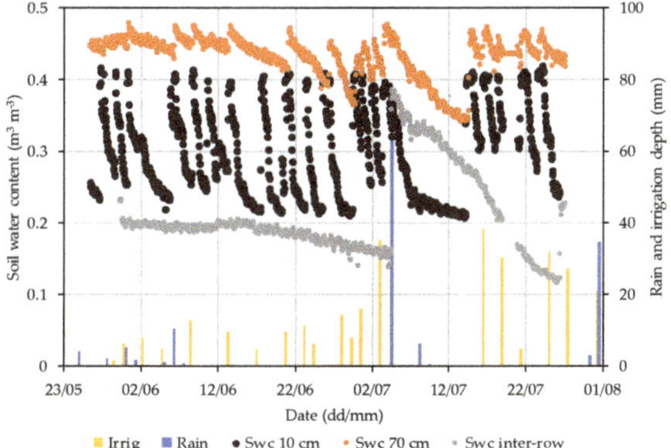

Figure 4. Hourly values of soil water content (SWC) by time domain reflectometry (TDR) probes horizontally and vertically installed at depths of 10 and 70 cm and in the middle of inter-row space. Daily rain and irrigation are also reported (from [36]).

3.2. Calibration of Hydrus-1D

Comparisons between SWC measured and simulated by Hydrus-1D for the five different approaches are shown in Figure 5. Simulated data show a significant gap with those measured by TDR, affecting the entire profile and especially the deepest layers. Simulated SWC of such a layer shows an almost constant trend except for the period when the irrigation was suspended, causing a reduction of SWC due to the root water uptake. This constant SWC trend in the deepest layer led us suppose the existence of an impervious layer, characterized by a low Ks value that prevented the flux of deep percolation. Therefore, a 6-cm layer characterized by a Ks equal to 0.5 cm d^{-1}, keeping the other parameters unchanged, was inserted into the soil profile at a depth of 83 cm as shown in Figure 6.

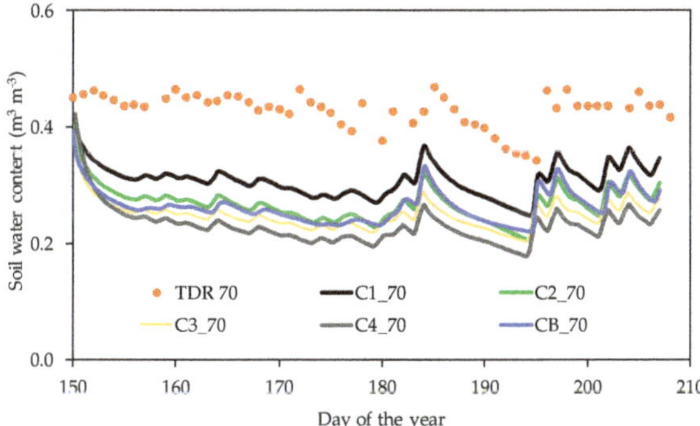

Figure 5. Comparison between measured and simulated soil water content at a 70-cm depth resulting from the five approaches.

The Table 3 reports 11 out of 24 the utilized statistical indicators and the ranking index calculated to compare the performances of the five approaches to simulate the measured SWC. C5, obtained from accurate lab measurements, provided the best values for all the indicators that were closest to the optimal values shown in the last row of the Table 3.

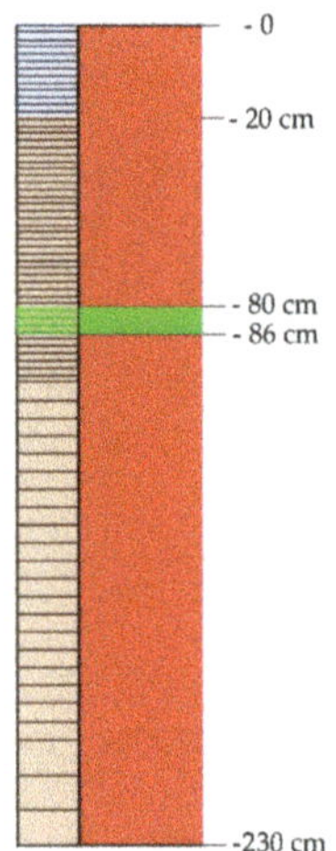

Figure 6. Soil profile discretization adopted in HYDRUS-1D. The blue, green, and brown colors indicate the first, impervious, and second layers, respectively.

Table 3. Statistical indicators comparing the performance of the five approaches in order to describe the measured soil water content.

Approach	RMSE *	rRMSE	ME	IA	CD	RM	AE	R^2	a	b	SE	RI
C1	0.064	0.169	−1.74	0.622	3.28	0.157	0.060	0.81	0.105	0.858	0.022	3.95
C2	0.079	0.209	−3.19	0.548	4.76	0.199	0.075	0.81	0.128	0.825	0.023	2.18
C3	0.073	0.194	−2.60	0.569	4.11	0.184	0.070	0.82	0.102	0.895	0.022	3.41
C4	0.093	0.247	−4.85	0.493	6.44	0.237	0.090	0.78	0.156	0.770	0.024	1.00
C5	0.035	0.093	0.16	0.814	1.43	0.067	0.028	0.80	0.097	0.798	0.023	4.45
Perfect fit	0	0	1	1	1	0	0	1	0	1	0	5

*: RMSE: root mean square error; rRMSE: root mean square error; ME: modelling efficiency; IA: index of agreement; CD: coefficient of determination; RM: coefficient of residual mass; AE: mean absolute error; a: regression intercept; b: regression slope; RI: ranking index.

Basically, for C5, the rRMSE value (less than 10%) was very favourable. Modelling efficiency, although significantly far from the optimal value of 1, was positive, unlike the other four approaches showing negative values (up to −4.85 of C4). The differences were high for index of agreement, coefficient of determination, residual mass, and mean absolute error, with values for C5 being 1.4, 3.2, 2.9 and 2.6 times better than the average values of the remaining approaches. However, the other approaches indicated a better regression coefficient and slope compared to C5.

The standard error was the indicator with the smallest differences between the 5 approaches, with values very close to 0.02, thus indicating a temporal variability that was independent of the set of adopted parameters. The ranking index (RI) equal to 4.45 indicated C5 as the approach that best described the SWC dynamic, followed by C1 and C3 (RI greater than 3), C2 (greater than 2) and finally C4, which had the worst performance (RI = 1). rRMSE, ME, and AE are aligned with the values obtained by Autovino et al. (2018) under olive cultivation simulated with two dimensional Hydrus-2D (in average for two years: 0.06, 0.4, and 0.02).

Figure 7 shows the time evolution of SWC as measured through TDR and simulated with the five parameter sets, at the four soil depths. The differences between simulated and measured values were quite narrowed in the shallowest layer and without relevant differences between the five approaches. Instead, such differences progressively increased with the probe installation depth. In the deepest layer, C5 was the only method to describe the SWC dynamic fairly well, unlike the other three approaches whose simulations were poor.

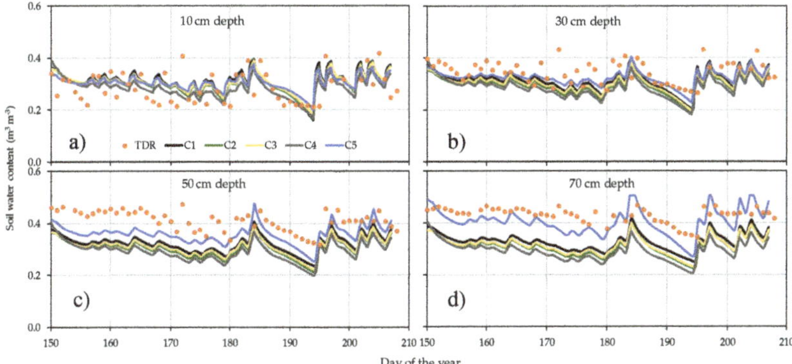

Figure 7. Comparison between measured (TDR) and simulated soil water content with the five approaches at five depths (from **a** to **d**) after inserting the impervious layer as specified in the text.

Figure 8 shows the comparison between the C5 approach and the TpNoAdj exercise in terms of their capacity to describe the transient soil water content of the cropping system in study. TpNoAdj does not take in account the particularity of our cropping system, which is the difference between soil surface covered by transpiring canopy, $cw(t)$, and surface of wetted soil (0.8 m), with $cw(t) >0.8$ m. Therefore, the time variable conditions of the top boundary of TpNoAdj were T_p (instead of T'_p) and irrigation depth considering A (Equation (1)) equal to 130×2.5 m (instead of 130×0.8 m). Figure 8 shows an evident larger disagreement between soil water content measured by TDR and simulated by TpNoAdj compared to C5, above all at depths of 50 and 70 cm. Moreover, TpNoAdj tends to dampen the time variability of SWC more than C5 and erroneously compared to TDR measurements. This result is expected considering that the no-adjusted potential transpiration and irrigation depth, imposed as top boundary condition for a one-dimensional model, failed to match the large SWC changes measured by TDR in the wetted soil volume. Therefore, under the experimental conditions characterized (clay soil cultivated with melon in an intensive cropping system), the direct measurement of soil parameters was useful in order to improve the simulation capabilities of Hydrus-1D in describing the water fluxes in the "soil–plant–atmosphere" continuum. These findings have obvious implications for simulating the water fluxes in semiarid Mediterranean cropping systems, because a great debate is still current on the best choice between standard and accurate (lab) or rapid and simplified (PTFs) methods for soil hydraulic characterization. Therefore, a more in-depth discussion on this main topic of soil hydrology is reported in Section 4, "Sensitivity analysis".

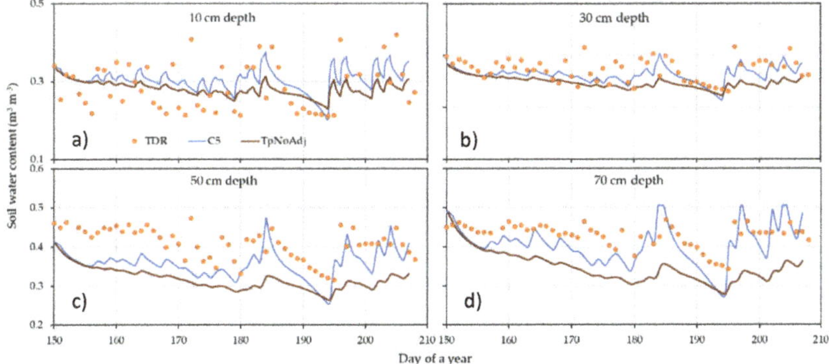

Figure 8. Comparison between measured (TDR) and simulated soil water content at five depths (from **a** to **d**) under the C5 approach and with the "TpNoAdj" exercise described in the text.

After evaluating the different capacities to simulate the soil water fluxes of the different parametrization approaches in study, we carried out a sensitivity analysis with the aim of identifying the most sensitive hydraulic parameters to describe the soil water balance. To do this, we perturbed each of the five parameters by increasing the corresponding values used in C5 by + 10% and running Hydrus-1D again. Figure 9 shows the SWC and soil pressure head (*h*) trends at two depths (10 and 70 cm). The perturbation of the hydraulic parameters had a significant impact in the calculation of the SWC both in the superficial and in the deep layer. θs and *n* were the most sensitive parameters and, for the shallowest layer, their +10% perturbation resulted in SWC variations that reached up to 45 and –30%, respectively. The impact on *h* occurred above all in the period with suspended irrigation (the first 10 days of July) which caused a strong reduction of SWC, especially in the 0–20 cm layer. The perturbation of θr and *Ks* had no effects; that of θs and *n* resulted in a 30% increase; and that of α in a reduction of 12%.

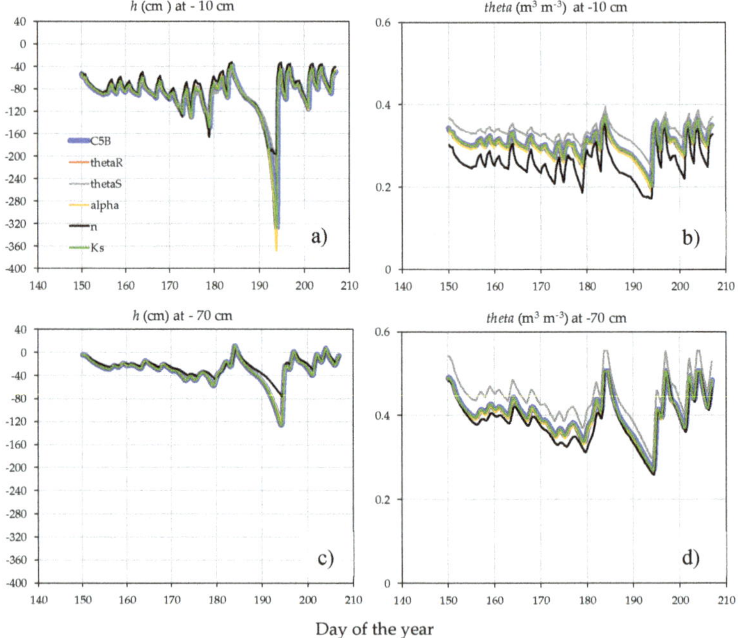

Figure 9. Temporal trend of soil pressure head (**a,c**) and soil water content (**b,d**) obtained by perturbating the van Genuchten parameters under the C5 approach.

Figure 10 shows the sensitivity of water fluxes of infiltration, percolation, and plant uptake to hydraulic parameters. The temporal evolution of soil pressure head explored by root apparatus (*hroot*) is also shown. These trends confirmed what was highlighted in the previous figure, with a greater weight of two parameters of water retention function, i.e., θs and *n*. The variables most affected by their variations were cumulative infiltration and *hroot*. The perturbation of θs and *n* caused changes in deep percolation of +14 and −11%, respectively. Also, for the volume explored by the roots, by increasing θs and *n*, the greatest variations of *h* were found in the period of July with a damping of the fall of 12 and 35%, respectively.

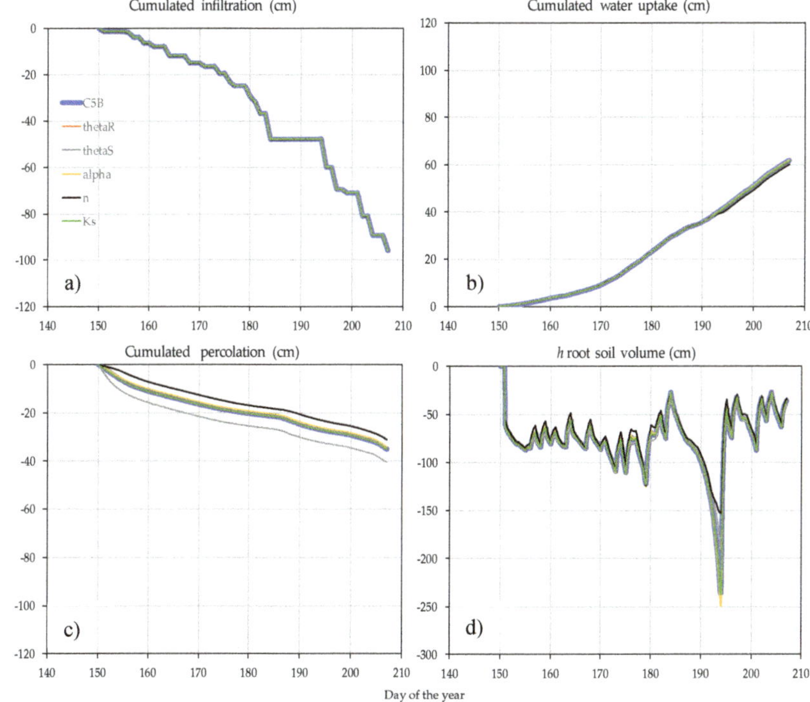

Figure 10. Temporal trend of soil infiltration (**a**), water uptake (**b**), percolation (**c**), and pressure head of soil root volume (**d**), obtained perturbating the van Genuchten parameters under the C5 approach.

As for the root water uptake, negligible variations were recorded and the most sensitive parameter was n, with a slight variation of 2.5%. Despite this small variation, the importance of two of the four parameters of the water retention function emerged, namely θs, the water content at saturation, and n, a fitting parameter that regulates the shape of the function itself. From an agronomic point of view, θs and n influence the irrigation scheduling variables, such as the irrigation height and the time interval between an irrigation and the next one.

Figure 10 also shows that, in the considered flow domain and in the two months of monitoring, against an actual infiltration of about 960 mm, there was a plant uptake of 600 mm and a percolation of 350 mm, highlighting an excess of irrigation carried out by the farmer. Finally, the 600 mm of uptake took place in a confined volume delimited by a smaller soil surface than the vegetable canopy area, as shown in Figure 1. Therefore, the uptake of 600 mm was greater than the evaporative demand of the environment of 400 mm (which refers to an entire evaporating surface) and the water absorbed by the roots was distributed over the area of vegetable canopy, corresponding to the evaporative demand and that was larger than the area of uptake.

3.3. Saturated Hydraulic Conductivity Measurement Impact

Compared to the other hydraulic parameter, Ks is characterized by a very large variability [37], it changes in space and in time and, for this reason, it is a key parameter for the hydrological models based on Richards equation, as reported in the large bibliography of Šimůnek et al. [7]. However, in agricultural cropping systems, due to the soil tillage carried out for seed bed preparation, this variability can be significantly reduced compared to non-agricultural systems. Indeed, Autovino et al. [2] showed Ks values for their olive orchard system between 20 and 70 cm d^{-1}. To take in account this peculiarity, we have expanded the range of variability in the Ks sensitivity analysis, utilizing the field measurement results for a soil with similar soil textural composition. The imposed variability for *Ks*, from 300 to almost 4000 cm d^{-1}, had no impact on temporal evolution of SWC and *h* at any depths (data not shown for −30 and −50 cm), as shown in Figure 11.

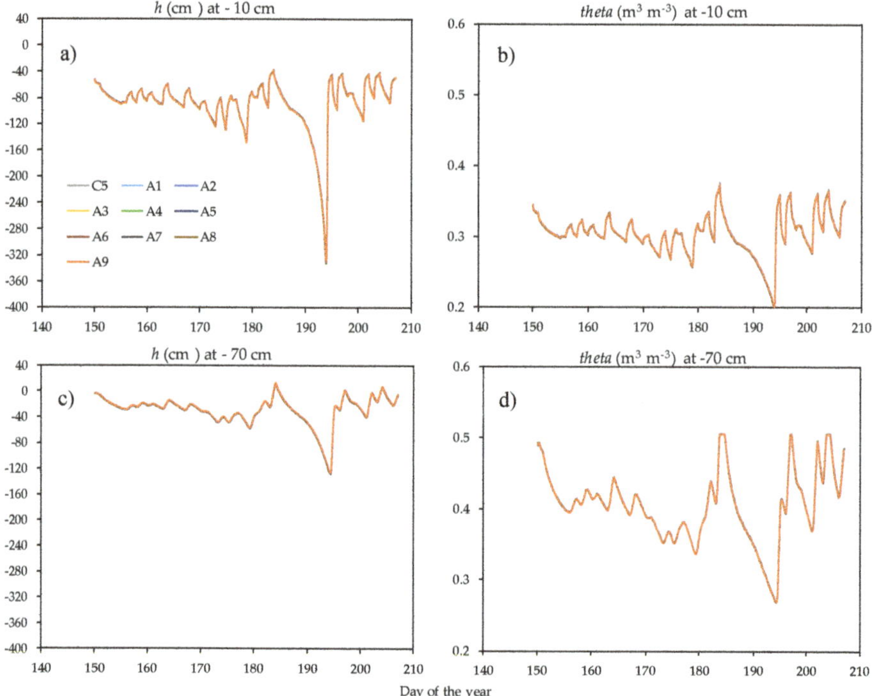

Figure 11. Temporal trend of pressure head (**a**,**c**) and soil water content (**b**,**d**) under the nine *Ks* scenarios.

With regard to the soil water fluxes, the variations were undetectable for the cumulative variables. For this reason, we considered the daily fluxes instead of the cumulative ones (Figure 12). The variations due to the *Ks* perturbation occurred following large rain or irrigation events and involved both infiltration and deep percolation, at the top and bottom boundary layer, respectively. In these same periods, also the plant uptake was irregularly and slightly affected. On the contrary, for *hroot* no impact was detected.

Therefore, with respect to the other parameters of the Mualem-van Genuchten model, *Ks* showed less sensitivity in influencing the state variables, *h* and SWC, and the main cumulative and daily fluxes, such as water infiltration, percolation, and uptake. This result should be put in relation to the fact that, in the climatic and agronomic context that we have considered, both the rains and the irrigations have not been so heavy as to create very high gradients of *h* along the soil profile.

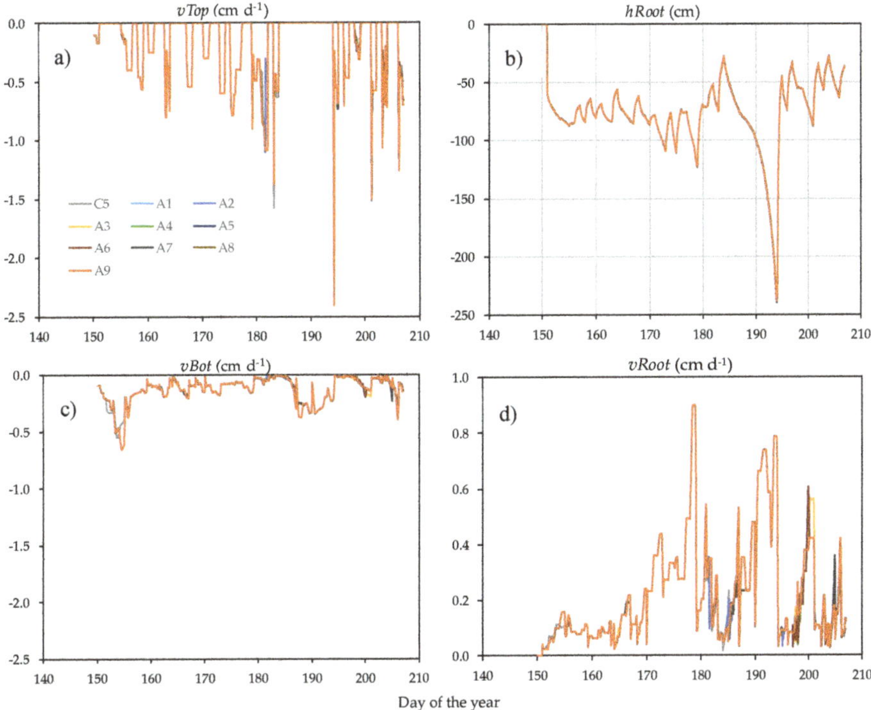

Figure 12. Dynamic of daily rates of soil infiltration (**a**), percolation (**c**), water uptake (**d**), and pressure head of soil root volume (**b**) under the nine *Ks* scenarios.

4. Discussion

Horticultural crops are widespread high-income crops in Mediterranean regions but need accurate and rational management for saving water resources [38]. In this investigation, we applied an integrated experimental approach in order to optimize the water use efficiency of a typical horticultural cropping system of Southern Italy, based on drip-irrigated watermelon cultivation. For this purpose, (1) accurate and simplified methods were selected and compared to measure and estimate, respectively, the physical and hydraulic properties of the soil, (2) the model by Coelho and Or [25] was applied to estimate root length density distribution starting from TDR measurements, and (3) the Hydrus-1D software was used to model the system under study, and pros and cons of the applied methods were critically discussed below in light of the available options.

Soil hydraulic properties may be responsible for soil water content field variability, which in turn affects the spatial distribution of root density [39]. Overall, in the literature it is agreed that irrigation scheduling based on soil properties knowledge allows management of the irrigation in a more efficient way [40]. For this, a great debate is still current on the best choice between accurate-lab or simplified methods for soil hydraulic characterization, and the pros and cons have been clearly summarized in the literature [41]. In fact, experimental efforts are often linked with the financial availability and the expected accuracy. Silva Ursulino et al. [5], for example, using TDR probes and Hydrus-1D for modelling one-dimensional flow in two plots in the Gameleira Experimental River Basin, Northeast Brazil, showed that simplified estimates of soil properties by the BEST method [27] may be considered adequate to estimate the hydraulic functions, even with the final aim of modelling the water flow processes and water budget for a soil profile. However, although BEST can be considered quicker and simpler than standard methods, it can be considered more accurate if compared with common

PTFs, so it could be suggested as a compromise between the two approaches. In this investigation, accurate measurements of soil hydraulic properties were obtained by the Wind's evaporation method, coupled with infiltrations of the Uhg type. Overall, Wind's approach is quite widespread in different experimental conditions, e.g., [42], because it provides multiple and simultaneous measurements of volumetric soil water content and soil pressure head. Moreover, it provides measurements of soil hydraulic properties from a depletion experiment involving an upward water flux that is more representative of natural hydrological processes [13]. Therefore, it is particularly suitable for agronomic applications. On the contrary, relatively worse results were obtained when PTFs of literature were used, confirming that they are useful tools, but specifically suitable for territorial-scale studies. Consequently, if the hydraulic characterization requires non-demanding soil sampling (i.e., few samples to analyze), the former approach should be suggested.

Another aspect that requires a choice a priori is linked to the approach to measure directly or to estimate quickly, and with little experimental burden, the root length density distribution. Overall, the option of digging a trench to establish the actual root development is not always feasible, especially in private farms. Application of the Coelho and Or [25] model and of the TDR probes system at multiple depths allowed us to estimate a deepening of the watermelon roots up to 70 cm of the soil profile, with about 70% of them developed in the first 50 cm. Therefore, this simplified solution seems to be applicable for estimating the watermelon root growth under drip irrigation and plastic mulch.

Finally, findings of this investigation show that the physically based Hydrus-1D model allows us to adequately simulate the water content within the framework of a typical drip-irrigated horticultural system. As a consequence, this free-software can be a viable alternative to more complex and paid license models. Moreover, since Hydrus is a very flexible type of software and allows an accurate spatial discretization of the soil, depending on the real field situations, it was possible to take into account a probable low permeability layer, which was set at an 80-cm depth according to the TDR measurements. This compact layer, which is probably linked to the pedological characteristics of the investigated soil profile, is plausible because the measured bulk density increased by about 10% from upper layers (0–20 cm) to deeper ones (21–40 cm). Consequently, considering this impervious layer was decisive for improving the model calibration performance. On the other hand, the presence of this low-permeability layer may have affected the soil pressure head distribution and, consequently, the relatively small hydraulic gradients reached. This hypothesis and the relatively low irrigation/rainfall events were addressed as the main factors for the observed low sensitivity of Ks to the main soil variables, namely soil pressure head or soil water content, and the main cumulative and daily fluxes, such as water infiltration, percolation, and water uptake. This unexpected finding deserves to be further investigated in the future with ad hoc studies.

5. Conclusions

The results of this study indicate that the physically based Hydrus-1D model allows for simulating fairly well the soil water contents measured at different distances under a typical cultivation scheme of a drip irrigated horticultural system, predicting the measured values with an average root mean square error lower than 9%. As expected: (1) the water flow domain was found to be characterized by two-dimensional pressure gradients, however (2) the fluxes of the "soil–plant–atmosphere" continuum were confined in a restricted soil volume, and (3) therefore it can be considered of one-dimensional type and can be described by a one-dimensional model such as Hydrus-1D.

Among the hydraulic parameters of the hydraulic retention and conductivity functions, it is convenient to directly determine the water content at saturation, θs, and the fitting parameters, α and n, with the laboratory or field methods reported in literature, while for the residual water content, θr, and saturated hydraulic conductivity, Ks, the pedotransfer functions can be suitable.

However, because the results obtained in this study refer to a specific cultivation system, for other physical and agronomical conditions (e.g., other textural and structural characteristics, irrigation methods with high irrigated flow rates), a direct determination of *Ks* with greater precision using the available measurement methods could be required for a better simulation of the soil water fluxes.

Author Contributions: D.V. conceptualized the research, carried out the simulation analysis, and wrote the paper. M.C. contributed to the methodology and carried out the retention function and hydraulic conductivity measurements. All authors contributed to critically discussing the results and reviewing the manuscript.

Funding: The work was supported and funded by the projects "STRATEGA, Sperimentazione e TRAsferimento di TEcniche innovative di aGricoltura conservativA" (Regione Puglia–Dipartimento Agricoltura, Sviluppo Rurale ed Ambientale, CUP: B36J14001230007) and "AQUATER", Decision Support Systems to Manage Water Resources for Irrigation at Districts Scale in Southern Italy Using Remote Sensing Information (Ministero delle politiche Agricole alimentari, forestali e del turismo (contract nr. 209/7303/05).

Conflicts of Interest: The authors declare no conflict of interest.

References

1. Ventrella, D.; Charfeddine, M.; Giglio, L.; Castellini, M. Application of DSSAT models for an agronomic adaptation strategy under climate change in Southern of Italy: Optimum sowing and transplanting time for winter durum wheat and tomato. *Ital. J. Agron.* **2012**, *7*, e16. [CrossRef]
2. Autovino, D.; Rallo, G.; Provenzano, G. Predicting soil and plant water status dynamic in olive orchards under different irrigation systems with Hydrus-2D: Model performance and scenario analysis. *Agr. Water Manag.* **2018**, *203*, 225–235. [CrossRef]
3. Kader, M.A.; Nakamura, K.; Senge, M.; Mojid, M.A.; Kawashima, S. Numerical simulation of water- and heat-flow regimes of mulched soil in rain-fed soybean field in central Japan. *Soil Till. Res.* **2019**, *191*, 142–155. [CrossRef]
4. Pinheiro, E.A.R.; de Jong van Lier, Q.; Inforsato, L.; Šimůnek, J. Measuring full-range soil hydraulic properties for the prediction of crop water availability using gamma-ray attenuation and inverse modeling. *Agr. Water Manag.* **2019**, *216*, 294–305. [CrossRef]
5. Silva Ursulino, B.; Maria Gico Lima Montenegro, S.; Paiva Coutinho, A.; Hugo Rabelo Coelho, V.; Cezar dos Santos Araújo, D.; Cláudia Villar Gusmão, A.; Martins dos Santos Neto, S.; Lassabatere, L.; Angulo-Jaramillo, R. Modelling soil water dynamics from soil hydraulic parameters estimated by an alternative method in a tropical experimental basin. *Water* **2019**, *11*, 1007. [CrossRef]
6. Di Prima, S.; Castellini, M.; Abou Najm, M.R.; Stewart, R.D.; Angulo-Jaramillo, R.; Winiarski, T.; Lassabatere, L. Experimental assessment of a new comprehensive model for single ring infiltration data. *J. Hydrol.* **2019**, *573*, 937–951. [CrossRef]
7. Šimůnek, J.; van Genuchten, M.T.; Šejna, M. Recent developments and applications of the HYDRUS computer software packages. *Vadose Zone J.* **2016**, *15*. [CrossRef]
8. Ventrella, D.; Moahanty, B.P.; Šimůnek, J.; Losavio, N.; van Genuchten, M.T. Use of HYDRUS-1D for simulating water and chloride transport in a bare clayey soil in presence of shallow groundwater. *Soil Sci.* **2000**, *165*, 624–631. [CrossRef]
9. Han, M.; Zhao, C.; Feng, G.; Yan, Y.; Sheng, Y. Evaluating the effects of mulch and irrigation amount on soil water distribution and root zone water balance using HYDRUS-2D. *Water* **2015**, *7*, 2622–2640. [CrossRef]
10. Yang, Q.; Zuo, H.; Xiao, X.; Wang, S.; Chen, B.; Chen, J. Modelling the effects of plastic mulch on water, heat and CO2 fluxes over cropland in an arid region. *J. Hydrol.* **2012**, *452–453*, 102–118. [CrossRef]
11. Zhang, H.; Huang, G.; Xu, X.; Xiong, Y.; Huang, Q. Estimating evapotranspiration of processing tomato under plastic mulch using the SIMDualKc model. *Water* **2018**, *10*, 1088. [CrossRef]
12. Ghazouani, H.; Autovino, D.; Rallo, G.; Rallo, G.; Provenzano, G. Using HYDRUS-2D model to assess the optimal drip lateral depth for Eggplant crop in a sandy loam soil of central Tunisia. *Ital. J. Agrometeorol.* **2016**, *1079*, 47–58.
13. Castellini, M.; Iovino, M. Pedotransfer functions for estimating soil water retention curve of Sicilian soils. *Arch. Agron. Soil Sci.* **2019**, *65*, 1401–1416. [CrossRef]
14. Schaap, M.G.; Leij, F.J.; van Genuchten, M.T. Rosetta: A computer program for estimating soil hydraulic parameters with hierarchical pedotransfer functions. *J. Hydrol.* **2001**, *251*, 163–176. [CrossRef]
15. Van Genuchten, M.T. A closed-form equation for predicting the hydraulic conductivity of unsaturated soils. *Soil Sci. Soc. Am. J.* **1980**, *44*, 892–898. [CrossRef]

16. Topp, G.C.; Davis, J.L.; Annan, A.P. Electromagnetic determination of soil water content: Measurements in coaxial transmission lines. *Water Resour. Res.* **1980**, *16*, 574–582. [CrossRef]

17. Allen, R.G.; Pereira, L.S.; Raes, D.; Smith, M. Crop evapotranspiration—Guidelines for computing crop water requirements. In *FAO Irrigation and Drainage*; Paper 56; Food and Agriculture Organization: Rome, Italy, 1998; p. 15.

18. Šimůnek, J.; Šejna, M.; Saito, H.; Sakai, M.; Van Genuchten, M.T. *The HYDRUS-1D Software Package for Simulating the One-Dimensional Movement of Water, Heat, and Multiple Solutes in Variably-Saturated Media*; Version 4.17; Department of Environmental Sciences University of California Riverside: Riverside, CA, USA, 2013.

19. Feddes, R.A.; Kowalik, P.J.; Zaradny, H. Simulation of field water use and crop yield. In *Simulation of Field Water Use and Crop Yield. Simul. Monogr.*; Pudoc: Wageningen, The Netherlands, 1978; p. 188.

20. Wind, G.P. Capillary conductivity data estimated by a simple method. In *Water in the Unsaturated Zone Proc Wageningen Symp*; Institute for Land and Water Management Research: Amsterdam, The Netherlands, 1969.

21. Klute, A.; Dirksen, C. Hydraulic conductivity and diffusivity: Laboratory methods. In *Methods of Soil Analysis: Part 1—Physical and Mineralogical Methods*; American Society of Agronomy-Soil Science Society of America: South Segoe Road, MA, USA, 1986; pp. 687–734.

22. Bagarello, V.; Castellini, M.; Iovino, M. Comparison of unconfined and confined unsaturated hydraulic conductivity. *Geoderma* **2007**, *137*, 394–400. [CrossRef]

23. Halbertsma, J.M.; Veerman, G.J. A new calculation procedure and simple set-up for the evaporation method to determine soil hydraulic functions. In *Report. 88*; Wageningen: Amsterdam, The Netherlands, 1994; p. 21.

24. Castellini, M.; Di Prima, S.; Iovino, M. An assessment of the BEST procedure to estimate the soil water retention curve: A comparison with the evaporation method. *Geoderma* **2018**, *320*, 82–94. [CrossRef]

25. Coelho, E.F.; Or, D. Root distribution and water uptake patterns of corn under surface and subsurface drip irrigation. *Plant Soil* **1999**, *206*, 123–136. [CrossRef]

26. Braud, I.; De Condappa, D.; Soria, J.M.; Haverkamp, R.; Angulo-Jaramillo, R.; Galle, S.; Vauclin, M. Use of scaled forms of the infiltration equation for the estimation of unsaturated soil hydraulic properties (the Beerkan method). *Eur. J. Soil Sci.* **2005**, *56*, 361–374. [CrossRef]

27. Lassabatere, L.; Angulo-Jaramillo, R.; Soria Ugalde, J.M.; Cuenca, R.; Braud, I.; Haverkamp, R. Beerkan estimation of soil transfer parameters through infiltration experiments—BEST. *Soil Sci. Soc. Am. J.* **2006**, *70*, 521. [CrossRef]

28. Stewart, R.D.; Abou Najm, M.R. A Comprehensive Model for Single Ring Infiltration I: Initial Water Content and Soil Hydraulic Properties. *Soil Sci. Soc. Am. J.* **2018**, *82*, 548–557. [CrossRef]

29. Bagarello, V.; Di Prima, S.; Iovino, M. Estimating saturated soil hydraulic conductivity by the near steady-state phase of a Beerkan infiltration test. *Geoderma* **2017**, *303*, 70–77. [CrossRef]

30. Fila, G.; Bellocchi, G.; Acutis, M.; Donatelli, M. Irene: A software to evaluate model performance. *Eur. J. Agron.* **2003**, *18*, 369–372. [CrossRef]

31. Kobayashi, K.; Salam, M.U. Comparing simulated and measured values using mean squared deviation and its components. *Agron J.* **2000**, *92*, 345–352. [CrossRef]

32. Greenwood, D.J.; Neeteson, J.J.; Draycott, A. Response of potatoes to N fertilizer: Dynamic model. *Plant Soil* **1985**, *85*, 185–203. [CrossRef]

33. Willmott, C.J.; Wicks, D.E. An Empirical method for the spatial interpolation of monthly precipitation within California. *Phys. Geogr.* **1980**, *1*, 59–73. [CrossRef]

34. Loague, K.; Green, R.E. Statistical and graphical methods for evaluating solute transport models: Overview and application. *J. Contam. Hydrol.* **1991**, *7*, 51–73. [CrossRef]

35. Shaeffer, D.L. A model evaluation methodology applicable to environmental assessment models. *Ecol. Model.* **1980**, *8*, 275–295. [CrossRef]

36. Ventrella, D.; Castellini, M.; Di Giacomo, E.; Giglio, L. Valutazione di diversi metodi di misura per il monitoraggio del contenuto idrico del suolo. In Proceedings of the XXXVII Conference of Italian Society of Agronomy, Il Contributo della Ricerca Agronomica all'Innovazione dei Sistemi Colturali Mediterranei, Catania, Italy, 13–14 September 2007; Cosentino, S.D., Tuttobene, R., Eds.; pp. 249–250.

37. Bagarello, V.; Baiamonte, G.; Castellini, M.; Di Prima, S.; Iovino, M. A comparison between the single ring pressure infiltrometer and simplified falling head techniques. *Hydrol. Process.* **2014**, *28*, 4843–4853. [CrossRef]

38. Fiorentino, C.; Ventrella, D.; Giglio, L.; Giacomo, E.D.; Lopez, R. Land use cover mapping of water melon and cereals in southern Italy. *Ital. J. Agron.* **2010**, *5*, 185–192. [CrossRef]

39. Rallo, G.; Provenzano, G.; Castellini, M.; Sirera, À.P. Application of EMI and FDR Sensors to assess the fraction of transpirable soil water over an olive grove. *Water* **2018**, *10*, 168. [CrossRef]

40. Liang, X.; Liakos, V.; Wendroth, O.; Vellidis, G. Scheduling irrigation using an approach based on the van Genuchten model. *Agr. Water Manag.* **2016**, *176*, 170–179. [CrossRef]

41. Vereecken, H.; Huisman, J.A.; Bogena, H.; Vanderborght, J.; Vrugt, J.A.; Hopmans, J.W. On the value of soil moisture measurements in vadose zone hydrology: A review. *Water Resour. Res.* **2008**, *44*, W00D06. [CrossRef]

42. Pirastru, M.; Niedda, M.; Castellini, M. Effects of maquis clearing on the properties of the soil and on the near-surface hydrological processes in a semi-arid Mediterranean environment. *J. Agric. Eng.* **2014**, *45*, 176. [CrossRef]

Article

Spatial Variability of Soil Physical and Hydraulic Properties in a Durum Wheat Field: An Assessment by the BEST-Procedure

Mirko Castellini [1,*], Anna Maria Stellacci [2], Matteo Tomaiuolo [3] and Emanuele Barca [4]

[1] Council for Agricultural Research and Economics-Research Center for Agriculture and Environment (CREA-AA), Via C. Ulpiani 5, 70125 Bari, Italy
[2] Department of Soil, Plant and Food Sciences, University of Bari "Aldo Moro", Via G. Amendola 165/a, 70126 Bari, Italy
[3] Council for Agricultural Research and Economics-Policies and Bioeconomy Research Centre (CREA-PB), Via C. Ulpiani 5, 70125 Bari, Italy
[4] Water Research Institute (IRSA)—National Research Council (CNR), Viale Francesco de Blasio 5, 70132 Bari, Italy
* Correspondence: mirko.castellini@crea.gov.it; Tel.: +39-080-5475039

Received: 10 June 2019; Accepted: 8 July 2019; Published: 12 July 2019

Abstract: Spatial variability of soil properties at the field scale can determine the extent of agricultural yields and specific research in this area is needed. The general objective of this study was to investigate the relationships between soil physical and hydraulic properties and wheat yield at the field scale and test the BEST-procedure for the spatialization of soil hydraulic properties. A simplified version of the BEST-procedure, to estimate some capacitive indicators from the soil water retention curve (air capacity, ACe, relative field capacity, $RFCe$, plant available water capacity, $PAWCe$), was applied and coupled to estimates of structure stability index (SSI), determinations of soil texture and measurements of bulk density (BD), soil organic carbon (TOC) and saturated hydraulic conductivity (Ks). Variables under study were spatialized to investigate correlations with observed medium-high levels of wheat yields. Soil physical quality assessment and correlations analysis highlighted some inconsistencies (i.e., a negative correlation between $PAWCe$ and crop yield), and only five variables (i.e., *clay* + *silt* fraction, BD, TOC, SSI and $PAWCe$) were spatially structured. Therefore, for the soil–crop system studied, application of the simplified BEST-procedure did not return completely reliable results. Results highlighted that (i) BD was the only variable selected by stepwise analysis as a function of crop yield, (ii) BD showed a spatial distribution in agreement with that detected for crop yield, and (iii) the cross-correlation analysis showed a significant positive relationship between BD and wheat yield up to a distance of approximately 25 m. Such results have implications for Mediterranean agro-environments management. In any case, the reliability of simplified measurement methods for estimating soil hydraulic properties needs to be further verified by adopting denser measurements grids in order to better capture the soil spatial variability. In addition, the temporal stability of observed spatial relationships, i.e., between BD or soil texture and crop yields, needs to be investigated along a larger time interval in order to properly use this information for improving agronomic management.

Keywords: spatial cross-correlation; saturated hydraulic conductivity; BEST-steady; durum wheat

---

1. Introduction

Soil physical and hydraulic properties can change drastically over space [1] and time [2] and their evaluation is essential for a rational soil management and, therefore, for increasing crop yields performance [3]. Moreover, they dynamically affect water balance components and crop yields by relating soil hydraulic functioning to climate patterns and crop water requirements [4,5].

Soil properties, such as saturated hydraulic conductivity, total organic carbon content, structure stability index, dry bulk density, as well as capacitive indicators obtained from the soil water retention curve (i.e., plant available water capacity, relative field capacity, air capacity) were widely and successfully applied to investigate soil management effects on soil physical and hydraulic properties [3,6–8]. Keller et al. [3], for example, have investigated the relationships between crop yield and soil structure in three Swedish fields, applying the simplified falling head (SFH) technique [9] to evaluate whether field-saturated hydraulic conductivity (Ks) could be used as a simple and quickly measurable indicator of crop yield. The main findings of their investigation showed that Ks may be a good indicator of low yielding zones, and degradation of soil structure has been indicated as the main reason for low yield. However, in other studies, spatial and temporal variation of soil water holding capacity was suggested to be a factor partly responsible for crop yield variation, regardless of the amount and timing of the rain contributions [10,11]. Consequently, investigations addressed at evaluating new experimental procedures for soil hydraulic characterization, and at establishing the usefulness and sensitivity of soil properties as predictors for crop yield, are needed. This is a current issue in Mediterranean agro-environments, where water resources are scarce and need to be optimally managed [12].

Soil hydraulic properties assessment, i.e., water retention curve and hydraulic conductivity function ($\theta(h)$ and $K(h)$ relationships, respectively), is expensive and time consuming, since standard methods need specific skills and their spatialization, i.e., prediction on a distributed large scale, which can be burdensome or wholly inapplicable [13].

Several quick methods are available to obtain $\theta(h)$ and $K(h)$ (or $K(\theta)$) relationships. Pedotransfer functions (PTFs), for example, allow for estimating soil hydraulic properties starting with basic soil variables such as soil texture, bulk density or organic matter or hydraulic conductivity [14]. However, PTFs are not able to accurately quantify the effects of different agronomic options for soil management, unless a site-specific calibration is performed [15]. On the other hand, the most widely applied laboratory methods as hanging water column apparatus [16], pressure cells [17] or evaporation method [18], may require up to a week (or more) to obtain a single soil water retention curve [19,20]. As a consequence, although reliable and accurate, they are not easily applicable for large-scale research and simplified techniques should be chosen for these purposes [21,22].

The BEST-procedure by Lassabatère et al. [23] allows for estimating hydraulic functions repeatedly over space and time with substantially limited experimental burdens. Basically, the procedure requires three sets of experimental information: (i) cumulative infiltration by a simple infiltrometric experiment (i.e., ring infiltration test of Beerkan type), (ii) soil bulk density and volumetric soil water content at the time of experiment by sampling few (generally two) undisturbed soil cores, and (iii) soil particle-size distribution or, alternatively, clay, silt and sand percentages according to the USDA classification. Specifically, the procedure makes use of well-known analytical solutions for $\theta(h)$ and $K(h)$ relationships and estimates (i) shape parameters, which are texture dependent, from particle-size analysis by physical-empirical PTFs, while (ii) scale parameters, which are structure dependent, by a three-dimensional field infiltration experiment at zero pressure head [23]. Therefore, BEST can be considered an adequate compromise between estimation accuracy and economic-experimental load. For example, some studies have applied BEST to establish the effects of droplet impact on soil sealing and crust formation [24,25], to carry out integrated soil physical quality assessment [26,27] or to identify the effects of tillage on some soil properties (i.e., Ks) under drip irrigation [28]. A main advantage of BEST is that it can be adopted when a large number of hydraulic measurements must be obtained at the field or at irrigation district scale, for example, for precision agriculture purposes. For instance, Mubarak et al. [29] assessed the temporal stability of both Ks and spatial structure of hydraulic properties of a loamy soil. Specifically, they compared Ks-BEST data with those estimated seventeen years earlier by applying the guelph permeameter method, under relatively comparable soil and agronomic conditions. Results showed that Ks changed significantly, but observed discrepancies were not higher than a factor of three or four. This suggests that BEST can represent an easy, robust, and inexpensive way for characterizing soil hydraulic behavior and its spatial [30] and temporal [29] variability at the field scale. The availability of a large number of georeferenced hydraulic measurements would allow delineating homogeneous sub-areas within the crop field to be submitted to uniform agronomic management [31,32]. This can lead to an increase in agricultural resources optimization [31]. In addition, dense spatial data can be used as auxiliary/covariate information in mixed effects models to improve the estimates of the target variables [33] or to improve the estimation of treatment significance reducing the risk of misleading or erroneous inferences in analysis of variance [34,35]. In other words, the potential application advantages seem attractive, but BEST has not yet been tested for soil hydraulic properties spatialization and the actual reliability for the mentioned purposes must be proven.

The general objective of this study was to test the BEST-procedure for the spatialization of soil hydraulic properties. In particular, the spatial distribution of the measured-estimated by BEST variables (soil texture, bulk density, saturated hydraulic conductivity, plant available water capacity, relative field capacity, air capacity) and ancillary soil properties (total organic carbon content, structure stability index) was investigated in a typical wheat cropping system in Southern Italy. Cross-correlation analysis was applied to establish strength and the extent of the spatial relationships between selected soil properties and crop yields.

2. Materials and Methods

2.1. Study Area

The research was carried out in the spring–summer period of 2016 at the experimental farm "Menichella" of Council for Agricultural Research and Economics, located near Foggia (41°27′02″ N, 15°30′36″ E), Southern Italy (Figure 1). The study was conducted in a field of about 5 ha (170 m × 250 m) conventionally cultivated with durum wheat (Figure 1). The field is located in a flat area characterized by Mediterranean climatic conditions and the soil was clay according to USDA classification. Additional information on the experimental site can be found in Cavallo et al. [31].

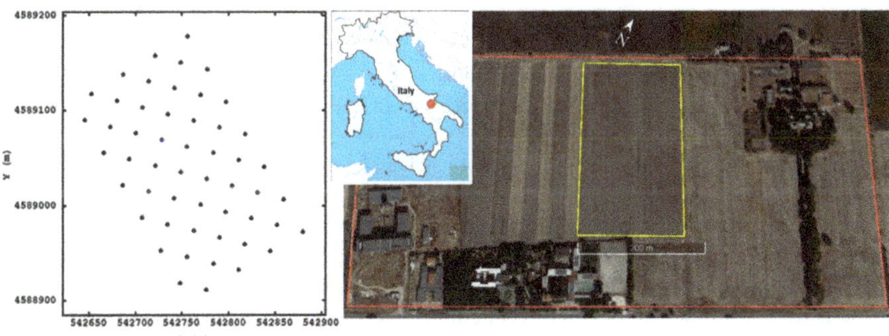

Figure 1. Geographical location, view of "Menichella" farm, and scheme of the fifty-two sampling points.

Fifty-two geo-referenced sampling points, located at the nodes of a regular grid (175 m × 250 m) with a mesh of 20 m × 40 m, were considered in this investigation (Figure 1). The amount of observations was checked to ensure a sufficient coverage of the study area, i.e., a number of pairs for each distance class larger than the minimum required for an effective spatial analysis, as reported by Myers [36] (minimum threshold equal to 25 pairs; actual pairs observed ranging between 66 and 239). At each selected point, the wheat yield and some physical, hydraulic and chemical soil properties (e.g., soil texture, dry bulk density, saturated hydraulic conductivity, total organic carbon content) were determined to establish possible spatial structures and correlations among variables. More details on the dataset composition are given in Section 2.2.

2.2. Lab and Field Measurements

For each of the selected fifty-two sampling points, a simplified version of the BEST-procedure by Lassabatère et al. [23] was applied. The saturated hydraulic conductivity (Ks) and soil bulk density (BD) were then measured, and three capacitive indicators were estimated from soil water retention curve of BEST, namely plant available water capacity ($PAWCe$), air capacity (ACe) and relative field capacity ($RFCe$) [8]. The subscript e is used for indicating variables estimated by BEST. In detail, as required for BEST application, a falling head infiltration experiment of Beerkan type was carried out in the third decade of April at each sampling point using a metal ring (15 cm inner diameter), and the cumulative infiltration curve as function of the time ($I(t)$) was obtained; eighteen water volumes of 200 mL each were used in this investigation and the BEST-steady algorithm by Bagarello et al. [37] was applied to estimate the aforementioned soil properties. A relatively higher number of water volumes than usual (eighteen instead of fifteen) was chosen to be sure of sampling a representative soil volume and be more confident to reach steady-state conditions of water flow. Two undisturbed soil cores (0.05 m in height by 0.05 m in diameter) were collected at the 0 to 0.05 m and 0.05 to 0.10 m depths close to the ring to determine the soil water content at the beginning of infiltration experiments, θ_i, and BD. A disturbed soil sample (0–0.10 m depth) was collected at each sampling point to quantify both the soil texture, i.e., clay (Cl), silt (Si) and sand (Sa) percentages, according to the USDA classification, and the soil total organic carbon content (TOC). Soil texture was obtained with the standard pipette method [38], and TOC was quantified through dry combustion using a TOC Vario Select analyzer [39]. Soil texture fractions, BD, θ_i and $I(t)$ were used to run BEST-steady and the infiltration constants, β and γ, were fixed at the reference values of literature [23,40]. More detail concerning the BEST-procedure application can be found in Castellini et al. [8]. At harvesting (end of June 2016), wheat grain yield data were collected on the fifty-two geo-referenced locations (i.e., nodes of a 20 m × 40 m grid). Yield data were recorded on sampling areas of 1 m^2 and normalized to 13% moisture content of grain.

2.3. Physical and Hydraulic Soil Properties

The saturated hydraulic conductivity (Ks) is the soil's ability to absorb and transmit soil water to the root zone, as well as drain excess water out of the root zone [41]. Since Ks is mainly controlled by soil structure and texture, e.g., [42], in the same soil or soil class it may be used as a measure of structural status of agricultural soils [3]. References of literature suggest optimal Ks values for agriculture soils within the range 0.005–0.05 mm s^{-1} to promote a rapid infiltration and redistribution of plant available water [41]. However, in order to select soil properties that directly (or indirectly) account for soil structure, we also considered (i) dry bulk density (BD), (ii) total organic carbon content (TOC) and (iii) structural stability index (SSI) by Pieri [43]. SSI is calculated from TOC and fine soil texture ($SSI = 1.724 \cdot TOC\%/(silt\% + clay\%) \cdot 100$) components [27]. For these soil indicators the following critical limits were considered: optimal BD values within the range 0.9–1.2 g cm^{-3}; optimal or poor TOC values, general for plant husbandry, were equal to 30–50 or <23 g kg^{-1}, respectively, although good values for agricultural fine-textured soils were reported in the range of 15–22 g kg^{-1} by Sequi and Nobili [44]. Therefore, a comparison between these two references has been made. SSI values >7% or ≤7% are representative of low or high risk of structural degradation, respectively [6].

Conversely, a second set of indicators that gives an account of the proportion between water and air into the soil was considered. Plant available water capacity ($PAWCe$) (cm^3 cm^{-3}), in fact, is the amount of water held in the soil and available for crop growth and obtained as the difference between the water contents at field capacity (at $h = -100$ cm) and at permanent wilting point (at $h = -15{,}300$ cm) [45]; according to the literature [6], the following $PAWCe$ limits were considered in this investigation: $PAWCe$ ≥ 0.20 ideal; 0.15 < $PAWCe$ < 0.20 good; 0.10 < $PAWCe$ < 0.15 limited; $PAWCe$ < 0.10 poor. Air capacity (ACe) (cm^3 cm^{-3}) provided information on the soil ability to store and transmit air (Reynolds et al. [6]). According to Castellini et al. [7], an optimal ACe value falls in the range 0.10–0.26 cm^3 cm^{-3}, while higher or lower values represent inadequate soil aeration conditions; this interval was optimal for a clay soil bordering with that studied [7]. Finally, the relative field capacity ($RFCe$), obtained as the ratio between the water contents at field capacity and at water saturation, was considered, since it partially combines *ACe* and *PAWCe*, thus expressing the soil capacity to store air and water relative to the soil's total pore volume [46]. Optimal values for *RFCe* were suggested within the range 0.6–0.7; consequently, lower or higher values are representative respectively of "water limited" or "aeration limited" (i.e., for a given soil texture, too porous or too compact) soil conditions [6].

2.4. Data Analysis

2.4.1. Preliminary Statistical Analysis

Descriptive statistics were computed in order to summarize the main features of data distribution for yield and the soil variables under study: bulk density (*BD*), capacitive indicators from the estimated soil water retention curve (*ACe*, *PAWCe*, *RFCe*), saturated soil hydraulic conductivity (*Ks*), soil total organic carbon content (*TOC*), fine soil texture components (*Cl* + *Si*) and soil structural stability index (*SSI*). In addition, hypothesis of normality was tested using the Kolmogorov–Smirnov test [47].

2.4.2. Correlation and Spatial Analysis

Relationships among studied variables were investigated using parametric correlation analysis, computing Pearson correlation coefficients.

The predisposition of the considered variables to be spatialized was investigated using Moran statistics. Moran's autocorrelation coefficient (often denoted as I) is an extension of the Pearson product–moment correlation coefficient to a univariate series [48]. Specifically, Moran statistics computes a weighted Pearson product–moment correlation of a variable against itself, where the weighting relates to the variable's spatial arrangement [49]. Moran's I allows for the investigation of correlation within a single variable due to the spatial relationship amongst its observations. The weights (w_{ij}) are a function of the distance between each pair of observations of the variable under study (x_i; x_j). In its simplest form, weights will take values 1 for close neighbours, and 0 otherwise; we also set $w_{ii} = 0$. These weights are sometimes referred to as a neighbouring function.

Moran's *I* statistics is represented by the following equations:

$$I = \frac{N}{S_0} \frac{\sum_{i=1}^{N} \sum_{j=1}^{N} w_{i,j} z_i z_j}{\sum_{i=1}^{N} z_i^2}$$
$$S_0 = \sum_{i=1}^{N} \sum_{j=1}^{N} w_{i,j} \tag{1}$$

where z_i is the deviation of an attribute from its mean $(x_i - \bar{x})$ and S_0 is the aggregate of all the spatial weights (w_{ij}).

2.4.3. Geostatistical Analysis

The geostatistical analysis is aimed at evaluating spatial heterogeneity of the studied variables by means of structural analysis (variography) and at producing maps of the collected data by means of spatial prediction approaches (kriging).

Variography

Under the name of (semi)variogram, two different types of functions related to geographically referenced data are usually denoted, namely the experimental variogram and the variogram model. The former is discrete and represents the half of the average squared difference between points separated by the distance h (the red-dotted line in Figure 2). The latter is parametric, continuous and a conditionally negative definite [50] (the solid line in Figure 2). The most important parameters for the model are the partial sill (σ^2) indicating the structured component of the variance, the nugget (σ_0^2), indicating the random uncorrelated component, and the range (α); this latter can be interpreted as the distance beyond which the spatial correlation becomes negligible (Figure 2). A list of the main variogram models is summarized in Figure 2.

Model	Variogram	Parameter Restrictions
Spherical	$\gamma(h) = \sigma^2[(1 - h/\alpha)^2 + (1 + h/2\alpha)]$	
Exponential	$\gamma(h) = \sigma^2(1 - exp\,(-h/\alpha))$	
Gaussian	$\gamma(h) = \sigma^2(1 - exp\,(-(h/\alpha)^2))$	
Power	$\gamma(h) = \sigma^2(h/\alpha)^a$	$0 < a \leq 2$
Matérn	$\gamma(h) = \sigma^2[1 - (\alpha h)^\nu \mathcal{K}_\nu(\alpha h)]$	$\nu > 0$

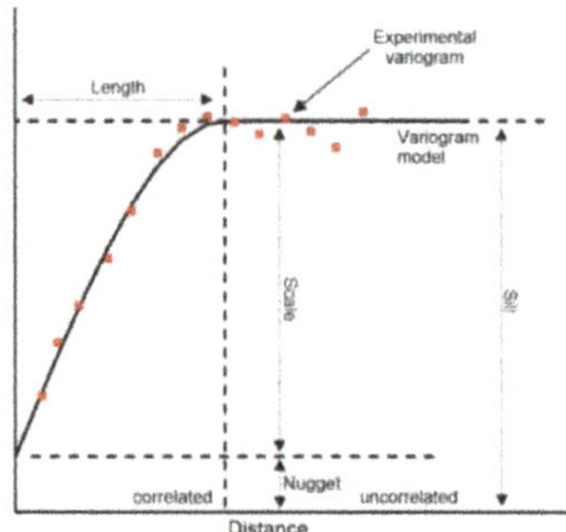

Figure 2. Examples of parametric families of theoretical variograms and experimental and model variograms (red squares and black line, respectively) with corresponding parameters, on the left and right respectively. The parameter σ^2 (the partial sill) identifies the variance of the process and is always positive; the parameter α (the range) identifies the length of the spatial process and is also positive. Other parameter restrictions are reported in the last column.

Variogram models were fitted to the experimental variograms of the variables under study. By virtue of the parsimony principle, the preferred spatial model should be isotropic. Regardless, a test has been applied to test for anisotropy. In detail, isotropic and anisotropic models were compared by means of a Likelihood Based Parameter Estimation for Gaussian Random Fields test, using REML, in order to estimate the two parameters characterizing anisotropy, namely the anisotropy angle and the ratio between the two ellipse axes [51].

Leave-one-out cross-validation was carried out and Pearson correlation coefficients (r) between predicted and observed data were used to quantify the goodness of model adaptation to the experimental variogram.

Spatial Interpolation

All the variables were interpolated using a univariate approach, the ordinary kriging (OK). Such a predictor is one of the most basic forms of kriging in which the unknown value $z(x_0)$ of a given realization of $Z(x_0)$ is predicted from the known values $z(x_i)$ $i = 1, 2, \ldots, N$, at the support points x_i. The ordinary kriging predictor can be written as:

$$z^*_{OK}(x_0) = \sum_{i=1}^{N} \lambda_i z(x_i) \tag{2}$$

where λ_i are weights associated with the N sampling points. The weights are chosen in such a way that the predictor is unbiased, the values are continuous and the estimation error is minimized:

$$E[Z^*(x_0) - Z(x_0)] = 0. \tag{3}$$

This ensures that kriging is an exact interpolator because the estimated values are identical to the observed values when a kriged location coincides with a sample location [33,52,53].

2.4.4. Cross-Correlogram

In order to compare prediction maps of different variables, cross-correlograms were computed. In detail, given the prediction maps of two different variables, M_A and M_B, there are several methods with which to compare them [33,54,55]. To account for the inherently spatial characteristics of map representation, the cross-correlogram, which measures the correlation as a function of the distance between observations, is particularly well-suited. The analytical formulation of the cross-correlogram is the following:

$$\rho_{A,B}(h) = \frac{E\left[z_{i,jA}, z_{i',j'B}\right] - m_A \cdot m_B}{s_A \cdot s_B} \tag{4}$$

where $z_{i,jA}$ and $z_{i',j'B}$ represent the values at locations (i, j) and (i', j') of the two maps separated by the h distance, respectively; $h = \sqrt{(i - i') + (j - j')}$ represents the distance between the two locations, E denotes the mathematical expectation, m_A and m_B represent the populations means and s_A and s_B represent the populations standard deviations. If patterns are completely similar, apart from a constant, $\rho_{A,B}(0)$ should be equal to 1. To estimate $\rho_{A,B}(h)$ from the available data the following equation can be used:

$$r_{A,B}(h) = \frac{\sum_{i,j=1}^{N(h)} z_{i,jA} \cdot z_{i',j'B} - \hat{m}_A \cdot \hat{m}_B}{\hat{s}_A \cdot \hat{s}_B} \tag{5}$$

To compute $r_{A,B}(h)$, the procedure is as follows: from both the maps, all the couples whose locations are separated by the distance h are collected. Indices $\hat{m}_A$ and $\hat{m}_B$ and $\hat{s}_A$ and $\hat{s}_B$ represent the mean and the standard deviation of mapped $z_{i,jA}$ and $z_{i',j'B}$, respectively. $N(h)$ is the total number of these pairs. If the methods being compared produce similar results, a decreasing cross-correlogram for increasing values of h is expected.

For summarizing the results and comparing different outcomes, results of cross-correlation analysis were reported in form of tables containing correlations at specified lag distances (0 m, 25 m, 50 m, 75 m, 100 m); in this way, practical information on the strength and extent of the spatial relationships between soil properties and crop yields was provided.

3. Results

3.1. Preliminary Statistical Analysis

The preliminary statistical analysis on the studied variables highlighted some slight departures from normal distribution (Table 1). In particular, Ks and TOC showed positive and negative longer tails due to a few larger and smaller observations. The Kolmogorov–Smirnov test results indicated a significant departure from normality only for ACe and Ks, although not extremely significant ($P = 0.0291$; $P = 0.0209$; Table 2). For this reason, it was not deemed necessary to transform the original data.

Table 1. Summary statistics for the studied variables.

Variable	n	Mean	st.dev.	Median	Range	Skewness	Kurtosis
$Yield$ (t ha^{-1})	52	3.86	0.77	3.84	3.86	0.38	0.18
BD (g cm^{-3})	52	1.06	0.09	1.07	0.36	−0.65	−0.32
$RFCe$ (-)	52	0.80	0.02	0.81	0.10	−0.81	0.04
ACe (cm^3 cm^{-3})	52	0.12	0.02	0.12	0.07	0.75	−0.12
$PAWCe$ (cm^3 cm^{-3})	52	0.22	0.02	0.22	0.06	−0.33	−0.79
Ks (mm s^{-1})	52	0.02	0.01	0.02	0.04	1.15	1.71
TOC (g kg^{-1})	52	15.95	1.16	16.14	6.97	−1.48	4.12
$Cl + Si$ (g 100g^{-1})	52	69.86	3.02	70.32	11.81	−0.27	−0.71
SSI (%)	52	3.94	0.36	4.03	1.81	−0.69	0.77

BD = soil bulk density; $RFCe$ = relative field capacity; Ace = air capacity; $PAWCe$ = plant available water capacity; Ks = saturated hydraulic conductivity; TOC = total organic carbon content; $Cl + Si$ = percentages of *clay + silt* content; SSI = structure stability index.

Table 2. Outcomes of the Kolmogorov-Smirnov test (D).

Variable	K-S (D)	*p*-Value
Yield (t ha^{-1})	0.08656	>0.1500
BD (g cm^{-3})	0.098441	>0.1500
RFCe (-)	0.114814	0.0860
ACe (cm^3 cm^{-3})	0.129142	0.0291
PAWCe (cm^3 cm^{-3})	0.081048	>0.1500
Ks (mm s^{-1})	0.133452	0.0209
TOC (g kg^{-1})	0.103982	>0.1500
Cl + Si (g 100g^{-1})	0.099201	>0.1500
SSI (%)	0.106077	0.1484

Note: That the acronyms of soil properties are specified in the Table 1.

3.2. Physical and Hydraulic Soil Properties

Physical and hydraulic properties of the investigated clay soil were summarized in Figure 3. Overall, according to the suggested criteria to detect relatively good or poor soil conditions of *BD*, *TOC*, *SSI*, *Ks*, *PAWCe*, *ACe*, *RFCe* and to obtain acceptable crop yields, results showed relatively good findings as only three of the seven soil properties indicated non optimal conditions. In further detail, the following was observed: (i) relatively low levels of organic carbon content (*TOC*) for threshold values reported by Reynolds [6], but good levels for ranges defined by Sequi and Nobili [44], (ii) risks of structural instability (*SSI*) and (iii) potential conditions of soil compaction (*RFCe*). In agreement with these findings, although *ACe* was within the suggested optimal range, it showed relatively low air capacity (Figure 3). Optimal hydrodynamic soil properties were also recognized for *Ks*, as the entire range of variation of the measurements fell within the limits defined by Reynolds et al. [41]. However, two clarifications should be provided because: (i) observed low *TOC* values are not entirely unexpected as they are quite common for a monoculture of wheat, in a typical Mediterranean environment, and (ii) *BD* evaluation showed optimal conditions, i.e., a soil compaction was not recognized, indicating a different assessment when compared with *RFCe* or *ACe* (Figure 3). In other words, soil physical and hydraulic assessment carried out using available references of literature has suggested optimal, or near optimal, conditions of investigated soil. This finding was in agreement with results of wheat yields, as obtained mean values equal to 3.86 t ha^{-1} may be considered as medium-high production levels for the investigated agro-environment. However, as *TOC* limits used do not appear to be entirely in line with the case under study, a specific discussion has been made in Section 4.

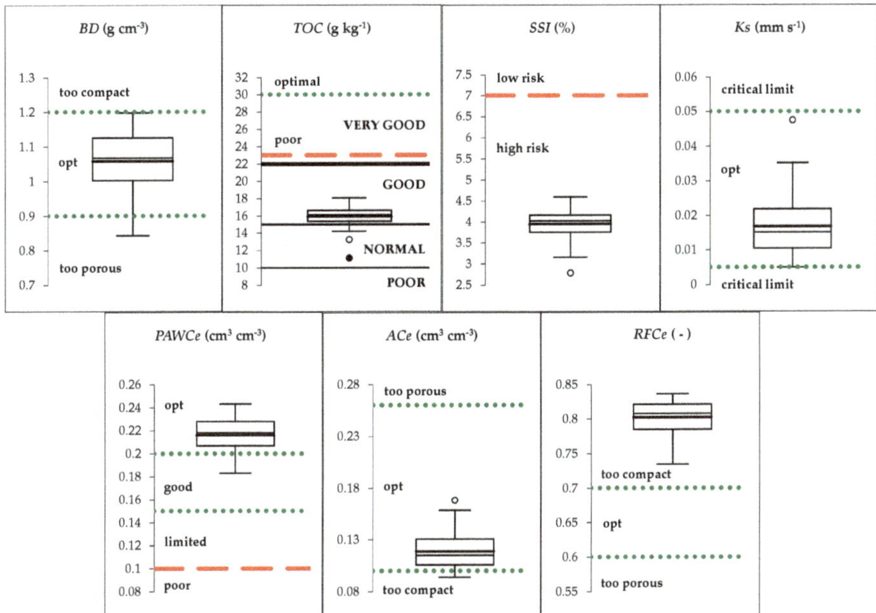

Figure 3. Box plots of dry bulk density (*BD*), total organic carbon content (*TOC*), structure stability index (*SSI*), saturated hydraulic conductivity (*Ks*), plant available water content (*PAWC*), air capacity (*AC*) and relative field capacity (*RFC*). The thick-black line within each box represents the mean value (the fine-black line, the median). Open circles represent outliers (closed circle, extreme outlier for *TOC*). Dot green-line or dashed red-line represents optimal or critical values, respectively, according to Section 2.3. For *TOC*, a further classification by Sequi and Nobili [43] was specifically reported (solid lines, which identify four areas, from poor to very good) for fine textured soils (clay, clay-loam, silty-clay and silty-clay-loam, according to USDA).

3.3. Correlation and Spatial Analysis

Results of correlation analysis are reported in Figure 4. Significant correlations were observed between crop yield and *BD* and *PAWCe*, with a positive and negative relationship respectively (r = 0.381, *P* = 0.005 and r = −0.400, *P* = 0.003); in addition, a positive correlation, although not significant (*P* = 0.063), was found with *Ks*. Overall, soil variables showed interesting relationships. As concerns capacitive indicators derived from the soil water retention curve of BEST, a strong negative correlation was found between *RFCe* and *ACe* (r = −0.922, *P* < 0.0001) as observed also in previous studies [7]. *PAWCe* was strongly negatively related to *BD* (r = −0.780, *P* < 0.0001). An interesting relationship was observed between *Ks* and the three capacitive indicators (*P* = < 0.0001, 0.001 and 0.009, respectively for *RFCe*, *ACe* and *PAWCe*). *TOC* was negatively correlated to *BD* (*P* < 0.006) and showed a weak relation with *PAWCe* (*P* < 0.065). Finally, *SSI*, synthetizing the information deriving from *TOC* and fine texture components (*Cl* + *Si*), showed to be an effective summary indicator highlighting good correlations also with *BD* (*P* = 0.005), *PAWCe* (*P* = 0.006) and *Ks* (*P* = 0.059). A more in-depth discussion on the correlations obtained is reported in the Discussion section.

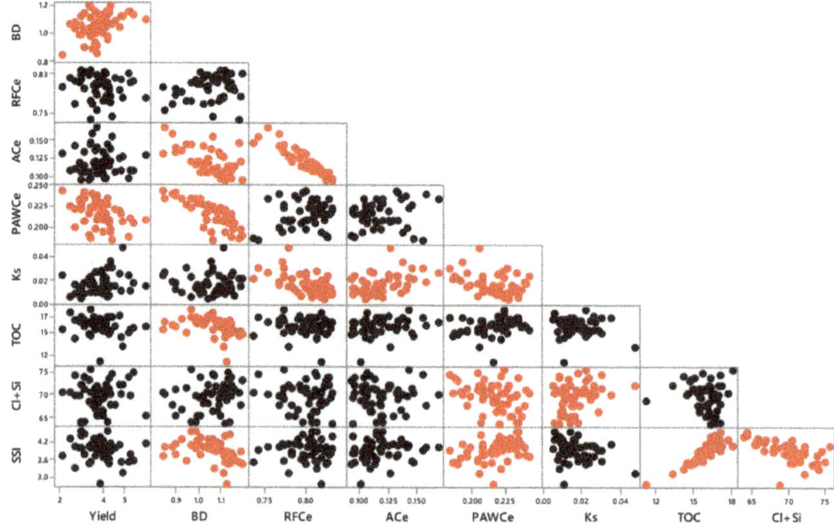

Figure 4. Bivariate scatterplots for the variables under study. Red dots highlight significant correlations at $P < 0.05$. *BD* is the bulk density (g cm^{-3}), *Cl + Si* is the sum of clay and silt content (%), *Ks* is the saturated soil hydraulic conductivity (mm s^{-1}), *TOC* is the soil total organic carbon (g kg^{-1}), *SSI* is the structure stability index (%), *PAWCe* is the plant available water capacity (cm^3 cm^{-3}), *ACe* is the air capacity (cm^3 cm^{-3}) and *RFCe* is the relative field capacity (dimensionless).

Moran's I spatial autocorrelation statistics gave a hint about the predisposition of the single variables to be spatialized (Table 3). For the case at hand, Moran values ranged from −0.2232 for *RFCe*, indicating a non-significant spatial correlation, to 0.3920, 0.4474, 0.4564 and 0.5889, for *BD*, *SSI*, *PAWCe* and *Cl + Si* respectively, indicating highly significant overall spatial dependence. Yield and *TOC* showed also a significant spatial correlation (Table 3). It is noteworthy that the variables correlated each other, showing also a similar overall spatial structure.

Table 3. Moran's I spatial autocorrelation statistics.

	Yield	*BD*	*RFCe*	*ACe*	*PAWCe*	*Ks*	*TOC*	*Cl + Si*	*SSI*
Moran I	0.2194 *	0.3920 **	−0.2232	−0.0394	0.4564 **	0.0121	0.2055 *	0.5889 **	0.4474 **
p-value	0.0312	0.0007	0.9435	0.5610	0.0001	0.4007	0.0331	0.001	0.0001

Note: * and ** indicate significance of the Moran I coefficient at probability equal or lower than 0.05 and 0.01, respectively.

Theoretical models were fitted to the experimental variograms of the variables under study, consisting generally of these nested models: a nugget effect and a spatial covariance function. Since the anisotropy ratio and the anisotropy angle parameters were not significant according to the Likelihood Based Parameter Estimation test, isotropic models were selected. In Table 4, the fitted variogram models are reported and described by the model name and the following parameters, namely nugget, partial sill and range. Actually, the Matérn model demonstrated to be the best suited theoretical function to describe spatial structure of the considered variables except for *BD* and *PAWCe*, since it was more flexible with respect to the classical models, having an additional shape parameter (*k*) (Figure 2). Following Cambardella et al. [56], the nugget-to-sill ratio can be a useful means to describe the spatial structure of the studied variables. In particular, the nugget semivariance expressed as a percentage of the total semivariance enables comparison of the relative size of the nugget effect among studied variables, with ratios <25% indicating strong spatial dependence, ratios between 25 and 75%

moderate spatial dependence, ratios >75% weak spatial dependence. The analysis of the nugget-to-sill ratios (Table 4) indicated almost strong spatial dependence for *TOC*, *Cl + Si* and *SSI* (with values of 0.264, 0.272 and 0.263, respectively); these results were confirmed by the high predicted vs. observed correlations (r = 0.42, 0.63 and 0.74). *RFCe*, together with *Ks*, showed a weak spatial dependence, with nugget to sill ratios of 0.847 and 0.910, results confirmed also by not significant cross-validation outcomes (predicted vs. observed correlations) and by the findings of overall spatial correlation analysis (Moran statistics, Table 3). Yield fell between the moderate to weak classes, with a nugget to sill ratio of 0.761. *PAWCe* and *BD* presented a moderate adaptation as showed by the predicted vs. observed correlation (r = 0.49 and 0.46); the nugget-to-sill ratio was not computed because the nugget was not significant different from zero. The variogram estimated parameters for *ACe* appeared to be not physically based (an estimated value of 773 m was observed for the range). All the remaining range parameters values were consistent and below the maximum distance between observations (Table 4).

Table 4. Parameters of the theoretical variograms fitted to the studied variables.

	Yield	*BD*	*RFCe*	*ACe*	*PAWCe*	*Ks*	*TOC*	*Cl + Si*	*SSI*
Nugget	0.48	0	0.00061	0.00028	0	6.59×10^{-5}	0.395	3.099	0.036
Partial sill	0.151	0.006	0.00011	0.00131	0.00012	6.55×10^{-6}	1.102	8.298	0.101
Nugget to sill ratio	0.761	–	0.847	–	–	0.910	0.264	0.272	0.263
Range	62	40.0	286.9	772.5	60.0	98.2	29.7	89.3	39.9
Model	Mat (k = 10)	Sph	Mat (k = 10)	Mat (k = 10)	Sph	Mat (k = 0.05)	Mat (k = 5)	Mat (k = 10)	Mat (k = 10)
Pred. vs. obs. (r)	0.31	0.46	–	–	0.49	–	0.42	0.74	0.63

Mat = Matérn model; Sph = Spherical model.

Spatial estimates of yield and spatially structured soil variables are reported in Figure 5. By visually inspecting the maps, a similar spatial behavior emerged for yield and *BD* with lower values in the left part of the experimental area and larger values in the lower right part. An inverse relationship was instead observed between yield and *PAWCe*. These results confirm the significant overall correlations already reported between yield and *BD* and *PAWCe* (Figure 4). *TOC* varied in a quite narrow range, apart from a few outliers, and the highest values were observed in the upper central part of the area. As expected, *SSI* map resembled *TOC* spatial behavior; at the same time, *SSI* brought the information related to the fine texture component (*Cl + Si*), showing as a consequence spatial similarity with the mentioned physical indicators maps.

To further deepen the *SSI* behavior observed, cross-correlograms were computed and cross-correlation coefficients at specific lags (0 m, 25 m, 50 m, 75 m, 100 m) were extracted (Table 5). *SSI* was strongly correlated with *TOC* and fine texture components, positively and negatively respectively, up to about 25 m distance for *TOC* and up to 50 m for *Cl + Si*. In addition, a moderate correlation was observed with *PAWCe* (positive) and *BD* (negative) up to 25 m, indicating the contribution of texture in driving this relationship. A weak correlation was observed between *SSI* and yield at lag 0, confirming that the behavior arose by the Pearson's overall correlation analysis (Figure 4). Finally, the cross-correlation between *SSI* and *RFCe* was negligible, as expected.

Table 5. Cross-correlation coefficients computed at different lags between *SSI* and other indicators maps.

	0 m	25 m	50 m	75 m	100 m
SSI vs. PAWCe	0.3805	0.2683	0.1448	0.0007	−0.0968
SSI vs. BD	−0.3825	−0.2554	−0.1249	0.0443	0.1092
SSI vs. Yield	−0.2582	−0.1791	−0.0329	0.0667	0.0591
SSI vs. TOC	0.8078	0.3554	0.0194	−0.1031	−0.0846
SSI vs. Cl + Si	−0.6202	−0.4288	−0.2312	−0.0173	0.1248
SSI vs. RFCe	−0.0168	−0.0341	−0.0546	0.0017	0.0465

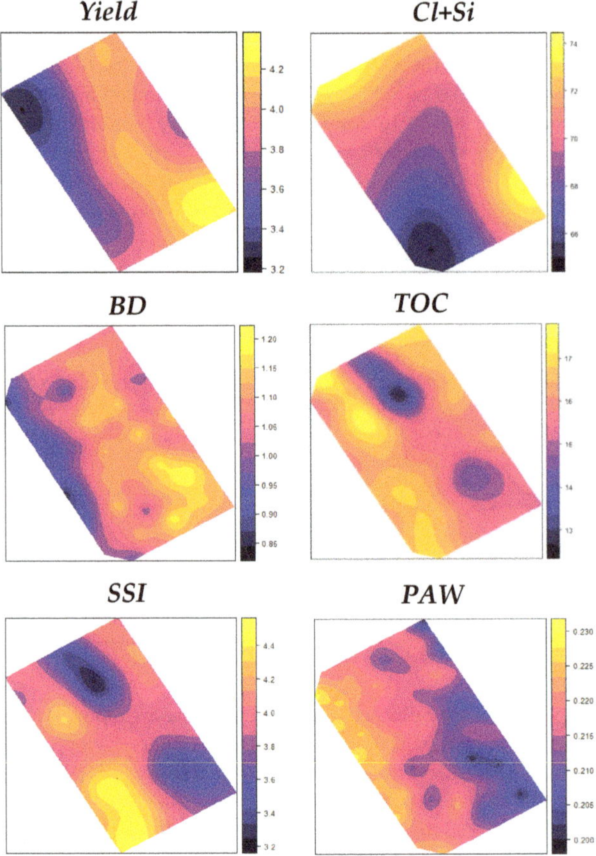

Figure 5. Spatial estimates of wheat yield (t ha^{-1}), clay and silt content *Cl* + *Si* (%), dry bulk density, *BD* (g cm^{-3}), soil total organic carbon, *TOC* (g kg^{-1}), structure stability index, *SSI* (%) and plant available water capacity, *PAWCe* (cm^3 cm^{-3}).

By considering the role of *BD* as key soil indicator, cross-correlograms between the maps of *BD* and the other indicators were also computed (Table 6). Strong negative correlations were observed with *PAWCe* up to about a 50 m distance. Moderate positive correlations were found with *RFCe* (up to 50 m) and yield (up to 25 m), whereas negative with *TOC* and *SSI*, as already observed, up to 25 m distance.

Table 6. Cross-correlation coefficients computed at different lags between *BD* and other indicators maps.

	0 m	25 m	50 m	75 m	100 m
BD vs. PAWCe	−0.7051	−0.4559	−0.2974	0.1381	0.0361
BD vs. SSI	−0.3825	−0.2554	−0.1249	0.0443	0.1092
BD vs. Yield	0.4219	0.3488	0.1937	0.0023	−0.0536
BD vs. TOC	−0.3924	−0.2814	−0.1355	0.0524	0.1030
BD vs. Cl + Si	0.1466	0.0650	0.0358	−0.0016	−0.0507
BD vs. RFCe	0.4643	0.3974	0.2894	0.0953	−0.0848

4. Discussion

Overall, the observed correlations showed physically plausible relationships among soil properties (Figure 4). In particular, inverse relationships between *ACe*, *TOC*, *SSI* and *BD* were detected, as well as between *ACe*, *Ks* and *RFCe*. As expected, *Ks* was positively correlated with *ACe*, as increasing values of soil aeration should match with increasing saturated hydraulic conductivity (*Ks*); a reasonable positive relationship between *Cl* + *Si* fraction and *Ks* was also detected. Expected correlations have been verified between *SSI* and *TOC* (positive) or between *SSI* and *Cl* + *Si* (negative), since *TOC* and *Cl* + *Si* terms appear, respectively, in the numerator and in the denominator of Pieri's equation [6]. However, uncertain results were obtained for the significant correlations of *PAWCe*, because indirect relationships between *PAWCe* and *Cl* + *Si*, as well as *BD*, were detected. In fact, although the latter was quite verified in the literature for medium-high *BD* values [6,57,58], the former relationship does not seem entirely plausible because higher *PAWCe* values should correspond to higher contents of fine particles of the soil. A similar explanation can also be given for the relationship (both general and spatial) between *PAWCe* and crop yield.

Agronomic results of this investigation showed medium-high wheat yields on average (Table 1). However, according with the hypothesis that good conditions of soil physical quality should correspond to good yields [3,59], findings showed non optimal soil conditions in terms of *TOC*, *SSI*, and *RFCe* (Figure 3), i.e., in 43% of cases, and some considerations are due, particularly for optimal *TOC* ranges. For instance, the suggested lower critical *TOC* limits for agricultural soils (i.e., about 20 g kg^{-1}) by Reynolds et al. [6] were obtained for urban soils, namely for sustainable establishment of plants in constructed landscaping soils used in urban parks, playing fields, curbside plantings, etc. [6,60]. Despite this critical threshold being applied in the literature e.g., [46,61,62], different limits should be considered for agricultural soils, and specifically for crops grown in Mediterranean environments. For example, *TOC* mean values observed in this investigation (i.e., about 16 g kg^{-1}) were relatively high as compared to those reported in Table 1 by Ventrella et al. [63], for eight experimental farms and soils with different texture, located from north to south of Italy (5.2 $\leq$ *TOC* $\leq$ 15.3 g kg^{-1}). In addition, in other investigations where a variety of plant species was considered, optimal *TOC* levels, as suggested in the literature (i.e., 30 < *TOC* < 50 g kg^{-1}), were seldom reached. For instance, for approximately two-hundred soil cores collected in three areas and 18 spot sites of agricultural and forest Sicilian environments, Castellini and Iovino ([14]; Table 2) reported *TOC* values equal to 10.8 g kg^{-1} on average (0.9–37 g kg^{-1} as a measured interval). To provide more concrete examples, Table 7 summarizes other results of some investigations carried out in the Mediterranean environments of Apulia, Sicily and Sardinia (south-central Italy). With reference to such investigations, listed *TOC* values were equal, at most, to 16.7 g kg^{-1} for conservation soil management of durum wheat, equal to 22.6 g kg^{-1} for a citrus groove or slightly higher for a grassland, suggesting that it is not common to find considerably higher *TOC* levels in Mediterranean agricultural environments. On the other hand, *TOC* values close to the upper limit suggested by Reynolds et al. [6], 50 g kg^{-1}, were only detected for a sandy loam soil under high maquis (holm oak), i.e., when organic matter accumulation and probably slow organic matter mineralization rate can lead to such high levels (Table 7). As an example, Figure 3 shows how the *TOC* classification by Sequi and Nobili [44], specifically referred to fine textured soils (clay, clay–loam, silty–clay and silty–clay–loam), is more suitable for the specific characteristics of Mediterranean agricultural soils. As a consequence, more accurate optimal or critical values should be provided for specific agro-environments, for example performing ad-hoc new research or by reviewing the available literature (e.g., through meta-analysis). Similar considerations can be made for *SSI*, as it is obtained from *TOC* and texture, and is not related directly to the soil structure, but is rather related to the soil resilience [6]. On the contrary, *RFCe* limits may be considered relatively more reliable as they were successfully verified in reference to other soil physical properties (i.e., *BD*, *TOC*, *ACe*, macroporosity) by Reynolds et al. [64] for a clay loam soil, as well as in comparison with literature guidelines.

Table 7. Mean values of soil total organic carbon contents (*TOC*) measured for different types of vegetation, Mediterranean agro-environments and management, and soils textural classes (USDA classification).

	Plant Species	*Agro-Environmental Management*	*Soil Texture*	*TOC (g kg^{-1})*
Castellini et al. [65]	durum wheat	burning of stubble and straw	clay	14.6–16.3
	durum wheat	incorporation of stubble and straw	clay	15.6–17.1
Iovino et al. [57]	vetch in annual rotation with durum wheat	shallow tillage (about 10–15 cm)	clay	8.9–7.1
	degraded olive orchard	shallow tillage (10 cm)	silty clay	5.3–5.7
	re-established grassland	undisturbed with moderate pasture	clay	6.5–25.9
	eucalyptus plantation	undisturbed with moderate pasture	clay	6.5–9.5
Castellini et al. [26]	holm oak forest	–	sandy loam	29
	branched asphodel grassland	–	sandy loam	36
	high maquis	–	sandy loam	50
	false yellowhead grassland	–	sandy loam	15
Ferrara et al. [66]	durum wheat	conventional tillage	silty clay	13.8
	durum wheat	no-tillage	silty clay	16.7
Castellini et al. [39]	corn	fertilized with sewage	sand	10.4
	durum wheat	conventional tillage	silty loam	15.7
	citrus groves	-	sandy loam	22.6
	citrus groves	-	sandy loam	11.6

Among those investigated, only five soil physical properties were found spatially structured, i.e., *Cl + Si*, *BD*, *TOC*, *SSI* and *PAWCe*, but their relationships with grain yield did not appear to be always convincing (Figure 5). In detail, in comparison to the yield map (i.e., lower or higher yielding zones), relatively coherent spatial distributions were identified in terms of fine soil texture components (*Cl + Si*) and *BD*; this suggests that, especially bulk density, can represent a reliable physical indicator to manage within-field sub-areas with different productivity. For this soil indicator, the map also shows that the sub-areas with relatively lower productivity correspond to those with very low *BD* values (as specified by the critical lower value of 0.9 g cm^{-3} of Figure 3), suggesting that this limit seems quite realistic for the case under study. On the contrary, conflicting information between yield and *PAWCe* maps were obtained (i.e., an inverse spatial structure), as already shown by the overall correlation. These results can be attributed to uncertainties of the *PAWCe* estimates obtained by BEST which would require further deepening. As expected, a relatively similar spatial distribution was observed for *TOC* and *SSI*, but not consistent with that of crop yield. In other words, findings provided by variables directly measured, both for overall correlation and for spatial analysis, were more convincing than those derived by variables estimated by BEST, for which further investigations are probably needed to quantify the accuracy degree of estimated soil water retention curve.

To deepen the aforementioned results and corroborate the relationships between yield and considered soil properties, a stepwise analysis was carried out to identify the variables most affecting wheat yield among those directly quantified (*TOC*, *BD*, fine textural components) or derived from laboratory measurements (*SSI*). Results of stepwise analysis showed that *BD* was the only variable selected ($P = 0.0054$), thus confirming the key role of soil porosity and compaction in affecting crop yield. Consequently, for investigated soil variables, only the BD map seems utilisable for agronomic management, i.e., for precision agriculture applications, because several factors may have contributed to produce unexpected or uncertain results for the other variables. Among these factors, particular relevance can be attributed to: (i) the uncertainties of the $PAWCe$ estimates obtained by BEST; (ii) the relatively poor correspondence between observed physical–chemical fertility of the soil and agricultural yields in the specific conditions investigated; (iii) a possible spatial variability at a scale smaller than that experimentally measured (i.e., lower than about the mesh side). About the last statement, as Ks showed no significant spatial structure, we could make a plausible conjecture suggesting that spatial variability of Ks occurred at a smaller scale than that investigated [67]. However, no significant relationship between Ks and crop yield (or BD) was detected. Overall, Ks is reported to be a better soil structure indicator [68] than BD, as the latter does not provide any information on soil pore distribution, i.e., architecture, connectivity and tortuosity of soil porosity [3]. In our case study, a very good spatial correspondence between crop yield and soil bulk density was detected, and the cross-correlation analysis also showed a significant positive relationship with the crop yield for the lower distance, of about 25 m. Although this result could be considered questionable from the perspective of precision agriculture application (i.e., for implementation on platforms with on-the-go soil sensors), it is quite significant as BD measurements can be obtained more easily if compared to Ks, and it can be easily related to penetrometric soil measurements.

As a final remark, it should be noticed that SSI showed promising characteristics to be used as a representative indicator to identify homogeneous within-field areas, because of its tight relationships with chemical (TOC) and hydrological (fine texture components, and in turn $PAWCe$ and BD) indicators and its predisposition to spatialization, thus suggesting a possible significant correlation with yield under a denser spatial sampling scheme. If this were the case, SSI would candidate itself as a key indicator for precision agriculture applications.

5. Conclusions

In this investigation a set of eight physical and hydraulic soil properties directly measured or estimated by BEST was obtained and spatialized to investigate correlations and identify intervals corresponding to medium-high levels of wheat yields, in order to provide useful information for site-specific agronomic management.

According to the guidelines of literature, a soil physical quality evaluation highlighted that soil under study had optimal bulk density, plant available water capacity, air capacity and saturated hydraulic conductivity values, but that the total organic carbon content, structure stability index and relative field capacity suggested too low levels of organic carbon or excessive soil compaction. However, both a literature analysis for different types of Mediterranean vegetation cover and the correlation analysis (overall and spatial correlation) suggested that a review of the optimal or critical TOC values for typical crops of Mediterranean environments should be made, as TOC value around 20 g kg^{-1} is hardly achievable even under conservation agriculture systems, and literature values are probably not entirely realistic for Southern Italy's cereal crops. In addition, capacitive indicators estimated from the soil water retention curve of BEST provided both expected and uncertain correlations, especially with regard to the inverse relationship between plant available water capacity and wheat yield. Therefore, for the soil–crop system studied, application of the simplified BEST-procedure did not return completely reliable results and further investigations are needed to quantify the accuracy and reliability of estimated soil water retention curve. These main results open up new research perspectives to improve our knowledge on this topic.

Among measured soil properties, BD showed a spatial distribution in agreement with that detected for crop yield, and the cross-correlation analysis also showed a significant positive relationship only for short lags. Finally, SSI showed promising characteristics suggesting a possible significant correlation with yield under a denser spatial sampling scheme, and a potentiality as a key indicator for precision agriculture applications.

Further research on this topic is needed for Mediterranean agro-environments, by deepening: (i) the reliability of available measurement methods for accurately estimating representative physical and hydraulic soil properties, and (ii) the temporal stability of observed spatial relationships between soil properties (soil bulk density or soil texture) and crop yields along a larger time interval.

Author Contributions: M.C. outlined the investigation. M.T., A.M.S. and E.B. have carried out data analysis. All authors contributed to critically discuss the results, to write and review the manuscript.

Funding: This research received no external funding.

Acknowledgments: The authors would like to thank Alessandro Vittorio Vonella for his valuable support in the field measurements.

Conflicts of Interest: The authors declare no conflict of interest.

References

1. Bagarello, V.; Baiamonte, G.; Castellini, M.; Di Prima, S.; Iovino, M. A comparison between the single ring pressure infiltrometer and simplified falling head techniques. *Hydrol. Process.* **2014**, *28*, 4843–4853. [CrossRef]
2. Alletto, L.; Coquet, Y. Temporal and spatial variability of soil bulk density and near-saturated hydraulic conductivity under two contrasted tillage management systems. *Geoderma* **2009**, *152*, 85–94. [CrossRef]
3. Keller, T.; Stutter, J.A.; Nissen, K.; Rydberg, T. Using field measurement of saturated soil hydraulic conductivity to detect low-yielding zones in three Swedish fields. *Soil Tillage Res.* **2012**, *124*, 68–77. [CrossRef]
4. Pinheiro, E.A.R.; de Jong van Lier, Q.; Šimůnek, J. The role of soil hydraulic properties in crop water use efficiency: A process-based analysis for some Brazilian scenarios. *Agric. Syst.* **2019**, *173*, 364–377. [CrossRef]
5. Ventrella, D.; Charfeddine, M.; Giglio, L.; Castellini, M. Application of DSSAT models for an agronomic adaptation strategy under climate change in Southern Italy: Optimum sowing and transplanting time for winter durum wheat and tomato. *Ital. J. Agron.* **2012**, *7*, 109–115. [CrossRef]
6. Reynolds, W.D.; Drury, C.F.; Tan, C.S.; Fox, C.A.; Yang, X.M. Use of indicators and pore volume-function characteristics to quantify soil physical quality. *Geoderma* **2009**, *152*, 252–263. [CrossRef]
7. Castellini, M.; Stellacci, A.M.; Barca, E.; Iovino, M. Application of multivariate analysis techniques for selecting soil physical quality indicators: A case study in long-term field experiments in Apulia (southern Italy). *Soil Sci. Soc. Am. J.* **2019**, *83*, 707–720. [CrossRef]
8. Castellini, M.; Fornaro, F.; Garofalo, P.; Giglio, L.; Rinaldi, M.; Ventrella, D.; Vitti, C.; Vonella, A.V. Effects of No-Tillage and Conventional Tillage on Physical and Hydraulic Properties of Fine Textured Soils under Winter Wheat. *Water* **2019**, *11*, 484. [CrossRef]
9. Bagarello, V.; Iovino, M.; Elrick, D. A simplified falling-head technique for rapid determination of field-saturated hydraulic conductivity. *Soil Sci. Soc. Am. J.* **2004**, *68*, 66–73. [CrossRef]
10. Blackmore, S.; Godwin, R.J.; Fountas, S. The analysis of spatial and temporal trends in yield map data over six years. *Biosyst. Eng.* **2003**, *84*, 455–466. [CrossRef]
11. Hakojärvi, M.; Hautala, M.; Ristolainen, A.; Alakukku, L. Yield variation of spring cereals in relation to selected soil physical properties in three clay soil fields. *Eur. J. Argon.* **2013**, *49*, 1–11. [CrossRef]
12. Garofalo, P.; Ventrella, D.; Kersebaum, K.C.; Gobin, A.; Trnka, M.; Giglio, L.; Dubrovský, M.; Castellini, M. Water footprint of winter wheat under climate change: Trends and uncertainties associated to the ensemble of crop models. *Sci. Total Environ.* **2019**, *658*, 1186–1208. [CrossRef]
13. Fernández-Gálvez, J.; Pollacco, J.A.P.; Lassabatere, L.; Angulo-Jaramillo, R.; Carrick, S. A general Beerkan Estimation of Soil Transfer parameters method predicting hydraulic parameters of any unimodal water retention and hydraulic conductivity curves: Application to the Kosugi soil hydraulic model without using particle size distribution data. *Adv. Water Resour.* **2019**, *129*, 118–130. [CrossRef]
14. Castellini, M.; Iovino, M. Pedotransfer functions for estimating soil water retention curve of Sicilian soils. *Arch. Agron. Soil Sci.* **2019**, *65*, 1401–1416. [CrossRef]
15. Zhang, X.; Zhu, J.; Wendroth, O.; Matocha, C.; Edwards, D. Effect of Macroporosity on Pedotransfer Function Estimates at the Field Scale. *Vadose Zone J.* **2019**, *18*, 180151. [CrossRef]
16. Burke, W.; Gabriels, D.; Bouma, J. *Soil Structure Assessment*; Balkema: Rotterdam, The Netherlands, 1986.

17. Dane, J.H.; Hopmans, J.W. Water retention and storage: Laboratory. In *Methods of Soil Analysis, Part 4, Physical Methods*; Dane, J.H., Topp, G.C., Eds.; SSSA: Madison, WI, USA, 2002; pp. 688–692.

18. Wind, G.P. Capillary conductivity data estimated by a simple method. In *Water in the Unsaturated Zone: Proceedings of Wageningen Syposium, June 1966*; Rijtema, P.E., Wassink, H., Eds.; IASAH: Gentbrugge, Belgium, 1968; Volume 1, pp. 181–191.

19. Hosseini, S.M.M.M.; Ganjian, N.; Pisheh, Y.P. Estimation of the water retention curve for unsaturated clay. *Can. J. Soil Sci.* **2011**, *91*, 543–549. [CrossRef]

20. Rustanto, A.; Booij, M.J.; Wösten, H.; Hoekstra, A.Y. Application and recalibration of soil water retention pedotransfer functions in a tropical upstream catchment: Case study in Bengawan Solo, Indonesia. *J. Hydrol. Hydromech.* **2017**, *65*, 307–320. [CrossRef]

21. Lozano-Baez, S.E.; Cooper, M.; Ferraz, S.F.B.; Ribeiro Rodrigues, R.; Pirastru, M.; Di Prima, S. Previous land use affects the recovery of soil hydraulic properties after forest restoration. *Water* **2018**, *10*, 453. [CrossRef]

22. Di Prima, S.; Castellini, M.; Abou Najm, M.R.; Stewart, R.D.; Angulo-Jaramillo, R.; Winiarski, T.; Lassabatere, L. Experimental Assessment of a New Comprehensive Model for Single Ring Infiltration Data. *J. Hydrol.* **2019**, *573*, 937–951. [CrossRef]

23. Lassabatère, L.; Angulo-Jaramillo, R.; Ugalde, J.M.S.; Cuenca, R.; Braud, I.; Haverkamp, R. Beerkan estimation of soil transfer parameters through infiltration experiments: BEST. *Soil Sci. Soc. Am. J.* **2006**, *70*, 521–532. [CrossRef]

24. Bagarello, V.; Castellini, M.; Di Prima, S.; Iovino, M. Soil hydraulic properties determined by infiltration experiments and different heights of water pouring. *Geoderma* **2014**, *213*, 492–501. [CrossRef]

25. Alagna, V.; Bagarello, V.; Di Prima, S.; Guaitoli, F.; Iovino, M.; Keesstra, S.; Cerdà, A. Using Beerkan experiments to estimate hydraulic conductivity of a crusted loamy soil in a Mediterranean vineyard. *J. Hydrol. Hydromech.* **2019**, *67*, 191–200. [CrossRef]

26. Di Prima, S.; Rodrigo-Comino, J.; Novara, A.; Iovino, M.; Pirastru, M.; Keesstra, S.; Cerda, A. Assessing soil physical quality of citrus orchards under tillage, herbicide and organic managements. *Pedosphere* **2018**, *28*, 463–477. [CrossRef]

27. Castellini, M.; Iovino, M.; Pirastru, M.; Niedda, M.; Bagarello, V. Use of BEST procedure to assess soil physical quality in the Baratz Lake catchment (Sardinia, Italy). *Soil Sci. Soc. Am. J.* **2016**, *80*, 742–755. [CrossRef]

28. Khaledian, M.R.; Shabanpour, M.; Alinia, H. Saturated hydraulic conductivity variation in a small garden under drip irrigation. *Geosyst. Eng.* **2016**, *19*, 266–274. [CrossRef]

29. Mubarak, I.; Angulo-Jaramillo, R.; Mailhol, J.C.; Ruelle, P.; Khaledian, M.; Vauclin, M. Spatial analysis of soil surface hydraulic properties: Is infiltration method dependent? *Agric. Water Manag.* **2010**, *97*, 1517–1526. [CrossRef]

30. Lassabatère, L.; Di Prima, S.; Angulo-Jaramillo, R.; Keesstra, S.; Salesa, D. Beerkan multi-runs for characterizing water infiltration and spatial variability of soil hydraulic properties across scales. *Hydrol. Sci. J.* **2019**. [CrossRef]

31. Cavallo, G.; De Benedetto, D.; Castrignanò, A.; Quarto, R.; Vonella, A.V.; Buttafuoco, G. Use of geophysical data for assessing 3D soil variation in a durum wheat field and their association with crop yield. *Biosyst. Eng.* **2016**, *152*, 28–40. [CrossRef]

32. De Benedetto, D.; Castrignanò, A.; Rinaldi, M.; Ruggieri, S.; Santoro, F.; Figorito, B.; Gualano, S.; Diacono, M.; Tamborrino, R. An approach for delineating homogeneous zones by using multi-sensor data. *Geoderma* **2013**, *199*, 117–127. [CrossRef]

33. Barca, E.; De Benedetto, D.; Stellacci, A.M. Contribution of EMI and GPR proximal sensing data in soil water content assessment by using linear mixed effects models and geostatistical approaches. *Geoderma* **2019**, *343*, 280–293. [CrossRef]

34. Ventrella, D.; Stellacci, A.M.; Castrignanò, A.; Charfeddine, M.; Castellini, M. Effects of crop residue management on winter durum wheat productivity in a long term experiment in Southern Italy. *Eur. J. Agron.* **2016**, *77*, 188–198. [CrossRef]

35. Campi, P.; Mastrorilli, M.; Stellacci, A.M.; Modugno, F.; Palumbo, A.D. Increasing the effective use of water in green asparagus through deficit irrigation strategies. *Agric. Water Manag.* **2019**, *217*, 119–130. [CrossRef]

36. Myers, D.E. Interpolation and estimation with spatially located data. *Chemom. Intell. Lab. Syst.* **1991**, *11*, 209–228. [CrossRef]

37. Bagarello, V.; Di Prima, S.; Iovino, M. Comparing alternative algorithms to analyze the Beerkan infiltration experiment. *Soil Sci. Soc. Am. J.* **2014**, *78*, 724–736. [CrossRef]
38. Gee, G.W.; Bauder, J. *Particle-size Analysis, In Methods of Soil Analysis, Part 1*, 2nd ed.; American Society of Agronomy/Soil Science Society of America: Madison, WI, USA, 1986.
39. Vitti, C.; Stellacci, A.M.; Leogrande, R.; Mastrangelo, M.; Cazzato, E.; Ventrella, D. Assessment of organic carbon in soils: A comparison between the Springer–Klee wet digestion and the dry combustion methods in Mediterranean soils (Southern Italy). *Catena* **2016**, *137*, 113–119. [CrossRef]
40. Castellini, M.; Di Prima, S.; Iovino, M. An assessment of the BEST procedure to estimate the soil water retention curve: A comparison with the evaporation method. *Geoderma* **2018**, *320*, 82–94. [CrossRef]
41. Reynolds, W.D.; Drury, C.F.; Yang, X.M.; Fox, C.A.; Tan, C.S.; Zhang, T.Q. Land management effects on the near-surface physical quality of a clay loam soil. *Soil Tillage Res.* **2007**, *96*, 316–330. [CrossRef]
42. Dexter, A.R. Soil physical quality. Part I. Theory, effects of soil texture, density, and organic matter, and effects on root growth. *Geoderma* **2004**, *120*, 201–214. [CrossRef]
43. Pieri, C.J.M.G. Fertility of soils: A Future for Farming in the West African Savannah. In *Fertility of Soils: A Future for Farming in the West African Savannah*; Springer: Berlin, Germany, 1992; p. 558.
44. Sequi, P.; De Nobili, M., VII. Carbonio organico. In *Metodi di Analisi Chimica del Suolo. Ministero per le Politiche Agricole e Forestali, Osservatorio Nazionale Pedologico e per la Qualità del Suolo*; Violante, P., Ed.; Franco Angeli Editore: Milano, Italy, 2000.
45. Castellini, M.; Pirastru, M.; Niedda, M.; Ventrella, D. Comparing physical quality of tilled and no-tilled soils in an almond orchard in southern Italy. *Ital. J. Agron.* **2013**, *8*, 149–157. [CrossRef]
46. Reynolds, W.D.; Drury, C.F.; Yang, X.M.; Tan, C.S.; Yang, J.Y. Impacts of 48 years of consistent cropping, fertilization and land management on the physical quality of a clay loam soil. *Can. J. Soil Sci.* **2014**, *94*, 403–419. [CrossRef]
47. Barca, E.; Bruno, E.; Bruno, D.E.; Passarella, G. GTest: A software tool for graphical assessment of empirical distributions' Gaussianity. *Environ. Monit. Assess.* **2016**, *188*, 138. [CrossRef]
48. Moran, P.A.P. Notes on continuous stochastic phenomena. *Biometrika* **1950**, *37*, 17–23. [CrossRef]
49. Rura, M.J.; Griffith, D.A. Spatial Statistics in SAS. In *Handbook of Applied Spatial Analysis*; Fischer, M., Getis, A., Eds.; Springer: Berlin/Heidelberg, 2010.
50. Barca, E.; Porcu, E.; Bruno, D.; Passarella, G. An automated decision support system for aided assessment of variogram models. *Environ. Model. Softw.* **2017**, *87*, 72–83. [CrossRef]
51. Ribeiro, P.J., Jr.; Diggle, P.J. geoR: A package for geostatistical analysis. *R News* **2001**, *1/2*, 14–18.
52. Vieira, S.R.; Hatfield, T.L.; Nielsen, D.R.; Biggar, J.W. Geostatistical theory and application to variability of some agronomical properties. *Hilgardia Berkeley.* **1983**, *51*, 1–75. [CrossRef]
53. Isaaks, E.H.; Srivastava, R.M. *An Introduction to Applied Geostatistics*; Oxford University press: New York, NY, USA, 1989.
54. Stein, A.; Brouwer, J.; Bouma, J. Methods for comparing spatial variability patterns of millet yield and soil data. *Soil Sci. Soc. Am. J.* **1997**, *61*, 861–870. [CrossRef]
55. Barca, E.; Passarella, G. Spatial evaluation of the risk of groundwater quality degradation. A comparison between disjunctive kriging and geostatistical simulation. *Environ. Monit. Assess.* **2008**, *137*, 261–273. [CrossRef]
56. Cambardella, C.A.; Moorman, T.B.; Novak, J.M.; Parkin, T.B.; Karlen, D.L.; Turco, R.F.; Konopka, A.E. Field-scale variability of soil properties in central Iova soils. *Soil Sci. Soc. Am. J.* **1994**, *58*, 1501–1511. [CrossRef]
57. Hoshino, A.; Tamura, K.; Fujimaki, H.; Asano, M.; Ose, K.; Higashi, T. Effects of crop abandonment and grazing exclusion on available soil water and other soil properties in a semi-arid Mongolian grassland. *Soil Tillage Res.* **2009**, *105*, 228–235. [CrossRef]
58. Iovino, M.; Castellini, M.; Bagarello, V.; Giordano, G. Using static and dynamic indicators to evaluate soil physical quality in a Sicilian area. *Land Degrad. Dev.* **2016**, *27*, 200–210. [CrossRef]
59. Topp, G.C.; Reynolds, W.D.; Cook, F.J.; Kirby, J.M.; Carter, M.R. Physical attributes of soil quality. In *Soil Quality for Crop Production and EcosystemHealth*; Developments in Soil Science, volume 25; Gregorich, E.G., Carter, M.R., Eds.; Elsevier: New York, NY, USA, 1997; pp. 21–58.
60. Reynolds, W.D.; Yang, X.M.; Drury, C.F.; Zhang, T.Q.; Tan, C.S. Effects of selected conditioners and tillage on the physical quality of a clay loam soil. *Can. J. Soil Sci.* **2003**, *83*, 318–393. [CrossRef]

61. Agnese, C.; Bagarello, V.; Baiamonte, G.; Iovino, M. Comparing physical quality of forest and pasture soils in a Sicilian Watershed. *Soil Sci. Soc. Am. J.* **2011**, *75*, 1958. [CrossRef]
62. Cullotta, S.; Bagarello, V.; Baiamonte, G.; Gugliuzza, G.; Iovino, M.; La Mela Veca, D.S.; Maetzke, F.; Palmeri, V.; Sferlazza, S. Comparing Different Methods to Determine Soil Physical Quality in a Mediterranean Forest and Pasture Land. *Soil Sci. Soc. Am. J.* **2016**, *80*, 1038–1056. [CrossRef]
63. Ventrella, D.; Virzì, N.; Intrigliolo, F.; Palumbo, M.; Cambrea, M.; Platania, A.; Sciacca, F.; Licciardello, S.; Troccoli, A.; Russo, M.; et al. Environmental effectiveness of GAEC cross-compliance standard 2.1 'Maintaining the level of soil organic matter through management of stubble and crop residues' and economic evaluation of the competitiveness gap for farmers. *Ital. J. Agron.* **2015**, *10*, 697.
64. Reynolds, W.D.; Drury, C.F.; Yang, X.M.; Tan, C.S. Optimal soil physical quality inferred through structural regression and parameter interactions. *Geoderma* **2008**, *146*, 466–474. [CrossRef]
65. Castellini, M.; Niedda, M.; Pirastru, M.; Ventrella, D. Temporal changes of soil physical quality under two residue management systems. *Soil Use Manag.* **2014**, *30*, 423–434. [CrossRef]
66. Ferrara, R.M.; Mazza, G.; Muschitiello, C.; Castellini, M.; Stellacci, A.M.; Navarro, A.; Lagomarsino, A.; Vitti, C.; Rossi, R.; Rana, G. Short-term effects of conversion to no-tillage on respiration and chemical-physical properties of the soil: A case study in a wheat cropping system in semi-dry environment. *Ital. J. Agrometeorol.* **2017**, *1*, 47–58.
67. Bagarello, V.; Di Stefano, C.; Ferro, V.; Iovino, M.; Sgroi, A. Physical and hydraulic characterization of a clay soil at the plot scale. *J. Hydrol.* **2010**, *387*, 54–64. [CrossRef]
68. Dexter, A.R. Advances in characterization of soil structure. *Soil Tillage Res.* **1988**, *11*, 199–238. [CrossRef]

Article

Modelling Soil Water Dynamics from Soil Hydraulic Parameters Estimated by an Alternative Method in a Tropical Experimental Basin

Bruno Silva Ursulino [1,*], Suzana Maria Gico Lima Montenegro [1], Artur Paiva Coutinho [1,2,*], Victor Hugo Rabelo Coelho [3], Diego Cezar dos Santos Araújo [4], Ana Cláudia Villar Gusmão [1], Severino Martins dos Santos Neto [1], Laurent Lassabatere [5] and Rafael Angulo-Jaramillo [5]

[1] Centro de Tecnologia e Geociências, Universidade Federal de Pernambuco, Recife 50670-901, Brazil; suzanam.ufpe@gmail.com (S.M.G.L.M.); villarelunagusmao@gmail.com (A.C.V.G.); martinsdsn@gmail.com (S.M.d.S.N.)

[2] Centro Acadêmico do Agreste, Universidade Federal de Pernambuco, Caruaru 55014-900, Brazil

[3] Departamento de engenharia Civil e Ambiental, Universidade Federal da Paraíba, João Pessoa 58051-900, Brazil; victor-coelho@hotmail.com

[4] Departamento de Engenharia Agrícola, Universidade Federal Rural de Pernambuco, Recife 52171-900, Brazil; diego@agro.eng.br

[5] Laboratoire d'Ecologie des Hydrosystèmes Naturels et Anthropisés, Université de Lyon, site ENTPE, 69120 Vaulx-en-Velin, France; laurent.lassabatere@entpe.fr (L.L.); rafael.angulojaramillo@entpe.fr (R.A.-J.)

[*] Correspondence: brunosenga@gmail.com (B.S.U.); arthur.coutinho@yahoo.com.br (A.P.C.)

Received: 27 February 2019; Accepted: 9 May 2019; Published: 14 May 2019

Abstract: Knowledge about soil moisture dynamics and their relation with rainfall, evapotranspiration, and soil physical properties is fundamental for understanding the hydrological processes in a region. Given the difficulties of measurement and the scarcity of surface soil moisture data in some places such as Northeast Brazil, modelling has become a robust tool to overcome such limitations. This study investigated the dynamics of soil water content in two plots in the Gameleira Experimental River Basin, Northeast Brazil. For this, Time Domain Reflectometry (TDR) probes and Hydrus-1D for modelling one-dimensional flow were used in two stages: with hydraulic parameters estimated with the Beerkan Estimation of Soil Transfer Parameters (BEST) method and optimized by inverse modelling. The results showed that the soil water content in the plots is strongly influenced by rainfall, with the greatest variability in the dry–wet–dry transition periods. The modelling results were considered satisfactory with the data estimated by the BEST method (Root Mean Square Errors, RMSE = 0.023 and 0.022 and coefficients of determination, R^2 = 0.72 and 0.81) and after the optimization (RMSE = 0.012 and 0.020 and R^2 = 0.83 and 0.72). The performance analysis of the simulations provided strong indications of the efficiency of parameters estimated by BEST to predict the soil moisture variability in the studied river basin without the need for calibration or complex numerical approaches.

Keywords: soil moisture content; vadose zone; soil properties; BEST model; Hydrus-1D

---

1. Introduction

Surface soil moisture plays a key role in the hydrological cycle as it controls the water fluxes between soil, vegetation, and atmosphere [1–3]. Moreover, knowledge about soil moisture is widely required in many agricultural studies and applications related to irrigation management [1,4–6]. Therefore, monitoring and understanding of soil moisture variability and its exchange relationships with the surface and atmosphere are essential to improve weather forecasting, flooding and drought predictions, and climate projections [7,8]. However, soil moisture is highly variable in space and time due to the combined influence of many factors in a nonlinear fashion such as the hydraulic properties of the soil, topographic characteristics, interaction with surface water systems, precipitation features, and additional meteorological conditions [3,9,10]. Because of its high variability in natural conditions, achieving accuracy in soil moisture estimation to obtain areal information at different spatial and temporal scales is still a challenging task, especially at high spatiotemporal scales [6,11]. In the last 40 years, the scientific community has clearly recognized the importance of soil moisture as input for earth science applications, developing new approaches and techniques for monitoring, modelling, and use of soil moisture data [12].

Currently, three main approaches are used to provide soil moisture estimates: in situ (generally point-scale) measurements, remote sensing observations, and hydrological modelling applications [12,13]. The most accurate and direct methods to determine soil moisture are related to experimental techniques using classical devices such as soil sampling and Time Domain Reflectometry (TDR) probes [14]. However, these measurements are expensive and time-consuming and provide point-scale information but are unable to detect the spatial heterogeneity of the soil moisture at wider scales [6]. The use of remote sensing and modelling techniques has become a viable solution to overcome the limitations of in situ data [15,16]. For instance, remote sensing techniques provide surface soil moisture estimates over large areas by using active sensors operating in microwave bands that are not influenced by solar radiation and cloud cover [6]. However, the spatial and temporal resolutions of the current remote sensing products are still insufficient for small-scale applications (<1 km^2) [17], with soil moisture data only available for the first few centimetres of the soil layer [14,18]. Consequently, they do not cover the soil water in the root zone or over the whole profile, which is most interesting for hydrologists [13]. The limitations imposed by the complex interactions between vegetation, soil, and climate make the use of models based on physical processes an effective and viable (low-cost) alternative for evaluating the spatiotemporal soil moisture trends at different depths [2,19,20].

Many models with different levels of complexity and accuracy have been proposed to simulate the dynamics of water in the unsaturated zone of soil; they include, among others, the Soil Water Atmosphere Plant (SWAP) [21], Soil-Vegetation-Atmosphere Transfer (SVAT) [22], Soil Water Infiltration and Movement (SWIM) [23], Groundwater Loading Effects of Agricultural Management Systems (GLEAMS) [24], Water and Agrochemicals in the soil, crop and Vadose Environment (WAVE) [25], and Hydrus models. Most of the abovementioned models use the Richards' equation to represent the movement of water in unsaturated soil [26], allowing a detailed description of the soil water content distribution and fluxes inside the soil domain [27]. Hydrus-1D model is one of the most common numerical solutions based on the Richards' equation [28]. Over the years, Hydrus-1D model has been successfully applied in various studies worldwide for predicting soil moisture content and water movement under different conditions (e.g., [13,29–33]). As a physically based model, Hydrus-1D requires some input data for simulations such as meteorological data for surface boundary conditions, soil physical conditions, and physical parameters (saturated hydraulic conductivity and soil water retention curve), which may be obtained through experimentation [27,34–36]. Moreover, according to Bordoni et al. [2], for a better implementation of this type of model, a preliminary calibration of some hydrological parameters used in the methodology is sometimes required.

When performed in the laboratory, the methods commonly used to determine soil physical parameters may not be accurate enough to represent the actual field conditions or may be tedious and time-consuming [36,37]. As an alternative, the scientific community has developed simplified methods that are very often able to provide more accurate results, such as Beerkan [38,39]. Beerkan is a simple and easy method based on in situ single-ring water infiltration experiments which can be carried out in the field at low cost [40]. BEST (Beerkan Estimation of Soil Transfer Parameters) is an algorithm presented by Lassabatere et al. [41] which makes it possible to process the infiltration tests carried out using the Beerkan method [42]. The Hydrus-1D model is also able to estimate the soil physical parameters in a simple way through the inverse method, which has been widely used by researchers (e.g., [37,43–45]), requiring the use of other types of data observed in the field.

However, there is a lack of studies on the performance of Hydrus-1D to simulate the temporal variability of soil moisture from soil physical parameters estimated by the BEST algorithm.

This study focuses on the use of the combination of Hydrus-1D model with BEST method to estimate the temporal variability of soil moisture in an important experimental river basin located near to a highly urbanized area in Northeast Brazil. Specifically, this study aims to: (i) quantify the temporal variability of soil moisture (storage and movement) by using rainfall as the only water input (without irrigation) of the system over two years (2015–2016); (ii) analyze the response of the Hydrus-1D model to simulate the soil moisture dynamics by implementing soil hydrodynamic parameters estimated in the field by the BEST method; and (iii) verify the performance of the Hydrus-1D model after the calibration (inverse modelling) of the soil hydrodynamic parameters. The comparison of modelled data obtained with BEST method and those optimized using numerical inversion allows the characterization of the consistency of BEST estimates for the modelling of water flow. To the authors' knowledge, such a validation of the use of BEST method has never been done before. Moreover, information about soil moisture dynamics in tropical regions where the monitoring network is scarce are relevant, especially the knowledge of water transfer processes.

2. Materials and Methods

2.1. Study Site Description

This study was carried out in Gameleira Experimental River Basin (GERB), Pernambuco state, Northeast Brazil (Figure 1). The experimental basin covers 17 km^2 between the coordinates 8°04' and 8°06' S and 35°17' and 35°20' W. GERB is a sub-basin of the Tapacurá Representative River Basin (TRRB), which plays an important role in water supply and flood control for the Recife Metropolitan Region (RMR) [46], the fourth largest urban area in Brazil, hosting four million inhabitants. The experimental and representative catchments have been monitored by the Network of Hydrology of the Semi-Arid Region (REHISA), which currently consists of eight federal universities in Brazil [47]. In the last years, the hydrology of the GERB has been studied with regard to surface runoff [48], sediment yield [49,50], hydrodynamic soil characterization [51], and evapotranspiration [52,53].

GERB is strategically located near the transition zone between the humid (Atlantic Forest, tropical woodland) and the semi-arid zone (Caatinga, scrub vegetation). The study site is mostly covered by irrigated sugarcane and vegetable crops, evidencing the necessity of practical and precise studies considering information on the water-soil system in the region. Secondary natural vegetation and grassland for extensive cattle-raising are also found in GERB. According to Köppen's classification proposed by Alvares et al. [54] for the Brazilian territory, the climate in GERB is tropical with dry summer (As), with an average temperature above 23 °C. The average annual rainfall in the experimental basin is approximately 1050 mm, most of which is concentrated from March to July, when the relative air humidity is higher than 70% [46].

The catchment altitude ranges between 140 and 430 m above sea level, with many points presenting high slopes. The dominant soil types in GERB are Acrisols and Gleysols, according to the WRB 2006 [55]. Acrisols are well drained, with structure developed in blocks and a large percentage of silt in their composition. On the other hand, Gleysols occur at the banks of the watercourses and are characerized by a low coefficient of infiltration when saturated and low hydraulic conductivity [56].

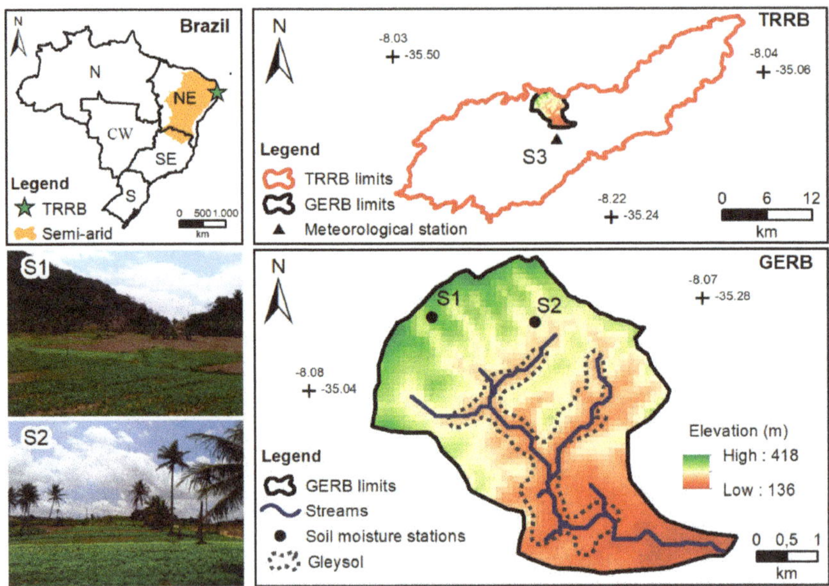

Figure 1. The location of Tapacaurá Representative River Basin (TRRB) and Gameleira Experimental River Basin (GERB), showing the soil moisture stations (S1 and S2) and meteorological monitoring station (S3).

2.2. Data Collection

Soil moisture data were automatically measured at two monitoring stations (S1 and S2, Figure 1) by using TDR probes (CS616, Campbell Scientific). Data from monitoring stations S1 and S2 were collected between March 2014 and December 2016. One TDR probe per site was vertically inserted in the soil to indirectly represent the average water content along the first 30 cm downward from the surface, i.e., over the depth of 0–30 cm. The two TDR probes were calibrated in the laboratory, with the reproduction of the field conditions in soil column tests. Table 1 shows some soil characteristics of monitoring stations S1 and S2, such as the soil type, textural characterization, bulk density, and particle density. Overall, these two sites present similar physical properties of the soil and homogeneity of the profiles along the depths of 0–10, 10–20, and 20–30 cm. During the rainy season, the groundwater levels are approximately 1 and 4 m above the soil surface in stations S1 and S2, respectively. Both monitored sites are covered by sparse grass frequently trimmed, with root depth inferior to 10 cm.

Table 1. Physical properties of the soil at the monitoring stations.

Site	Soil Type	Depth (cm)	Soil Particle Composition			Bulk Density	Particle Density
			Sand (%)	Silt (%)	Clay (%)	(g cm^{-3})	
		0–10	71.61	20.06	8.33	1.54	2.62
S1	Acrisol	10–20	71.72	18.68	9.60	1.50	2.60
		20–30	76.14	14.11	9.75	1.51	2.58
		0–10	67.56	24.01	8.43	1.58	2.58
S2	Acrisol	10–20	60.40	21.26	18.34	1.62	2.56
		20–30	65.88	23.10	11.02	1.64	2.58

Daily meteorological data were also used in this study to calculate the potential evapotranspiration (ET_p) by Penman's formulation [57] modified by Shuttleworth [58]. This formulation is a physically-based form of ET_p, which means that all key variables that govern the evaporative process are explicit in the equation [59]. Donohue et al. [60] demonstrated that this form of ET_p is the most appropriated when considering a changing climate, which was also used by Coelho et al. [61] in a study carried out in the Brazilian Northeast region. The data necessary to calculate the ET_p by the Penman's formulation were acquired from a meteorological station (named S3) operated by Pernambuco Water and Climate Agency (APAC). Meteorological station S3 is approximately 6 km from stations S1 and S2 and automatically monitors the following variables: rainfall, air temperature, relative humidity, atmospheric pressure, wind speed, and solar radiation. In addition, rainfall data were also obtained by using two automatic rain gauges installed in stations S1 and S2.

2.3. Modelling the Unsaturated Water Flow

Unsaturated water flow in GERB was modelled using the Hydrus-1D package, a free modelling environment that allows the simulation of water, heat, and solute transport in variably saturated media under permanent or transient regimes [28]. The governing equation of one-dimensional water flow for a partially saturated porous medium can be described by a modified form of the Richards equation [26], as described in Equation (1). The application of this modified form assumes that the air phase plays an insignificant role in the liquid flow process and the water flow due to thermal gradients, and therefore they can be neglected [62].

$$\frac{\partial \theta}{\partial t} = \frac{\partial}{\partial z}\left[K(h)\left(\frac{\partial h}{\partial z}+1\right)\right]-S(h) \tag{1}$$

where h is the water pressure head (cm), θ is the volumetric water content ($cm^3\ cm^{-3}$), t is time (day), x is the vertical coordinate axis (cm) (positive upward), $K(h)$ is the hydraulic conductivity ($cm\ day^{-1}$), and $S(h)$ is the sink term in the flow equation ($cm^3\ cm^{-3}\ day^{-1}$), which represents the root water uptake (actual transpiration), obtained using the Feddes equation (more information can be found in the manual [28]). $K(h)$ and $\theta(h)$ are obtained from the following functions according to van Genuchten [63]:

$$\theta(h) = \theta_r + \frac{\theta_s - \theta_r}{[1+|\alpha h|^n]^m}, \quad h < 0 \tag{2}$$

$$\theta(h) = \theta_s, \quad h \geq 0 \tag{3}$$

$$K(h) = K_s S_e^l \left[1-\left(1-S_e^{1/m}\right)^m\right]^2 \tag{4}$$

where θ_r is the residual water content ($cm^3\ cm^{-3}$), θ_s is the saturated water content ($cm^3\ cm^{-3}$), K_s is the saturated hydraulic conductivity ($cm\ day^{-1}$), l is the shape factor in the hydraulic conductivity function (assumed to be equal to 0.5), m, α, and n are empirical shape factors in the water retention function ($m = 1 - 1/n$), and S_e is the relative saturation, calculated as follows:

$$S_e = \frac{\theta - \theta_r}{\theta_s - \theta_r} \tag{5}$$

2.4. Estimation of Soil Hydraulic Parameters Using BEST Methods

The soil hydraulic parameters required for the Hydrus-1D simulation were obtained by the BEST method [64]. The BEST method has wide advantages when compared to other classical methods because it provides a complete characterization of both soil water retention and hydraulic conductivity functions. BEST is based on the van Genuchten [63] relationship for the water retention curve, by using the Burdine [65] condition and the Brooks and Corey [66] relationship for hydraulic conductivity

(Equation (6) [65]) and Equation (7). These relationships were used because Fuentes et al. [67] found they gave more accuracy when describing the hydraulic behaviour of most soil types analyzed.

$$\frac{\theta - \theta_r}{\theta_s - \theta_r} = \left[1 + \left(\frac{h}{h_g} \right)^n \right]^{-m} \text{ with } m = 1 - \frac{2}{n} \tag{6}$$

$$\frac{K(\theta)}{K_s} = \left(\frac{\theta - \theta_r}{\theta_s - \theta_r} \right)^\eta \tag{7}$$

where h_g is the scale parameter (cm) of $\theta(h)$, and η is the shape parameter of the $K(\theta)$ relationship. Usually, θ_r is very low and is assumed to be zero in BEST. The saturated water content (θ_s) is derived from the value of bulk density, being assumed to be equal to the porosity. This set of formulae for water retention and hydraulic conductivity functions is referred to as the vGBC functions (van Genucthen–Brooks and Corey functions).

BEST methods require two sets of data to estimate all the hydraulic parameters: (i) the Particle-Size Distribution (PSD) of the soil and bulk density and (ii) cumulative infiltration along with the initial and final soil water contents of the infiltration experiment [40,41]. Three algorithms were developed to analyze the infiltration data: (i) the original version, namely BEST-slope [41], (ii) an improved method for coarse media, namely BEST-intercept [68], and (iii) a third method, BEST-steady [69], based on the analysis of the last part of the cumulative infiltration [40]. The difference between the three methods lies in the way in which the cumulative infiltration data are processed. In this study the BEST-slope and BEST-intercept algorithms were used; however, the results of the first algorithm were discarded due to divergent values of K_s. According to Yilmaz et al. [68], the BEST-slope may lead to erroneous values of K_s, especially when a very high level of precision relative to the steady-state infiltration rate cannot be obtained.

In this study, the infiltration experiments were performed at depths of approximately 10 cm from the soil surface in order to represent the range of soil moisture obtained by the probes installed in the field. For the experiments, fixed water volumes (70 mL) were poured into a 75-mm-diameter ring, leading to a ponded water thickness of a few millimetres. Three infiltration experiments were performed at each of the monitoring stations S1 and S2 to improve the representativeness of the site. The time required for the infiltration of each water volume was measured until steady state was reached and the surface soil was sampled at the end of each experiment. These collected samples were used to determine the final profiles of θ_s and to analyze the particle size of the soils, as previously done by Coutinho et al. [40]. Although BEST assumes that θ_r is equal to zero, the value of this parameter was determined in the laboratory and fixed for the initial model simulations.

The shape parameters n, m, and η are derived from the pedotransfer functions detailed in BEST [41]. Then, the cumulative infiltrations are treated and fitted to the models developed for water infiltration into the disc source by Haverkamp et al. [38] to derive the field-saturated soil hydraulic conductivity (K_s) and soil sorptivity (S). The scale parameter for the water retention curve, hg, is then obtained from the estimates of K_s and S. In this step, all unsaturated hydraulic parameters, namely θ_r, θ_s, n (and thus m), η, h_g, and K_s, are fully estimated and the complete water retention and hydraulic conductivity functions are determined [40,41].

The retention curve parameters obtained by the BEST method were adjusted in this study because the Hydrus-1D model uses the van Genuchten equation with the Mualem [70] pore distribution condition. Therefore, the Retension Curve (RETC) program elaborated by van Genuchten et al. [71] was used. RETC uses the parametric models of vGBC functions to fit the hydraulic curves previously determined with BEST. After adjusting the retention curve, the parameters $\alpha = 1/h_g$ (cm^{-1}) and a new value of n were obtained.

2.5. Boundary Condition

For the atmosphere boundary condition, the variable flow of daily precipitation and potential evapotranspiration were used, considering free drainage for the lower boundary condition. Hydrus-1D requires separated input values of both potential evaporation (the upper boundary condition) and potential transpiration (the sink term in the Richards equation) fluxes at a daily time step to simulate the influence of soil water on transpiration (root water uptake) [62]. Then, potential evaporation (E_p) and potential transpiration (T_p) were separated according to the model based on the Leaf Area Index (LAI) using Beer's law [72] as follows:

$$T_p = ET_p\left[1 - e^{(-k\, \text{LAI})}\right] \tag{8}$$

$$E_p = ET_p e^{(-0.463\, \text{LAI})} \tag{9}$$

where k is an extension coefficient for global solar radiation and depends on the angle of the sun, the distribution of plants, and the arrangement of leaves [62,73]. In this study, we adopted the same k-value of 0.463 that was considered by Chen et al. [13] and Šimůnek [28]. The value of LAI was assumed to be equal to 1.0 and constant for the whole period of simulation because S1 and S2 are installed in a sparse lawn area. This constant LAI with time is the same used by Chen et al. [13], which selected values in a feasible range for grasses and found LAI = 1 the more consistent after optimization and calibration of the parameters. The partitioned daily values of ET_p were then used as input data for the model (Figure 2).

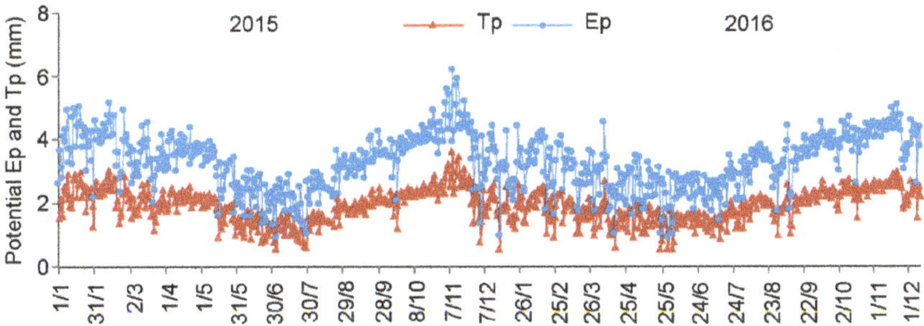

Figure 2. Daily time series of evaporation (E_p) and transpiration (T_p) for S3 estimated using the Penman-Monteith equation from 2015 to 2016.

2.6. Time and Space Discretisation

Regarding the geometry of the numerical domain, we considered a profile with a depth of 60 cm composed of a single control layer (corresponding to the physical and hydrodynamic properties obtained in the infiltration test performed for the 10 cm layer). This assumption of a homogeneous soil along the first 60 cm of the profile was considered after the collection of samples for soil characterization, where the same structural class was found. The numerical domain is made of a compact mesh of 201 nodes. Daily time discretisation was used in this study, with the initial time considered as 0 (the day before the simulation) and the final time referring to the days for calibration and validation. The soil moisture measured on the day before the simulation start was used as the initial condition. For the convergence criteria, the maximum number of 10 iterations for the resolution of the Richards non-linear equation was chosen, with a value of 0.001 for the tolerance of the water content in the unsaturated region of the nodes, in agreement with the default values [28]. As mentioned above, the hydraulic curves were described using the van Genuchten model for the water retention curve [63], with the Mualem condition, and the Mualem model for hydraulic conductivity. This set of formulae for water retention and hydraulic conductivity functions is referred to as the vGM functions. We did not consider the hysteresis phenomenon.

2.7. Simulations of Volumetric Water Content and Model Evaluation

The simulations were started with the hydraulic parameters obtained by the BEST method and adjusted to the vGM formulations for the period of two years. Then, optimization of the hydraulic parameters was performed by inverse modelling with the model parameters adjusted to minimize the difference between observed and simulated values [74]. The parameter algorithm employed by the Hydrus-1D model for optimization purposes is based on the Marquardt-Levenberg method [28]. The optimization was performed by using 365 and 268 days throughout 2015 for S2 and S1, respectively. A reduced number of days was used for this purpose at S1 because the station only started to operate at the end of March 2015.

On the other hand, the year 2016 was used for the validation of the model, performed soon after the calibration of the hydraulic parameters. Three statistical criteria were adopted to analyze the performance of the model: (i) the root mean square error (RMSE); (ii) the Willmott index (d); and (iii) the Nash-Sutcliffe model efficiency coefficient (NSE).

$$\text{RMSE} = \sqrt{\frac{1}{n}\left[\sum_{i=1}^{n}(\theta_o - \theta_m)^2\right]} \tag{10}$$

$$d = 1 - \frac{\sum_{i=1}^{n}(\theta_m - \theta_o)^2}{\sum_{i=1}^{n}\left(\left|\theta_m - \overline{\theta}\right| + \left|\theta_o - \overline{\theta}\right|\right)} \tag{11}$$

$$\text{NSE} = 1 - \frac{\sum_{i=1}^{n}(\theta_o - \theta_m)^2}{\sum_{i=1}^{n}\left(\theta_o - \overline{\theta}\right)^2} \tag{12}$$

where θ_o and θ_m are the observed and modelled soil moisture values, respectively, $\overline{\theta}$ is the mean of the observed moisture values, and n is the number of comparable paired points.

3. Results and Discussion

3.1. Rainfall and Soil Moisture Dynamics

Figure 3 shows the differences between monthly rainfall regimes in the two studied hydrological years and the average monthly rainfall time series from 1970 to 2000 in the Tapacurá representative river basin, TRRB. In 2015, the rainy period occurred between March and July as well as in December. On the other hand, the rainfall was more concentrated from January to May in 2016, which does not correspond to the rainy period registered in the monthly long-term time-series for TRRB [75]. In the two analyzed years, the precipitation in GERB was around 20% less than the annual average for the region, with values of 1821 and 2060 mm registered for S1 and S2, respectively.

The time series of volumetric water content (Figure 4) were obtained based on the calibration curves of CS616 probes performed in the laboratory. Analysing the soil moisture behaviour as a function of precipitation, it is observed that rainfall events above 15 mm generally provide rapid responses in the two stations. The highest increases in soil water content occurred during the rainfall events observed from May to July 2015, with maximum values of 0.32 and 0.33 cm^3 cm^{-3} in S1 and S2, respectively. After the abrupt increase of soil moisture at the end of May 2015, which was influenced by a rainfall event of 61 mm, the water content in the soil remained slightly stable in S1 and S2 until the end of the rainy period in July. In 2016, the soil moisture variability was higher than in 2015 during the rainfall events, exhibiting many abrupt increases and decreases in soil water content. These different soil moisture temporal patterns are probably associated with the divergent behaviour of rainfall between the two analyzed years. While in 2015, the rainy season was more concentrated and without the presence of abrupt decreases of soil moisture, throughout the year 2016, the rain was sparser during the rainy season and consequently depletions in the soil moisture were smoother in the two stations. After the rainy period, the water content in the soil presented a mild depletion during the first days of August in both analyzed years, that is, at the beginning of the dry season. These decreases after the rainy season are more pronounced at S2, probably due to the surrounding vegetation slightly lower, and tend to stabilize at values close to 0.20 cm^3 cm^{-3}.

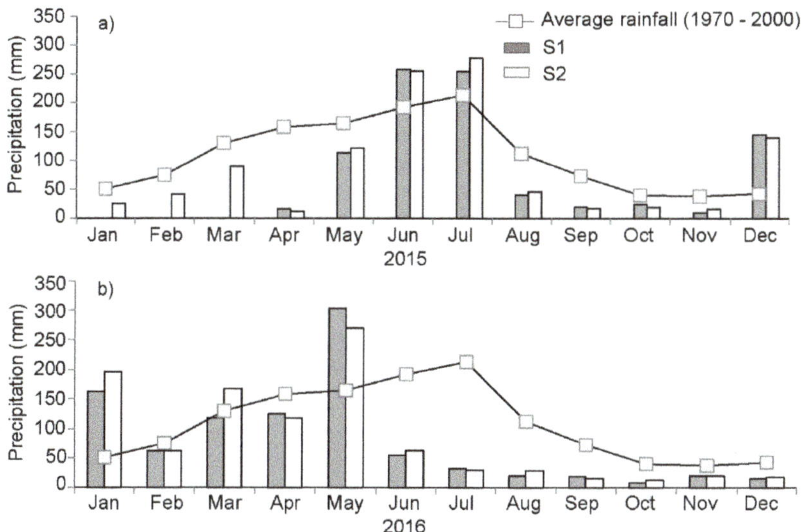

Figure 3. Distribution of total monthly rainfall during the observed period at S1 and S2 in comparison to rainfall monthly historical series (1970–2000) in the Tapacurá representative river basin, TRRB, for 2015 (**a**), and 2016 (**b**).

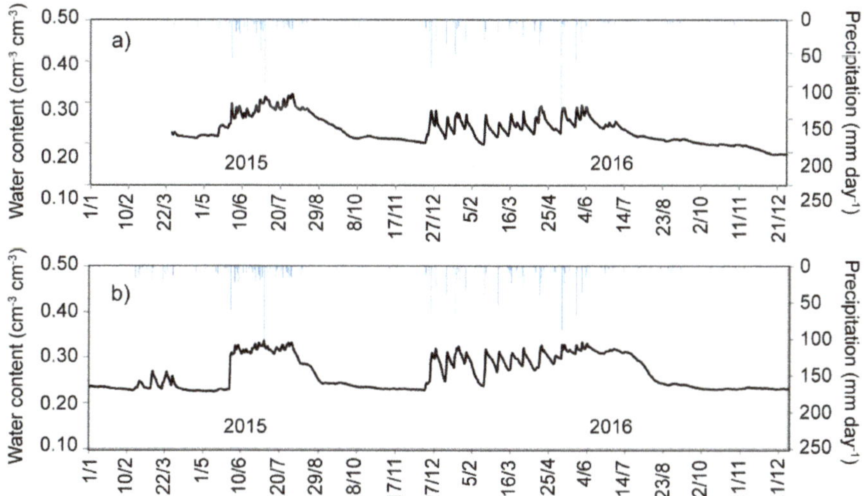

Figure 4. Temporal distribution of soil moisture and precipitation at sites S1 (**a**) and S2 (**b**) during the studied period (2015–2016).

Figure 5 shows box plots with the variability and distribution of the observed mean daily soil moisture (the dot inside the box) for each month at the two monitoring stations, and the median (central marks), minimum, and maximum are also presented. The box delimits the lower and upper quartiles and outliers. The presence of outliers was only noted in 2015 due to the higher values of soil moisture, probably caused by isolated rainfall events. The isolated rainfall events observed in 2015 generated more accentuated increases in soil water content, differing from the other events. The presence of these atypical values can also be associated with the beginning and ending of the rainy season, as observed in May and August.

The highest dispersion of data is observed in the transition months between dry–wet and wet–dry periods, such as August and December 2015 and the first semester of 2016, and during the wetter periods. This variability can be noted from the difference between maximum and minimum values as well as the distance between the quartiles and the median. On the other hand, the driest months presented the lowest variability of data, as expected due to the absence of high rainfall volumes. These lowest variabilities were observed from September to October 2015 and after the end of the rainy season in 2016 (starting in August).

3.2. Hydraulic Characterization of the Sites

Table 2 presents the results of the mean values of the shape (m, n, and η) and normalisation (S, K_s, θ_s, and h_g) parameters of the soils generated by the BEST method for stations S1 and S2. Adjusted parameters (α and n) for the vGM functions using the RETC program are also exhibited in Table 2. The parameters m and n, which depend on the soil texture, presented the highest values in the soil with the sandiest fraction, that is, at station S1. The shape parameters strongly depend on the type of soil, with higher values of parameters n and m for the coarser materials (sand and silty sand) and smaller values of the parameter η for coarse soils [40,76]. The soil at station S1 also exhibits slightly higher values of S and K_s. The estimated values are similar to the results found by Furtunato et al. [51], who also used the BEST method at 102 points distributed throughout the GERB.

Figure 6 shows the hydraulic conductivity and water retention curves of the soils at stations S1 and S2 obtained from the mean values of the hydraulic parameters. The water retention curves were fitted to the vGM functions, generating the new parameters α and n (Table 2).

Accordingly, a similar hydraulic behaviour was verified between the soils present at the studied stations. Both hydraulic curves present an accentuated format typical of coarser materials, with a very marked inflexion point before saturation. This similarity of shapes is in agreement with the proximity of the adjusted values of the parameter n in the two soils (Table 2), as this parameter mainly determines the slope of the retention curve at larger potential values [77]. In this case, water retention is mainly affected by the soil texture [78], which is similar in the two stations (see Table 1). It is possible to observe that the Ks values at station S1 are higher than those noted at S2 for almost all ranges of soil water content. The values of K at the two stations only reach similar amounts when the soils are near to saturation ($\sim$160 cm day^{-1}).

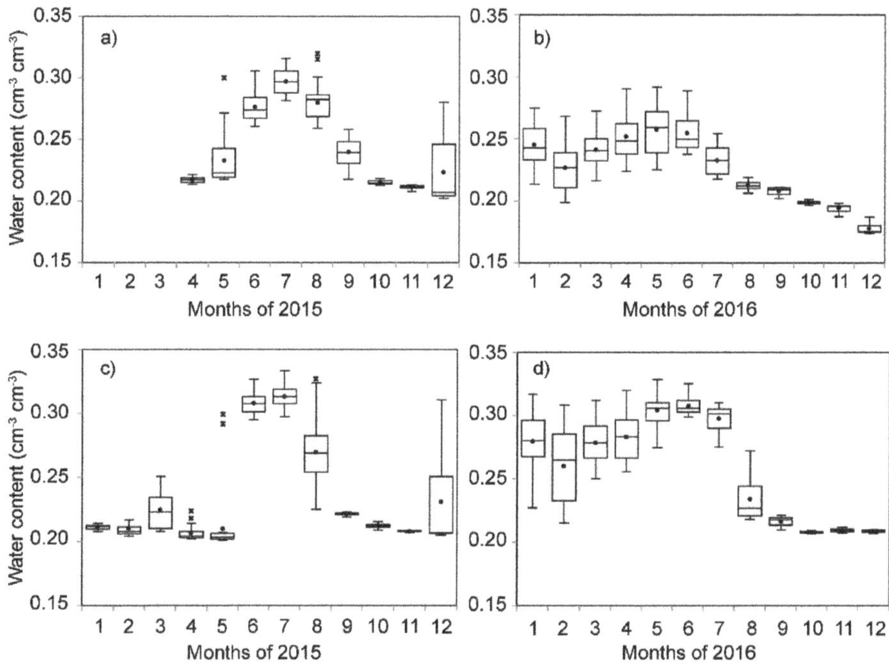

Figure 5. Box plot of monthly soil moisture variability at sites S1 (**a,b**) and S2 (**c,d**) during the studied period (2015–2016, respectively). The dot represents the mean. The box plots present the 10, 25, 50, 75, and 90% quantiles, with the marks after that representing the outliers.

Table 2. Mean values of the hydraulic parameters derived from Beerkan Estimation of Soil Transfer (BEST) and Retension Curve (RETC) of the three replicates performed at each station.

Site	BEST Output							RETC Output		
	m	n	η	S (mm s$^{-1/2}$)	K_s (cm day^{-1})	θ_r [1] (cm^3 cm^{-3})	θ_s (cm^3 cm^{-3})	h_g (cm)	A (cm^{-1})	n
S1	0.102	2.223	11.482	1.31	159.97	0.063	0.377	−18.5	0.041	1.245
S2	0.094	2.204	12.682	1.20	155.52	0.026	0.379	−16.7	0.020	1.273

[1] The value of θ_r was measured in the laboratory.

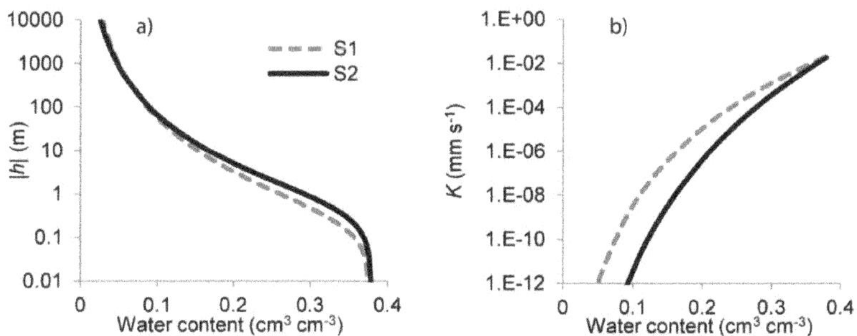

Figure 6. Soil hydraulic characteristics curves: water retention curve (**a**) and hydraulic conductivity (**b**) for the soil of the sites as estimated using BEST and adjusted by the RETC procedure.

3.3. Simulation

The first simulations were performed with the hydraulic parameters estimated by the BEST method and adjusted to vGM functions (Table 2). The results of the simulation at both station S1 and station S2 present similar tendencies of variation, with estimated soil moisture values close to the observed data in almost the entire series throughout the studied period (Figure 7). Exceptions are noted during the dry periods, especially in S2, when the simulated data cannot effectively capture the observed recession curves. This may have been influenced by the lower boundary condition adopted in the simulation, which in this case was free drainage disregarding the depth of the groundwater. It is possible that in the field there is a small contribution of capillarity to soil moisture. Another possible explanation for the underestimated results of the model at the two sites may be the constant value considered for LAI over the whole simulation, promoting periods with higher evaporation rates and consequently reducing soil moisture rates.

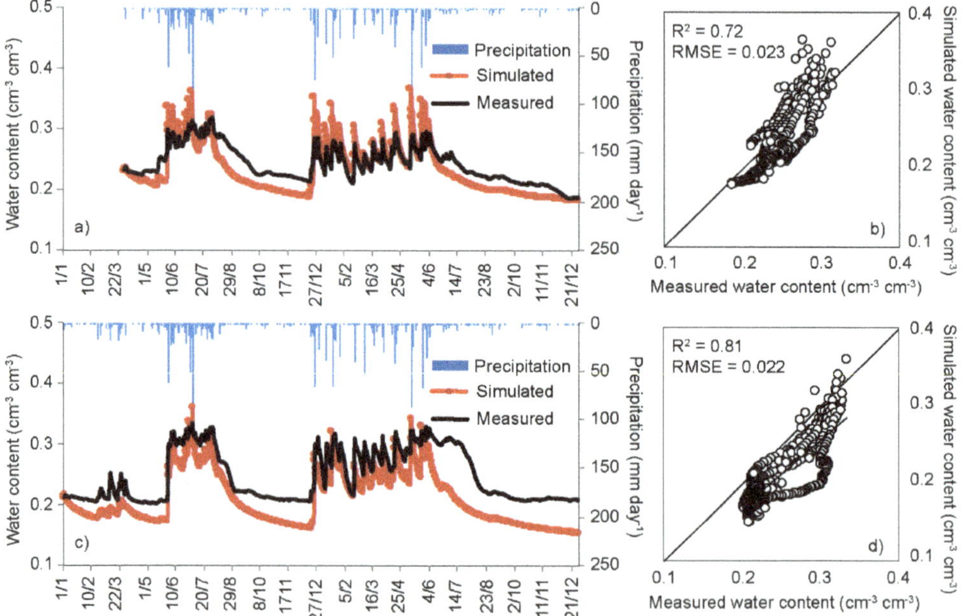

Figure 7. Measured and simulated water contents from the hydraulic parameters estimated by the BEST methods during the studied period (2015–2016) in stations S1 (**a,b**) and S2 (**b,c**). Scatter plots show the simulated versus observed water contents with the indication of the coefficient of determination (R^2) and the root mean square error (RMSE).

The scatter plots also show the good agreement between observed and estimated data, with values of R^2 ranging from 0.72 (S1) to 0.81 (S2) and RMSE below 0.024 cm^3 cm^{-3} (Figure 7b,d). Honari et al. [36] used the soil parameters estimated by the BEST method to evaluate the performance of Hydrus-3D model in simulating 3D water flow and water content in a short time period and at different depths of a soil cultivated with corn and durum in Montpellier, France. They also obtained good results without parameter calibration, with RMSE values between 0.0174 and 0.019 cm^3 cm^{-3}. The authors emphasized that the quality of hydrodynamic property estimates is fundamental for accuracy in simulations.

Some studies used hydraulic parameters estimated by the Rosetta Lite pedotransference function [79] to start the simulations of the behaviour of water in the vadose zone using Hydrus-1D [30,32,33,73]. By analysing the efficiency of this function, Okamoto et al. [80] observed that the hydraulic parameters measured in the field presented better results than those estimated by Rosetta Lite. They highlighted the importance of the use of measured parameters when simulating the water flux in the unsaturated zone. The results found at this step demonstrate the good efficiency of hydrodynamic parameters determined in situ with the BEST model, adjusted by the RETC, to characterize the temporal variability of soil moisture in the GERB.

Table 3 presents the hydraulic parameters of the soil calibrated by inverse modelling. After calibration, residual moisture values increased and saturated moisture decreased in the two stations. The saturated hydraulic conductivity in the two stations increased approximately 50% of predicted values using the BEST methods (Table 2 versus Table 3). Chen et al. [13] used Hydrus-1D to simulate soil moisture in two sub-basins of the Goulburn River, Australia, for a period of three years. They observed that inverse modelling modified the soil properties, reducing the values of θ_r while θ_s and K_s increased. According to the aforementioned authors, the calibrated hydraulic parameters represent, in addition to soil texture, factors such as the number of macropores caused by roots or cracks, vegetation type, organic matter, and anthropic activities.

Table 3. Calibrated soil hydraulic parameters by using the inverse solution for the monitoring stations.

Site	θ_r $(cm^3\ cm^{-3})$	θ_s $(cm^3\ cm^{-3})$	α (cm^{-1})	n	K_s $(cm\ day^{-1})$	l
S1	0.107	0.342	0.037	1.23	234.2	0.5
S2	0.056	0.369	0.018	1.17	243.7	0.5

Figure 8 shows the results of the modelling validation. It is noticeable that the estimated soil water content is better adjusted to the observed values obtained in situ. However, a discrepancy between the observed and estimated data remains in both stations after the end of the rainy season. At S1, the simulated data follow the same pattern of the measured soil moisture variation; however, the model slightly overestimated the measured values for some rainfall events (Figure 8a). For instance, the response of the model to the rainfall of 133.8 mm that occurred on 9 May 2016 was 0.33 $cm^3\ cm^{-3}$, while the observed value registered was 0.27 $cm^3\ cm^{-3}$. For the same station, the model was unable to detect the rainfall events that occurred from 13 June 2016 to 10 August 2016, underestimating the measured soil moisture data.

The simulated data for S2 also presented overestimated values for some rainfall events (Figure 8). Similar underestimation was also observed after the rainy season from 10 July 2016 to 9 August 2016, with a considerable difference between the observed and estimated data. This behaviour of the model in underestimating the observed data is probably linked to the high infiltration rate considered by Hydrus-1D, which is larger than that which actually occurs with the antecedent soil moisture.

The correlation between estimated and simulated values for the two monitoring stations is shown in Figure 8b,d. The dispersion between the data indicates that the observed soil water content is represented well by the modelling results, with R^2 values equal to 0.83 (S1) and 0.72 (S2). Silva et al. [29] obtained similar values of R^2 (0.79 for natural cover and 0.78 for bare soil) by using the Hydrus-1D model to simulate soil moisture in an Acrisol in Brazil using 151 days of the year 2013 for validation. The authors attributed the results obtained to the different hydrological behaviours between the years used for calibration and validation, particularly the changes in interception rate and evapotranspiration and the scarcity of rainfall events in the calibration period. Any change in vegetation or soil structure at the surface may impact the water budget, mainly the runoff and infiltration processes.

Figure 8b,d also shows that some points are still far from the straight 1:1 line for the two stations, even though R^2 presents good values. This occurred because of the response of the model to some isolated rainfall events, which raised the soil moisture to values higher than those measured in the field. Another divergent behaviour is observed below the 1:1 line, where the model failed to respond to the small rainfall events that occurred in June and July after the rainy season, underestimating the soil moisture values measured in the field.

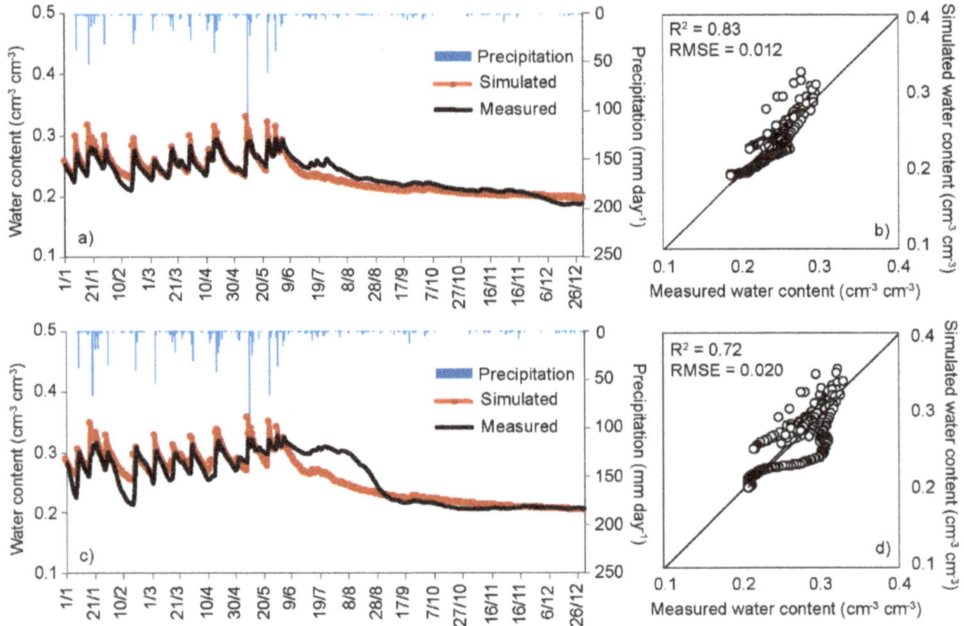

Figure 8. Measured and simulated water content after calibration for the year 2016 at S1 (**a,b**) and S2 (**c,d**). Scatter plots show the simulated versus observed water contents with the indication of the coefficient of determination (R^2).

Another possible cause of the observed divergence between simulated and observed values is associated with some possible small errors that accumulated throughout the data acquisition process, such as: (i) characterization of soil properties, (ii) calibration of the sensors in the laboratory, and (iii) some criteria adopted for simulation of the model (e.g., LAI and lower boundary condition), as previously described.

Table 4 presents the results of the performance evaluation of the model. The RMSE for the simulated and observed soil moisture data exhibited good absolute values equal to 0.012 and 0.020 cm^3 cm^{-3} for S1 and S2, respectively. The Willmott coefficient of agreement (d) close to 1 also evidences the good accuracy of the soil moisture simulations under the investigated conditions. The NSE values for the simulations at both station S1 and station S2 were classified as being "very good" (0.75 < NSE < 1) and "good" (0.65 < NSE ≤ 0.75), respectively. These results are similar to those found by Silva et al. [29] using Hydrus-1D for simulations, with d equal to 0.93 and 0.94 and NSE equal to 0.72 and 0.74 for natural and bare soils, respectively.

Table 4. Hydrus-1D model goodness of fit and accuracy. RMSE: Residual Mean Square Errors; d: Willmott coefficient of agreement; NSE: Nash-Sutcliffe model efficiency coefficient.

Site	RMSE (cm^3 cm^{-3})	d	NSE
S1	0.012	0.95	0.783
S2	0.020	0.93	0.740

The comparison between simulated and validated results shows that the RMSE values were very close, with a small reduction of 0.009 cm^3 cm^{-3} at S1 and a of 0.002 cm^3 cm^{-3} at S2 (see Figures 7 and 8). Considering the aforementioned results and the entire time series used in both the initial simulation and the validation (2015 and 2016), it can be noticed that the parameters estimated by the BEST method are efficient for use in the Hydrus-1D model and describe the soil conditions in the GERB well, without the need for calibration. Therefore, simplified estimates of soil properties by the BEST method, associated with meteorological data and using Hydrus-1D model, can provide acceptable predictions of soil water content in GERB.

4. Conclusions

This study was conducted in an experimental basin in Brazil to investigate the behaviour and prediction of soil water content dynamics in the region using field measurements and numerical methods. Soil moisture presented the greatest and the lowest variability during the rainy transitions (dry–rainy–dry) and dry periods, respectively. The soil water content did not approximate the residual and saturation contents throughout the period. The Hydrus-1D model adequately simulated the soil moisture variability both with the parameters estimated by the BEST model and after their calibration (numerical optimization of parameters by fitting experimental measures). Indeed, these optimized parameters reduced the differences between the measured and simulated values, as expected. However, the inverse modelling requires a considerable number of observations, which means it is limited to well monitored basins. Meanwhile, the gain in accuracy was not drastic, suggesting that BEST estimates were sufficient. Thus, the performance analysis of the simulations provided strong indications of the efficiency of parameters estimated by BEST to predict the soil moisture variability in the studied region, without the need for calibration or complex numerical approaches. This result indicates that the BEST method can be adequate for depicting the water retention and hydraulic conductivity functions, even with the final aim of modelling the water flow processes and water budget for a soil profile. It must be mentioned that the approach was based on the modelling of a soil profile, considering a uniform soil. The impact of soil heterogeneity and spatial variability of soil hydraulic characteristics in different layers of the soil profile should be investigated and the resulting flow heterogeneity should be accounted for. This issue will be the object of further investigations.

Author Contributions: B.S.U., S.M.G.L.M., and A.P.C. developed the study conception and methodology; B.S.U., A.P.C., D.C.d.S.A., S.M.d.S.N., and A.C.V.G. were responsible for the construction and validation of the dataset; B.S.U., S.M.G.L.M., A.P.C., V.H.R.C., and L.L. analyzed and interpreted the dataset; B.S.U., S.M.G.L.M., A.P.C., V.H.R.C., L.L., and R.A.-J. wrote and edited the manuscript.

Funding: This research was funded by the Brazilian Innovation Agency (FINEP) trhough the REHIDRO 1830 project. The National Council for Scientific and Technological Development (CNPq) funded the Universal MCTI/CNPq No. 14/2014 (448236/2014-1) and MCTI/CNPq/ANA No. 23/2015 (446254/2015-0) projects. This study was financed in part by the Coordenação de Aperfeiçoamento de Pessoal de Nível Superior (CAPES)–Finance Code 001 and process No. 88881.159246/2017-01. The Foundation for Science and Technology of Pernambuco State (FACEPE) is also acknowledged for funding the UNIVERSITAS consortium (APQ-0300-5.03/17) and the projects APQ-0885-3.01/16 and BCT-0240-3.01/17.

Acknowledgments: Special thanks are offered to Pernambuco Water and Climate Agency (APAC) for providing the meteorological data. We also acknowledge the anonymous reviewers for their comments and suggestions that allowed improve the quality of the manuscript.

Conflicts of Interest: The authors declare no conflict of interest.

References

1. Šípek, V.; Tesar, M. Soil moisture simulations using two different modelling approaches. *Die Bodenkult.* **2013**, *64*, 99–103.
2. Bordoni, M.; Bittelli, M.; Valentino, R.; Chersich, S.; Persichillo, M.G.; Meisina, C. Soil water content estimated by support vector machine for the assessment of shallow landslides triggering: The role of antecedent meteorological conditions. *Environ. Model. Assess.* **2018**, *23*, 333–352. [CrossRef]
3. Zarlenga, A.; Fiori, A.; Russo, D. Spatial variability of soil moisture and the scale issue: A geostatistical approach. *Water Resour. Res.* **2018**, *54*, 1765–1780. [CrossRef]
4. Rodriguez-Iturbe, I.; Porporato, A. *Ecohydrology of Water-Controlled Ecosystems: Soil Moisture and Plant Dynamics*, 1st ed.; Cambridge University Press: Cambridge, UK, 2005. [CrossRef]

5. Sheikh, V.; van Loon, E.E. Comparing performance and parameterization of a one-dimensional unsaturated zone model across scales. *Vadose Zone J.* **2007**, *6*, 638–650. [CrossRef]

6. Ojha, R.; Morbidelli, R.; Saltalippi, C.; Flammini, A.; Govindaraju, R.S. Scaling of surface soil moisture over heterogeneous fields subjected to a single rainfall event. *J. Hydrol.* **2014**, *516*, 21–36. [CrossRef]

7. Dirmeyer, P.A.; Gao, X.; Zhao, M.; Guo, Z.; Oki, T.; Hanasaki, N. GSWP-2: Multimodel analysis and implications for our perception of the land surface. *Bull. Am. Meteorol. Soc.* **2006**, *87*, 1381–1397. [CrossRef]

8. Li, S.; Liang, W.; Zhang, W.; Liu, Q. Response of soil moisture to hydro-meteorological variables under different precipitation gradients in the Yellow River basin. *Water Resour. Manag.* **2016**, *30*, 1867–1884. [CrossRef]

9. Melo, O.R.; Montenegro, A.A.A. Dinâmica temporal da umidade do solo em uma bacia hidrográfica no semiárido pernambucano. *Rev. Bras. Recur. Hídr.* **2015**, *20*, 430–441. [CrossRef]

10. Huang, X.; Shi, Z.H.; Zhu, H.D.; Zhang, H.Y.; Ai, L.; Yin, W. Soil moisture dynamics within soil profiles and associated environmental controls. *Catena* **2016**, *136*, 189–196. [CrossRef]

11. Negm, A.; Capodici, F.; Ciraolo, G.; Maltese, A.; Provenzano, G.; Rallo, G. Assessing the performance of thermal inertia and Hydrus models to estimate surface soil water content. *Appl. Sci.* **2017**, *7*, 975. [CrossRef]

12. Brocca, L.; Ciabatta, L.; Massari, C.; Camici, S.; Tarpanelli, A. Soil moisture for hydrological applications: Open questions and new opportunities. *Water* **2017**, *9*, 140. [CrossRef]

13. Chen, M.; Willgoose, G.R.; Saco, P.M. Spatial prediction of temporal soil moisture dynamics using HYDRUS-1D. *Hydrol. Process.* **2014**, *28*, 171–185. [CrossRef]

14. Tavakoli, M.; De Smedt, F. Validation of soil moisture simulation with a distributed hydrologic model (WetSpa). *Environ. Earth Sci.* **2013**, *69*, 739–747. [CrossRef]

15. Legates, D.R.; Mahmood, R.; Levia, D.F.; DeLiberty, T.L.; Quiring, S.M.; Houser, C.; Nelson, F.E. Soil moisture: A central and unifying theme in physical geography. *Prog. Phys. Geogr.* **2011**, *35*, 65–86. [CrossRef]

16. Escorihuela, M.J.; Quintana-Seguí, P. Comparison of remote sensing and simulated soil moisture datasets in Mediterranean landscapes. *Remote Sens. Environ.* **2016**, *180*, 99–114. [CrossRef]

17. Espejo-Pérez, A.J.; Brocca, L.; Moramarco, T.; Giráldez, J.V.; Triantafilis, J.; Vanderlinden, K. Analysis of soil moisture dynamics beneath olive trees. *Hydrol. Process.* **2016**, *30*, 4339–4352. [CrossRef]

18. Zribi, M.; Le Hégarat-Mascle, S.; Ottlé, C.; Kammoun, B.; Guerin, C. Surface soil moisture estimation from the synergistic use of the (multi-incidence and multi-resolution) active microwave ERS Wind Scatterometer and SAR data. *Remote Sens. Environ.* **2003**, *86*, 30–41. [CrossRef]

19. She, D.; Liu, D.; Xia, Y.; Shao, M. Modeling effects of land use and vegetation density on soil water dynamics: Implications on water resource management. *Water Resour. Manag.* **2014**, *28*, 2063–2076. [CrossRef]

20. Chen, B.; Liu, E.; Mei, X.; Yan, C.; Garré, S. Modelling soil water dynamic in rain-fed spring maize field with plastic mulching. *Agric. Water Manag.* **2018**, *198*, 19–27. [CrossRef]

21. Van Dam, J.C.; Huygen, J.; Wesseling, J.G.; Feddes, R.A.; Kabat, P.; van Walsum, P.E.V.; Groenendijk, P.; van Diepen, C.A. *Theory of SWAP Version 2.0: Simulation of Water Flow, Solute Transport and Plant Growth in the Soil-Water-Atmosphere-Plant Environment*; Technical Document 45; Wageningen Agricultural University and DLO Winand Staring Centre: Wageningen, The Netherlands, 1997.

22. Avissar, R. Which type of soil-vegetation-atmosphere transfer scheme is needed for general circulation models: A proposal for a higher-order scheme. *J. Hydrol.* **1998**, *212–213*, 136–154. [CrossRef]

23. Verburg, K.; Ross, P.J.; Bristow, K.L. *SWIM v2.1 User Manual*; Divisional Report No. 130; CSIRO Division of Soils: Canberra, Australia, 1996.

24. Leonard, R.A.; Knisel, W.G.; Still, D.A. GLEAMS: Groundwater loading effects of agricultural management systems. *Trans. ASAE* **1987**, *30*, 1403–1418. [CrossRef]

25. Vanclooster, M.; Viaene, P.; Diels, J.; Christiaens, K. WAVE: A mathematical model for simulating water and agrochemicals in the soil and vadose environment. In *Reference and User's Manual, Release 2.1*; Institute for Land and Water Management, Katholieke Universiteit Leuven: Leuven, Belgium, 1996.

26. Richards, L.A. Capillary conduction of liquids through porous mediums. *J. Appl. Phys.* **1931**, *1*, 318–333. [CrossRef]

27. Tinet, A.J.; Chanzy, A.; Braud, I.; Crevoisier, D.; Lafolie, F. Development and Evaluation of an Efficient Soil-Atmosphere Model (FHAVeT) based on the Ross fast solution of the Richards equation for bare soil conditions. *Hydrol. Earth Syst. Sci.* **2015**, *19*, 969–980. [CrossRef]

28. Šimůnek, J.; Šejna, M.; Saito, H.; Sakai, M.; van Genuchten, M.T. *The HYDRUS-1D Software Package for Simulating the One-Dimensional Movement of Water, heat, and Multiple Solutes in Variably-Saturated Media Version 4.17*; University of California Riverside: Riverside, CA, USA, 2013; pp. 1–342.

29. Da Silva, J.R.L.; Montenegro, A.A.A.; Monteiro, A.L.N.; Silva, P.V. Modelagem da dinâmica de umidade do solo em diferentes condições de cobertura no semiárido pernambucano. *Rev. Bras. Cienc. Agrar.* **2015**, *10*, 293–303. [CrossRef]

30. Lai, X.; Liao, K.; Feng, H.; Zhu, Q. Responses of soil water percolation to dynamic interactions among rainfall, antecedent moisture and season in a forest site. *J. Hydrol.* **2016**, *540*, 565–573. [CrossRef]

31. Li, Y.; Šimůnek, J.; Wang, S.; Yuan, J.; Zhang, W. Modeling of soil water regime and water balance in a transplanted rice field experiment with reduced irrigation. *Water* **2017**, *9*, 248. [CrossRef]

32. Wang, H.; Tetzlaff, D.; Soulsby, C. Modelling the effects of land cover and climate change on soil water partitioning in a boreal headwater catchment. *J. Hydrol.* **2018**, *558*, 520–531. [CrossRef]

33. Gabiri, G.; Burghof, S.; Diekkrüger, B.; Leemhuis, C.; Steinbach, S.; Näschen, K. Modeling spatial soilwater dynamics in a tropical floodplain, East Africa. *Water* **2018**, *10*, 191. [CrossRef]

34. Šimůnek, J.; van Genuchten, M.T. Using the HYDRUS-1D and HYDRUS-2D codes for estimating unsaturated soil hydraulic and solute transport parameters. In *Characterization and Measurement of the Hydraulic Properties of Unsaturated Porous Media*; van Genuchten, M.T., Leij, F.J., Wu, L., Eds.; University of California: Riverside, CA, USA, 1999; pp. 1523–1536.

35. Kato, C.; Nishimura, T.; Imoto, H.; Miyazaki, T. Predicting soil moisture and temperature of andisols under a monsoon climate in Japan. *Vadose Zone J.* **2011**, *10*, 541–551. [CrossRef]

36. Honari, M.; Ashrafzadeh, A.; Khaledian, M.; Vazifedoust, M.; Mailhol, J.C. Comparison of HYDRUS-3D soil moisture simulations of subsurface drip irrigation with experimental observations in the south of France. *J. Irrig. Drain. Eng.* **2017**, *143*, 1–8. [CrossRef]

37. Siltecho, S.; Hammecker, C.; Sriboonlue, V.; Clermont-Dauphin, C.; Trelo-Ges, V.; Antonino, A.C.D.; Angulo-Jaramillo, R. Use of field and laboratory methods for estimating unsaturated hydraulic properties under different land uses. *Hydrol. Earth Syst. Sci.* **2015**, *19*, 1193–1207. [CrossRef]

38. Haverkamp, R.; Ross, P.J.; Smettem, K.R.J.; Parlange, J.Y. Three-dimensional analysis of infiltration from the disc infiltrometer: 2. Physically based infiltration equation. *Water Resour. Res.* **1994**, *30*, 2931–2935. [CrossRef]

39. Braud, I.; De Condappa, D.; Soria, J.M.; Haverkamp, R.; Angulo-Jaramillo, R.; Galle, S.; Vauclin, M. Use of scaled forms of the infiltration equation for the estimation of unsaturated soil hydraulic properties (the Beerkan method). *Eur. J. Soil Sci.* **2005**, *56*, 361–374. [CrossRef]

40. Coutinho, A.P.; Lassabatere, L.; Montenegro, S.; Antonino, A.C.D.; Angulo-Jaramillo, R.; Cabral, J.J.S.P. Hydraulic characterization and hydrological behaviour of a pilot permeable pavement in an urban centre, Brazil. *Hydrol. Process.* **2016**, *30*, 4242–4254. [CrossRef]

41. Lassabatère, L.; Angulo-Jaramillo, R.; Soria Ugalde, J.M.; Cuenca, R.; Braud, I.; Haverkamp, R. Beerkan estimation of soil transfer parameters through infiltration experiments—BEST. *Soil Sci. Soc. Am. J.* **2006**, *70*, 521. [CrossRef]

42. Kanzari, S. Spatio-Temporal variability of the soil hydraulic properties—Effect on modelling of water flow and solute transport at field-scale. In *Recent Advances in Environmental Science from the Euro-Mediterranean and Surrounding Regions*; Kallel, A., Ksibi, M., Ben Dhia, H., Khélifi, N., Eds.; Springer: Cham, Switzerland, 2018; pp. 1279–1281. [CrossRef]

43. Lai, J.; Ren, L. Estimation of effective hydraulic parameters in heterogeneous soils at field scale. *Geoderma* **2016**, *264*, 28–41. [CrossRef]

44. Nascimento, Í.V.; de Assis Júnior, R.N.; de Araújo, J.C.; de Alencar, T.L.; Freire, A.G.; Lobato, M.G.R.; da Silva, C.P.; Mota, J.C.A.; Nascimento, C.D.V. Estimation of van Genuchten equation parameters in laboratory and through inverse modeling with Hydrus-1D. *J. Agric. Sci.* **2018**, *10*, 102. [CrossRef]

45. Graham, S.L.; Srinivasan, M.S.; Faulkner, N.; Carrick, S. Soil hydraulic modeling outcomes with four parameterization methods: Comparing soil description and inverse estimation approaches. *Vadose Zone J.* **2018**, *17*. [CrossRef]

46. Montenegro, S.M.G.L.M.; Ragab, R. Impact of possible climate and land use changes in the semi arid regions: A case study from north eastern Brazil. *J. Hydrol.* **2012**, *434–435*, 55–68. [CrossRef]

47. Montenegro, A.A.A.; Ragab, R. Hydrological response of a brazilian semi-arid catchment to different land use and climate change scenarios: A modelling study. *Hydrol. Process.* **2010**, *24*, 2705–2723. [CrossRef]

48. Araújo Filho, P.; Cabral, J. Modelagem hidrológica da bacia do riacho Gameleira (Pernambuco) utilizando TOPSIMPL, uma versão simplificada do modelo TOPMODEL. *Rev. Bras. Recur. Hídr.* **2005**, *10*, 61–72. [CrossRef]

49. Paiva, F.M.L.; Montenegro, S.M.G.L.; Salcedo, I.H.; Araújo Filho, P.F.; Srinivasan, V.S.; Silva Filho, S.L.; Azevedo, J.R.G.; Silva, R.M.; Silva, L.P. Análise do transporte de sedimentos em suspensão num pequeno curso d'água na bacia experimental do riacho Gameleira, PE. In Proceedings of the XIX Simpósio Brasileiro de Recursos Hídricos (SBRH), Maceió, Brazil, 27 November–1 December 2011; Available online: https://abrh.s3.sa-east-1.amazonaws.com/Sumarios/81/e7457e425ff7a2508f2b4bfb293cfbd9_e6315e8645bda4bec9188b7bdc137669.pdf (accessed on 17 December 2018).

50. Da Silva, R.M.; Paiva, F.M.D.L.; Montenegro, S.M.G.D.L.; Augusto, C.; Santos, G. Aplicação de Eqsuações de Razão de Transferência de Sedimentos na Bacia do rio Gameleira com Suporte de Sistemas de Informação Geográfica. In Proceedings of the XVIII Simpósio Brasileiro de Recursos Hídricos (SBRH), Campo Grande, Brazil, 22–26 November 2009; Available online: https://abrh.s3.sa-east-1.amazonaws.com/Sumarios/110/c9e2c8a28e3027a52b5b34e91b6f9b14_705324c4dd596d1379d42fc4fe927d8f.pdf (accessed on 17 December 2018).

51. Furtunato, O.M.; Montenegro, S.M.G.L.; Antonino, A.C.D.; de Oliveira, L.M.M.; de Souza, E.S.; Moura, A.E.S.S. Variabilidade espacial de atributos físico-hídricos de solos em uma bacia experimental no estado de Pernambuco. *Rev. Bras. Recur. Hídr.* **2013**, *18*, 135–147. [CrossRef]

52. Oliveira, L.M.M.; Montenegro, S.M.G.L. Evapotranspiração de referência na bacia experimental do riacho Gameleira, PE, utilizando-se lisímetro e métodos indiretos. *Rev. Bras. Ciênc. Agrár.* **2008**, *81*, 58–67. [CrossRef]

53. Moura, A.R.C.; Montenegro, S.M.G.L.; Antonino, A.C.D.; de Azevedo, J.R.G.; da Silva, B.B.; de Oliveira, L.M.M. Evapotranspiração de referência baseada em métodos empíricos em bacia experimental no estado de Pernambuco - Brasil. *Rev. Bras. Meteorol.* **2013**, *28*, 181–191. [CrossRef]

54. Alvares, C.A.; Stape, J.L.; Sentelhas, P.C.; de Moraes Gonçalves, J.L.; Sparovek, G. Köppen's climate classification map for Brazil. *Meteorol. Z.* **2013**, *22*, 711–728. [CrossRef]

55. WRB, I.W.G. *World Reference Base for Soil Resources 2006: A Framework for International Classification, Correlation and Communication*; World Soil Resources Reports; Food and Agriculture Organization: Rome, Italy, 2006.

56. Braga, R.A.P. *Gestão Ambiental Da Bacia Do Rio Tapacurá - Plano de Ação*; Universitária-UFPE: Recife, Brazil, 2001; p. 101.

57. Penman, H.L. Natural evaporation from open water, bare soil, and grass. *Proc. R. Soc. Lond.* **1948**, *193*, 120–146. [CrossRef]

58. Shuttleworth, W.J. Evaporation. In *Handbook of Hydrology*; Maidment, D.R., Ed.; McGraw Hill: New York, NY, USA, 1993; pp. 4.1–4.53.

59. McVicar, T.R.; Roderick, M.L.; Donohue, R.J.; Li, L.T.; Van Niel, T.G.; Thomas, A.; Grieser, J.; Jhajharia, D.; Himri, Y.; Mahowald, N.M.; et al. Global review and synthesis of trends in observed terrestrial near-surface wind speeds: Implications for evaporation. *J. Hydrol.* **2012**, *416–417*, 182–205. [CrossRef]

60. Donohue, R.J.; McVicar, T.R.; Roderick, M.L. Assessing the ability of potential evaporation formulations to capture the dynamics in evaporative demand within a changing climate. *J. Hydrol.* **2010**, *386*, 186–197. [CrossRef]

61. Coelho, V.H.R.; Montenegro, S.; Almeida, C.N.; Silva, B.B.; Oliveira, L.M.; Gusmão, A.C.V.; Freitas, E.S.; Montenegro, A.A.A. Alluvial Groundwater Recharge Estimation in Semi-Arid Environment Using Remotely Sensed Data. *J. Hydrol.* **2017**, *548*, 1–15. [CrossRef]

62. Li, Y.; Šimůnek, J.; Jing, L.; Zhang, Z.; Ni, L. Evaluation of water movement and water losses in a direct-seeded-rice field experiment using Hydrus-1D. *Agric. Water Manag.* **2014**, *142*, 38–46. [CrossRef]

63. Van Genuchten, M.T. A closed-form equation for predicting the hydraulic conductivity of unsaturated soils. *Soil Sci. Soc. Am. J.* **1980**, *44*, 892–898. [CrossRef]

64. Angulo-Jaramillo, R.; Bagarello, V.; Iovino, M.; Lassabatère, L. *Infiltration Measurements for Soil Hydraulic Characterization*; Springer International Publishing: New York, NY, USA, 2016; ISBN 978-3-319-31786-1. [CrossRef]

65. Burdine, N.T. Relative permeability calculations from pore size distribution Data. *J. Pet. Technol.* **1953**, *5*, 71–78. [CrossRef]

66. Brooks, R.H.; Corey, A.T. *Hydraulic Properties of Porous Media*; Hydrology Paper 3; Colorade State University: Fort Collins, CO, USA, 1964.

67. Fuentes, C.; Haverkamp, R.; Parlange, J.Y. Parameter constraints on closed-form soilwater relationships. *J. Hydrol.* **1992**, *134*, 117–142. [CrossRef]

68. Yilmaz, D.; Lassabatère, L.; Angulo-Jaramillo, R.; Deneele, D.; Legret, M. Hydrodynamic characterization of basic oxygen furnace slag through an adapted BEST method. *Vadose Zone J.* **2010**, *9*, 107–116. [CrossRef]

69. Bagarello, V.; Di Prima, S.; Giordano, G.; Iovino, M. A test of the Beerkan Estimation of Soil Transfer Parameters (BEST) procedure. *Geoderma* **2014**, *221–222*, 20–27. [CrossRef]

70. Mualem, Y. A new model for predicting the hydraulic conductivity of unsaturated porous media. *Water Resour. Res.* **1976**, *12*, 513–522. [CrossRef]

71. Van Genuchten, M.T.; Leij, F.J.; Yates, S.R. *The RETC Code for Quantifying the Hydraulic Functions of Unsaturated Soils*; Research Report n. EPA/600/2-91/065; U.S. Salinity Laboratory, USDA-ARS: Riverside, CA, USA, 1991; 93p. [CrossRef]

72. Ritchie, J.T. Model for predicting evaporation from a row crop with incomplete cover. *Water Resour. Res.* **1972**, *8*, 1204–1213. [CrossRef]

73. Qu, W.; Bogena, H.R.; Huisman, J.A.; Martinez, G.; Pachepsky, Y.A.; Vereecken, H. Effects of soil hydraulic properties on the spatial variability of soil water content: Evidence from sensor network data and inverse modeling. *Vadose Zone J.* **2014**, *13*. [CrossRef]

74. Vereecken, H.; Huisman, J.A.; Bogena, H.; Vanderborght, J.; Vrugt, J.A.; Hopmans, J.W. On the value of soil moisture measurements in vadose zone hydrology: A review. *Water Resour. Res.* **2008**, *44*. [CrossRef]

75. Silva, R.M.; Silva, L.P.; Montenegro, S.M.G.L.; Santos, C.A.G. Análise da variedade espaço-temporal e identificação do padrão da precipitação na bacia do rio Tapacurá, Pernambuco. *Soc. Nat.* **2010**, *22*, 357–372. [CrossRef]

76. Souza, E.S.; Antonino, A.C.D.; Angulo-Jaramillo, R.; Netto, A.M. Caracterização hidrodinâmica de solos: Aplicação do método Beerkan. *Rev. Bras. Eng. Agrícola Ambient.* **2008**, *12*, 128–135. [CrossRef]

77. Pirastru, M.; Castellini, M.; Giadrossich, F.; Niedda, M. Comparing the hydraulic properties of forested and grassed soils on an experimental hillslope in a mediterranean environment. *Procedia Environ. Sci.* **2013**, *19*, 341–350. [CrossRef]

78. Van Genuchten, M.T.; Nielsen, D. On describing and predicting the hydraulic properties of unsaturated soils. *Ann. Geophys.* **1985**, *3*, 615–627. [CrossRef]

79. Schaap, M.G.; Leij, F.J.; van Genuchten, M.T. Rosetta: A computer program for estimating soil hydraulic parameters with hierarchical pedotransfer functions. *J. Hydrol.* **2001**, *251*, 163–176. [CrossRef]

80. Okamoto, K.; Sakai, K.; Nakamura, S.; Cho, H.; Nakandakari, T.; Ootani, S. Optimal choice of soil hydraulic parameters for simulating the unsaturated flow: A case study on the island of Miyakojima, Japan. *Water* **2015**, *7*, 5676–5688. [CrossRef]

 *water*

Article

Time Scale Effects and Interactions of Rainfall Erosivity and Cover Management Factors on Vineyard Soil Loss Erosion in the Semi-Arid Area of Southern Sicily

Giorgio Baiamonte *, Mario Minacapilli, Agata Novara and Luciano Gristina

Department of Agricultural, Food and Forest Sciences (SAAF). University of Palermo, Viale delle Scienze, 90128 Palermo, Italy; mario.minacapilli@unipa.it or mminacapilli@gmail.com (M.M.); agata.novara@unipa.it (A.N.); luciano.gristina@unipa.it (L.G.)
* Correspondence: giorgio.baiamonte@gmail.com; Tel.: +39-091-238-97054

Received: 23 April 2019; Accepted: 4 May 2019; Published: 9 May 2019

Abstract: Several authors describe the effectiveness of cover crop management practice as an important tool to prevent soil erosion, but at the same time, they stress on the high soil loss variability due to the interaction of several factors characterized by large uncertainty. In this paper the Revised Universal Soil Loss Equation (RUSLE) model is applied to two Sicilian vineyards that are characterized by different topographic factors; one is subjected to Conventional Practice (CP) and the other to Best Management Practice (BMP). By using climatic input data at a high temporal scale resolution for the rainfall erosivity (R) factor, and remotely sensed imagery for the cover and management (C) factor, the importance of an appropriate R and C factor assessment and their inter and intra-annual interactions in determining soil erosion variability are showed. Different temporal analysis at ten-year, seasonal, monthly and event scales showed that results at events scales allow evidencing the interacting factors that determine erosion risk features which at other temporal scales of resolution can be hidden. The impact of BMP in preventing soil erosion is described in terms of average saved soil loss over the 10-year period of observation. The evaluation of soil erosion at a different temporal scale and its implications can help stakeholders and scientists formulate better soil conservation practices and agricultural management, and also consider that erosivity rates are expected to raise for the increase of rainfall intensity linked to climate change.

Keywords: soil erosion; RUSLE model; rainfall erosivity factor; cover management factor; NDVI

---

1. Introduction

Soil erosion in vineyards represent several environmental issues. In the semi-arid environment, where the soil is maintained bare, to limit water competition with weeds and rainfall trend is variable and considerable mainly in the winter period, protective practices are also supported from an economic point of view both for soil erosion control and soil organic matter improvement.

Commonly, the Revised Universal Soil Loss Equation (RUSLE) and its family of models have been widely applied in order to evaluate the soil erosion risk [1]. The RUSLE model integrates a number of sub-factors describing the main components of the erosion processes, including rainfall erosivity, soil erodibility, topographic factors, and cover and practices management.

Very important information about rainfall erosivity, which in the RUSLE approach is considered by the R sub-factor, is the time or month of the year when erosivity is at its maximum as well as when it is at its minimum, but also its variability within the considered time or period. The high level of risk linked to the well-known uncertainty of rainfall events allows us to mainly consider the probability of erosivity events.

Despite several studies highlighting that in the semi-arid environment the rainfall intensity is higher in winter or autumn [2–5], in Sicily, D'Asaro et al. [6] estimated the summer erosivity to be equal to or slightly higher than the winter one. The same intensity of rainfall erosivity can result in more or less effectiveness on soil erosion according to the time of year when it occurs depending on different factors, like crop cover, that change during the year. Therefore, knowledge about the time of year when the highest erosivity occurs is critical for management practices, as it allows to optimize mitigation and adaptation procedures. But for these reasons, also probability of the cover factor in the same period or month of the erosivity factor, must be taken into account together with its occurrence probability.

At the same time, a high erosivity factor can occur together with an excellent crop or residue cover not affecting soil erosion processes. On the contrary, serious problems result if a bare soil period matches with a sporadic high erosivity event.

The best management practices (BMP) aimed at soil erosion control use cover cropping as the main tool to achieve the goal, but the presence of soil covering is affected by a high intra-annual and inter-annual variability due to climatic and environmental factors (seedling date, emergence, drought during spring and cover crops burying period, etc.).

In the RUSLE model, the inter-annual variability of crop management is described by the C factor. In particular, the C factor monthly variability needs to be accounted for. At this aim, the potentiality of remote sensing can be exploited using the well-known Normalized Difference Vegetation Index (NDVI) estimated from time-series satellite imagery. NDVI quantifies vegetation by measuring the difference between near-infrared (which vegetation strongly reflects) and red radiation (which vegetation absorbs). In particular, using data from the Moderate Resolution Imaging Spectroradiometer (MODIS), the National Aeronautics and Space Administration Earth Observing System (NASA/EOS) produced a global dataset of NDVI time-series at 8 and 16-day time steps and characterized by a pixel resolutions ranging from 250–1000 m, which can be easily retrieved from the Land Processes Distributed Active Archive Center [7]. The inter-annual variability of observed NDVI can be transformed in a corresponding C factor variability using one of the various approaches proposed in literature through regression correlation analysis [8–10].

According to Panagos [11], monthly erosivity probability allows imposing and/or recommend sustainable agricultural management practices (crop residues, reduced tillage) in regions with high erosivity probability.

If the spatiotemporal rainfall erosivity analysis is a first step towards developing dynamic maps of soil loss by water erosion, the same approach must be applied to the other factor characterized by inter-annual and intra-annual variability such as the cover factor. For these reasons, the distribution over time of the factors involved in the RUSLE equation (R and C factor) must be known. Additionally, their interaction with the goal to identify the riskiest period.

The consequent study of probability is important to support flood prevention, hazard mitigation, ecosystem services, land use change, agricultural production and food security, in general.

For the reasons in this paper, the interactions of rainfall erosivity and cover management factors, as well as their time scale effects, for the vineyard crop, were studied. In particular, the following main objectives of the research were pursued:

(i)   Modeling inter-annual variability of C factor using time-series products of NDVI provided by MODIS satellite platform;

(ii) Assessing the importance of monthly C and R factors and their temporal interactions in comparison to the corresponding annual values, and;

(iii) Assessing soil erosion risk in relation to BMP adoption in the semi-arid vineyard.

2. Materials and Methods

2.1. Study Area

The study area is located in southern Sicily (Figure 1) and is one of the 18 DOC (Controlled Denomination of Origin) vineyards areas on the island. For the studied area, the mean annual precipitation is 516 mm. On average, 3% of the mean annual rainfall occurs during summer (June, July, and August) while 42% occurs during November, December, and January. The minimum, maximum and mean of annual average incoming global solar radiation and vap pressure deficit (VPD) of the investigated areas are equal to 21, 353 and 223 W m^{-2} and 0.03, 2.41 and 0.7 kPa, respectively. Two vineyards with different soil management were selected (Figure 1): One vineyard was managed with at least five shallow tillages per year (Conventional Practice farm—CP farm) to control weeds and reduce water competition, by using a field cultivator with a working width equal to 1.5 m and equipped with seven steel shanks; the other vineyard was managed (Best Management Practice—BMP farm) with agro-environmental measure management involving annual cover cropping using legumes like faba bean (Vicia faba) were investigated. In both surveyed areas, the soil was classified as Vertic Haploxerept according to Soil Taxonomy, with 58.3% and 51.2% sand, 11% and 13.5% silt, and 30.7% and 35.3% clay in the topsoil (0–20 cm) in BMP and CP farm, respectively.

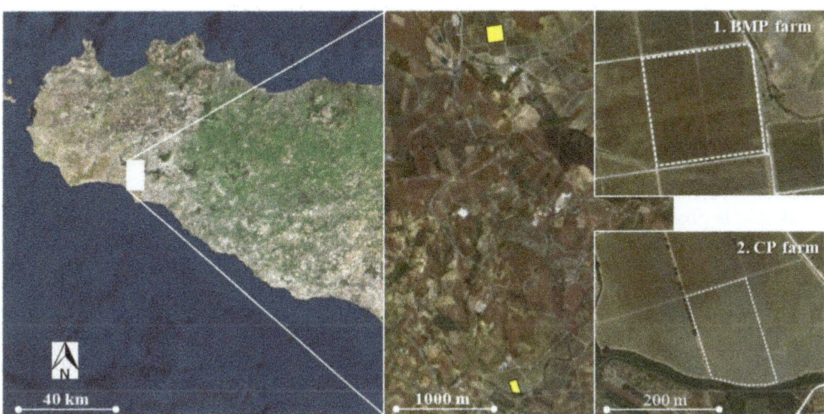

Figure 1. Geographic location of the study area in the Sicily region. The two different investigated vineyards (Best Management Practices, BMP, and Conventional Practice farm, CP, farms) are illustrated.

The two vineyards were around 3.6 km from each other. The main features driving the choice of the two farms were (i) a sufficient uniform area matching the spatial resolution of MODIS data, (ii) BMP adoption under temporary cover crop during the winter period. The two farms are characterized by different slope and length (7.8% and 301 m for BMP and 0.6% and 250 m for CP farm).

2.2. The RUSLE Model

Following the original USLE model, the RUSLE computes the yearly soil loss erosion according to the equation:

$$A = R \times K \times L \times S \times C \times P \tag{1}$$

where A is the mean annual soil loss (t ha^{-1} year^{-1}), R is the rainfall erosivity factor (MJ mm ha^{-1} h^{-1} year^{-1}), K is the soil erodibility factor (t ha h MJ^{-1} h^{-1} mm^{-1}), L and S are the slope-length and the slope-steepness factors (dimensionless), C is the cover and management factor (dimensionless) and P is the support practice factor (dimensionless).

With respect to the original USLE formulation, the revised USLE equation includes improvements in the sub-factors estimation, which were tested in a number of applications and experimental studies performed in many countries [12]. The detailed description of RUSLE factors algorithm is reported in the handbook edited by Renard et al. [1].

Recently, Panagos et al. [11] applied the RUSLE model for all of Europe providing a soil loss map with a detailed spatial resolution, although limited to an average annual soil loss estimation. However, uncertainties from soil erosion modeling arose from applications focused on intra-annual soil loss variability investigations that the RUSLE is also able to consider. In fact, although Equation (1) is usually applied for estimating yearly soil loss, by using the annual R-factor retrieved by iso-erodent maps or annual rainfall data [13,14], the RUSLE model can be also applied at a time scale shorter than the year, up until the event scale. In this case, sub-daily rainfall data is required in order to compute the single event rainfall erosivity index, R_e:

$$R_e = EI_{30} = \left(\sum_{i=1}^{k} e_i^i h_i \right) I_{30} \tag{2}$$

where e_i is the unit rainfall energy (MJ ha^{-1} mm^{-1}) and h_i the rainfall volume (mm) during the ith time period of a rainfall event divided in k-parts. I_{30} is the maximum 30-minutes rainfall intensity (mm h^{-1}). The unit rainfall energy, e_i', is calculated for each time interval as follows:

$$e_i' = 0.29[1 - 0.72exp[-0.05I_i]] \tag{3}$$

where I_i (mm h^{-1}) is the rainfall intensity during the time interval, h_i/t_i.

According to the RUSLE handbook [1], the erosive rainfall events were computed based on the following criteria: (i) The cumulative rainfall of an event is greater than 12.7 mm, (ii) the event has at least one peak that is greater than 6.35 mm during a period of 15 min (or 12.7 mm during a period of 30 min), and (iii) a rainfall accumulation of less than 1.27 mm during a period of six hours splits a longer storm period into two storms.

In our case, in the 2007–2016 period, rainfall events partition according to (iii) was first detected by calculating the sequence of waiting times, also denoted as dry spells [15], between two consecutive rainfall events recorded at 10 min time-scale by using a MATLAB code. The described procedure was applied to the rainfall data-series collected by a single station of the Agro-Meteorological Information Service of Sicily (SIAS) covering both vineyards (Figure 2). Figure 2, also reports the different NVDI time-series collected for the two farms, which were used to derive the cover management C factor.

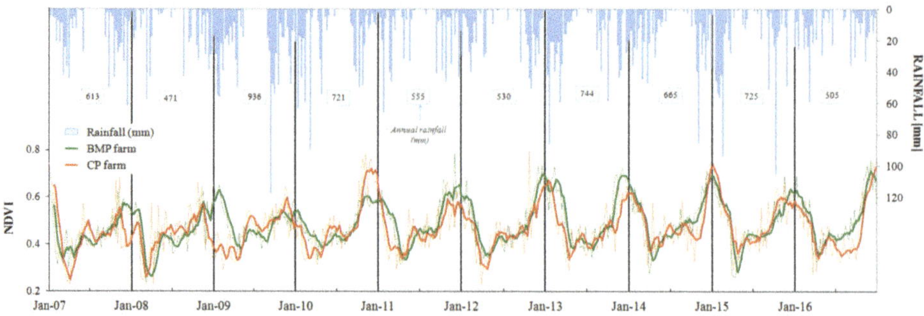

Figure 2. Rainfall data and Modis NDVI time-series used in the investigation.

The C factor accounts for the influence of vegetation upon the soil erosion process, ranging from 0 to 1, where higher values indicate no cover effects, whereas lower values mean a strong cover effect resulting in very limited soil erosion. According to the RUSLE model, a combination of sub-factors, such as impacts of previous management, canopy height, surface cover and roughness, etc., can be used to estimate the C factor [1]. This method requires extensive knowledge of the study area's cover characteristics and maybe not be suitable without detailed monitoring of agricultural management. To address this, many authors suggested evaluating the C-factor through the Normalized Difference Vegetation Index (NDVI) obtained from satellite imagery [10,16–18]. This index has been widely used to map, at various observation scales, crop variables like biomass Leaf Area Index, LAI, plant coverage, etc. [19,20]. NDVI can be obtained from multispectral sensors able to capture reflectance images in red (RED) and infrared (NIR) spectral regions and was computed using the following expression:

$$NDVI = \left(\frac{NIR - RED}{NIR + RED} \right) \tag{4}$$

An advantage of using NDVI consists of the capability to determine sub-annual C values, if satellite images are available, which can lead to understanding the cover contribute to seasonal soil erosion and to identify critical periods within the year where soil erosion could be considered a risk [4]. Following this approach, we used the global dataset of NDVI time-series at 8 and 16-day time-steps for the 2007–2016 period, provided by NASA using data from MODIS platforms (satellites Terra and Aqua), which can be retrieved from the Land Processes Distributed Active Archive Center [7]. The use of the MODIS NDVI time-series product was also supported considering its spatial resolution equal to 250 m which resulted appropriate for representing the two vineyard plots.

In order to transform the observed NDVI in a corresponding C factor, the following equation suggested and validated by Van der Knijff et al. [10], was applied:

$$C = exp\left(\frac{-2NDVI}{1 - NDVI} \right) \tag{5}$$

Application of the RUSLE model also required the K factor estimation, which represents the influence of different soil properties on the slope's susceptibility to erosion [1]. K factor also defines the "mean annual soil loss per unit of rainfall erosivity for a standard condition of bare soil, recently tilled up down slope with no conservation practice" [21]. In the RUSLE, Renard et al. [1] proposed an equation that relates textural and organic matter characteristics, soil structure and profile-permeability with the K factor or soil erodibility factor.

Recently, the new soil erodibility map for Europe generated by Panagos et al. [22] using the Land Use/Cover Area Frame Survey (LUCAS) soil European data-set, allowed a rough evaluation of the K factor in absence of the required soil textural information for the K factor RUSLE estimation. Following this approach, a value of 0.03 t ha^{-1} ha h MJ^{-1} mm^{-1} was estimated for both considered vineyards (Figure 3).

As far as the LS factor, which accounts for the influence of the slope's length and steepness of soil erosion processes, it is defined as the ratio of expected soil loss from a field slope relative to the original USLE unit plot [23]. The USLE method of calculating the slope length and steepness factor was originally applied at the unit plot and field scale, and the RUSLE extended this to the one-dimensional hillslope scale, with different equations depending on the slope gradient [1].

The LS factor was extended to topographically complex units using a method that incorporates contributing area and flow accumulation [24] derived from DEM and GIS approaches. In our case, a high-resolution DEM was used to assess LS factor of the two vineyard plots. However, the almost planar topography of the two plots suggested not using the contributing area and flow accumulation algorithms, restricting the use of DEM for an accurate estimation of the planar slope of the two plots (Figure 4).

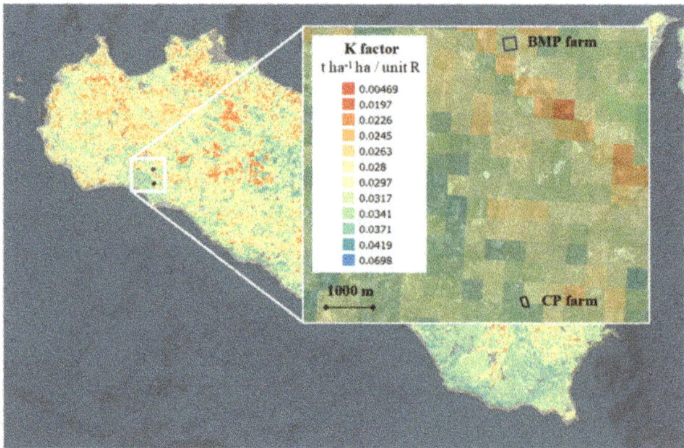

Figure 3. Spatial variability of soil erodibility K factor over the study area [22].

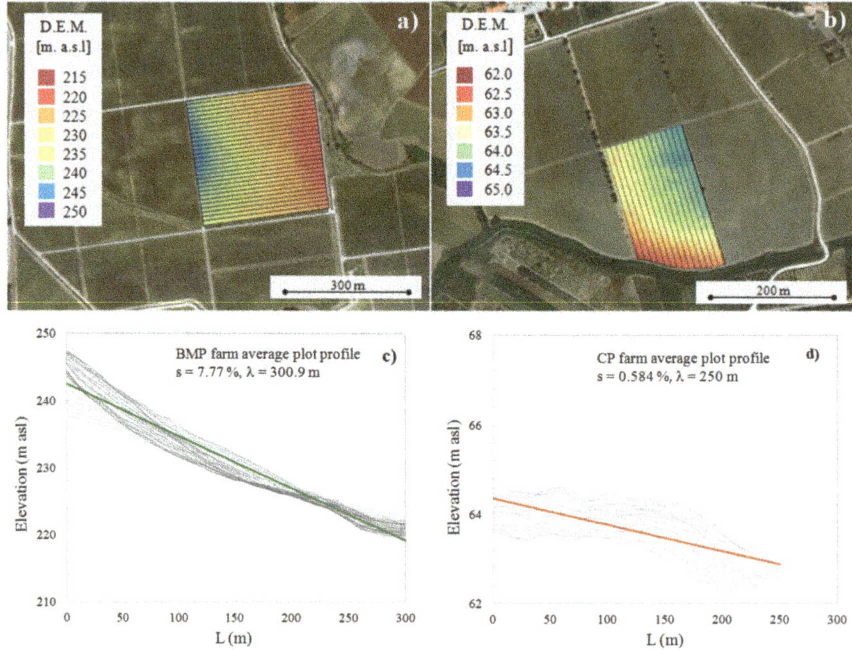

Figure 4. Elevation and topographic attributes of the two vineyards plots. (**a**) BMP farm, (**b**) CP farm, (**c**) BMP farm average profile, (**d**) CP farm average profile.

Therefore, according to Renard et al. [1] the following L and S relationships were applied:

$$L = \left(\frac{\lambda}{22.1} \right)^m \tag{6}$$

$$S = a\,sen\beta + b \tag{7}$$

where λ (m) is the plot length, β is the plot angle on the horizontal whereas the coefficients m, a, b, F, and p have the following expressions and values:

$$m = \left(\frac{pF}{1 + pF}\right) \tag{8}$$

$$F = \left(\frac{sen\beta/0.0896}{3sen^{0.8}\beta + 0.56}\right) \tag{9}$$

$$p = 0.5 - 2 \, (interrill - rill) \tag{10}$$

$$a = 10.8 \text{ (if } \tan\beta < 0.09) \text{ or } 16.8 \text{ (if } \tan\beta \geq 0.09) \tag{11}$$

$$b = 0.03 \text{ (if } \tan\beta < 0.09) \text{ or } -0.50 \text{ (if } \tan\beta \geq 0.09) \tag{12}$$

The P factor of Equation (1) accounts for management practices that affect soil erosion by modifying the flow pattern, such as contouring, strip-cropping, or terracing. The P factor is defined as the ratio between the soil loss under a specific soil conservation practice and that corresponding to a field with upslope and downslope tillage [1]. In our case, since no support practices were adopted in the two vineyard plots, the P factor was fixed equal to the unity [25].

3. Results and Discussion

Using the input data-set described in the previous section, the RUSLE model was applied at different time-scales of aggregations. The weekly data, which corresponds to the minimum C factor temporal scale of observation, were aggregated in annual, seasonal, monthly and the RUSLE input factors were accordingly analyzed.

With the aim to evaluate the C and R factor interactions, disaggregated data in the two vineyards were analyzed. In this case, in order to evaluate soil erosion independently from the highly different topographic characteristics of the two plots, the product of R, K and C factor, RKC, which refers to soil loss (t ha^{-1}) per LS unit, was also investigated.

3.1. Yearly and Monthly Average Soil Loss

The ten-year calculated R factor (Figure 5a) shows a high interannual variability ranging from 661–4377 MJ mm ha^{-1} h^{-1}. The R factor variability appears not correlated to the number of erosive events, N_{ev}, (Figure 5a), whereas it seems dependent on the annual rainfall, H_y (Figure 2).

As an example, the yearly R factor value in 2015 (R = 2122 t ha MJ mm ha^{-1} h^{-1}, H_y = 705 mm, N_{ev} = 13) resulted much higher than 2016 (R = 661 t ha MJ mm ha^{-1} h^{-1}, h_{an} = 505 mm, N_{ev} = 18), evidencing for the latter case the occurrence of lower energy rainfall values than the former one.

Differences between the two vineyards characterized by different soil management, in terms of yearly C values, appear almost slight with a greater variability for CP farm than BMP farm (Figure 5a), although differences in terms of ten-year average C values are appreciable (see dash-dot lines). The higher variability of the C factor for CP than BMP farm can be ascribed to the unpredictable rainfall regime that affects the time scheduling of soil practices (see box plots Figure 5a).

The yearly soil loss calculated by the RUSLE model (Figure 5b) mainly reflects the variability of the R factor, as a consequence of the slight aforementioned C factor variability and of the lumped K and LS factors assumed for the two vineyards. Generally, yearly soil loss between the two vineyards are characterized by the same interannual variability (see box plot, Figure 5b). The large differences in yearly soil loss between the two vineyards are mainly due to the different LS factors (LS = 1.9, for BMP farm and LS = 0.1 for the CP farm). Results illustrated in Figure 5b highlight the prevailing role played by the R factor rather than the C factor in determining soil erosion when input data at a higher temporal scale are analyzed, and suggested that inter-annual and intra-annual soil erosion variability needs considering.

Event scale input data and the corresponding results of the RUSLE model in the two vineyard farms are reported in Figures 6 and 7, respectively.

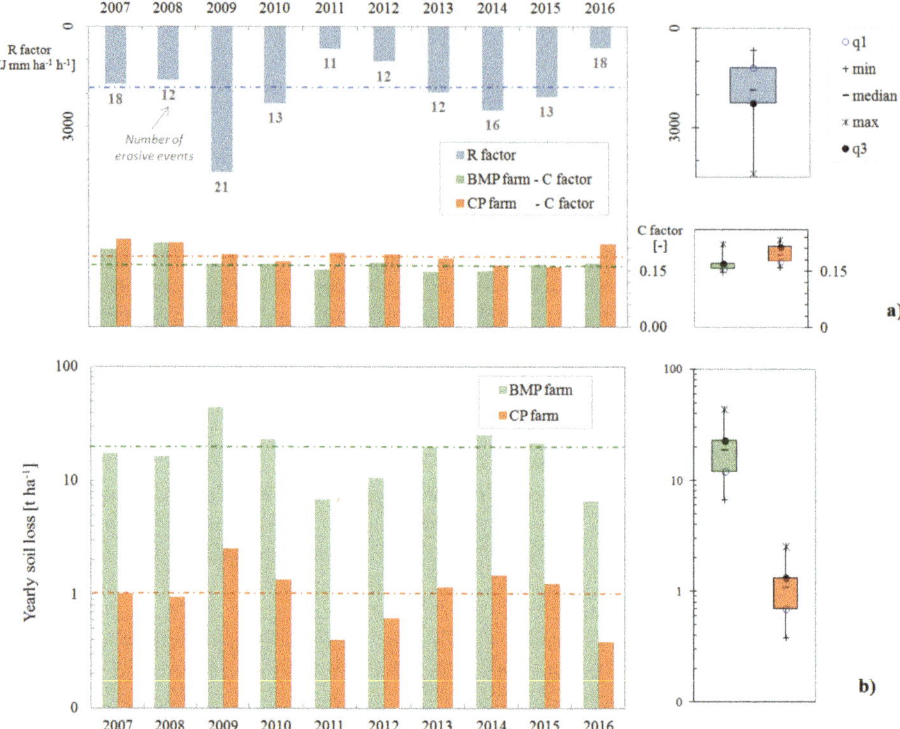

Figure 5. (a) Rainfall erosivity (R) and cover management (C) factors and (b) output results of the RUSLE model applied at average yearly and annual time scale. In the right of both figures the box-plots of the corresponding variables are also reported (q1 and q3 are the first and the third quartiles).

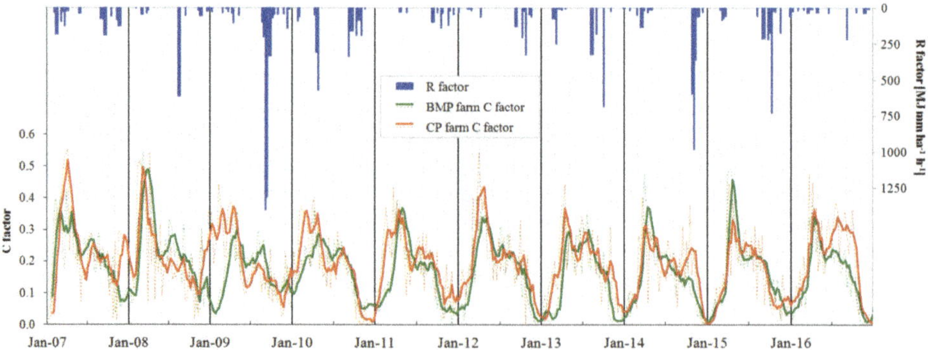

Figure 6. R and C factors calculated at event time-scale for the two vineyard plots during the investigated period.

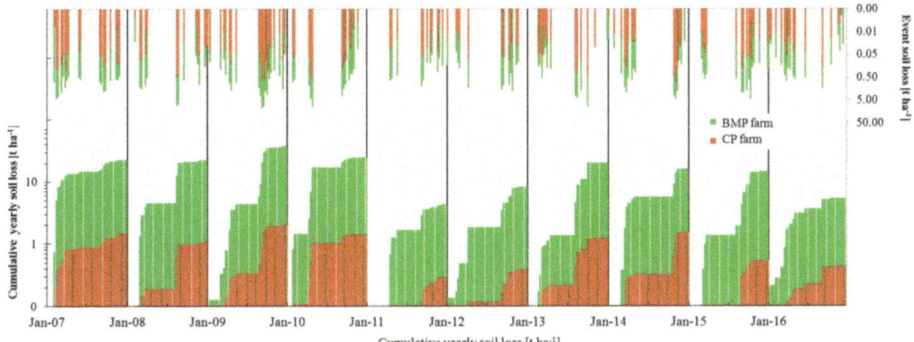

Figure 7. Event and cumulative yearly soil loss estimated using RUSLE model for the two vineyard plots during the investigated period.

Figure 6 indicates that generally high values of R factors occur when the two plots are characterized by low C values, suggesting the need of cover crop soil management in the corresponding periods. Analysis at the event scale allows detecting R and C factor interactions evidencing the occurrence of peak R factor that could also be associated with high differences in C values between BMP and CP farms. See, as an example, the difference in the peaks of R values in 2013 and 2014 and the corresponding C factors (Figure 6) that on a yearly scale was not possible to detect. In particular, the greater peak of the R value in 2014 rather than in 2013 is associated with higher differences in C values.

The cumulative yearly soil loss generally shows a drop occurring in the second semester of the year (Figure 7), which corresponds to high R values that in turn are affected by high rainfall intensities, concentrated after the summer season, typical of the Mediterranean region [26]. These results suggest that the analysis at the event scale could help in identifying the most critical period for erosion risks and the importance of alternative soil management, including BMP or spontaneous vegetation cover. Differences in cumulative yearly soil loss between the two plots of around one order of magnitude are of course due to the LS factor, making it difficult to discuss the effect of the cover crop factor alone and its interactions with the R factor in terms of soil loss.

3.2. C and R Factors Seasonal Interactions and Impact of Crop Management

Results described in Section 3.1 suggest exploring the C and R factors intra-annual interactions, which rather than a simple analysis of means (Figure 5), allow emphasizing the impact of soil management practice and at the same time, the extremes and intra-annual dynamics.

From summer to autumn, seasonal R factor increases, with the median peak (R = 173 MJ mm h⁻¹ ha⁻¹) in September (Figure 8). High R factor values also occurred in the winter season with the median value (R = 37 MJ mm h⁻¹ ha⁻¹) in February. During these periods, different C factor distributions between the considered plots could be observed. In particular, C factor values corresponding to BMP farm generally resulted lower than the CP farm, as a consequence of the different cover cropping management. On the contrary, during the April–October period, a C factor invariance between BMP and CP management due to the dominant vigor of vineyard canopy was observed.

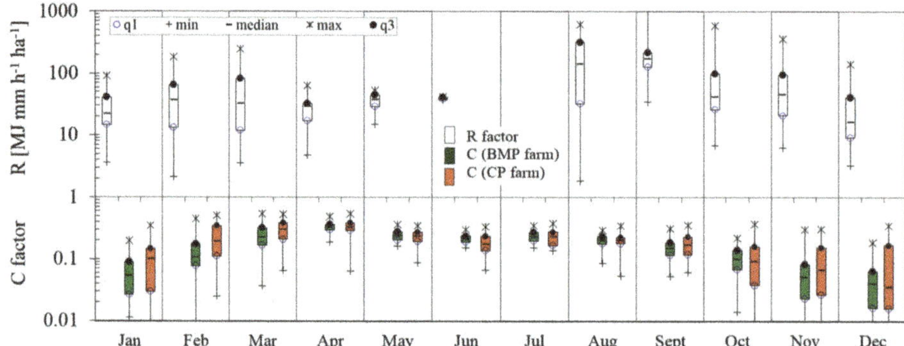

Figure 8. Box-plots of the monthly variability of R and C factors for the two investigated plots (q1 and q3 are the first and the third quartiles).

Differences in the C factor between the two plots, and its interactions with the R factor, can be better evidenced if referring to soil loss per unit of the LS factor. Moreover, in order to assess the importance of seasonal C and R factors and their temporal interactions, compared to the corresponding annual values, for the periods January–March, April–October and November–December, in Figure 9 frequency distributions of soil loss per unit LS are plotted. For the January–March period (Figure 9a), the lower RKC values of the BMP farm provide lower soil loss values than the CP farm and a similar behavior occurs during the November–December period (Figure 9c), especially for high frequencies. Contrarily, during the April–October period (Figure 9b), no significant differences between the two vineyards can be observed. This result could be ascribed to the temporary cover crop and to conventional soil management, which characterizes BMP and CP farms, respectively.

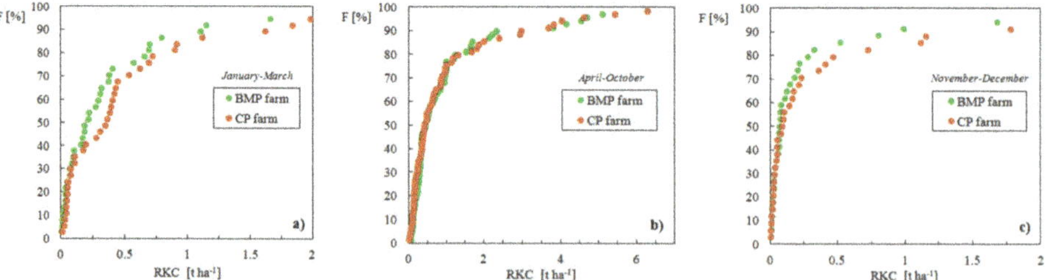

Figure 9. Seasonal distributions of soil loss per unit LS and P, RKC (see RUSLE model), for the two investigated plots. F is the frequency of non-exceedance. (**a**) January–March period, (**b**) April–October period, (**c**) November–December period.

Figure 10 shows the distributions of yearly soil loss per unit LS and P (RKC) for BMP and CP farms. At low frequencies there is no relevant effect of management practice, whereas for high frequencies the corresponding RKC value is generally lower for the BMP than the CP farm, indicating the positive impact of BMP cover management independently from the slope of the two plots. These results indicate that despite the seasonal R and C factors interactions and the differences between the two farms could not appear relevant, it seems they play an important role at the annual scale.

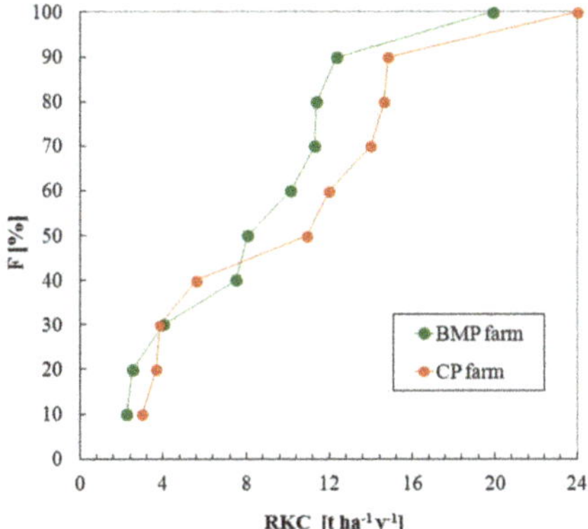

Figure 10. Distributions of yearly soil loss per unit LS and P, RKC (see RUSLE model), for the two investigated vineyards. F is the frequency of non-exceedance.

Finally, in order to assess the effectiveness of cover crop management in controlling soil loss, yearly soil loss assuming a different scenario of management practice consisting of no cover cropping was assumed. In particular, C values corresponding to CP management were applied to the BMP farm. This choice was justified by considering that the observed BMP and CP C factors are almost invariant during the period covering phenological vineyard cycle (April–October, see Figure 9), while they vary during the remaining periods according to the different management practice.

In Figure 11, the difference between soil loss calculated for the actual BMP farm and those calculated with the different management practice mentioned above denoted as saved soil loss is plotted for the ten-year investigated period.

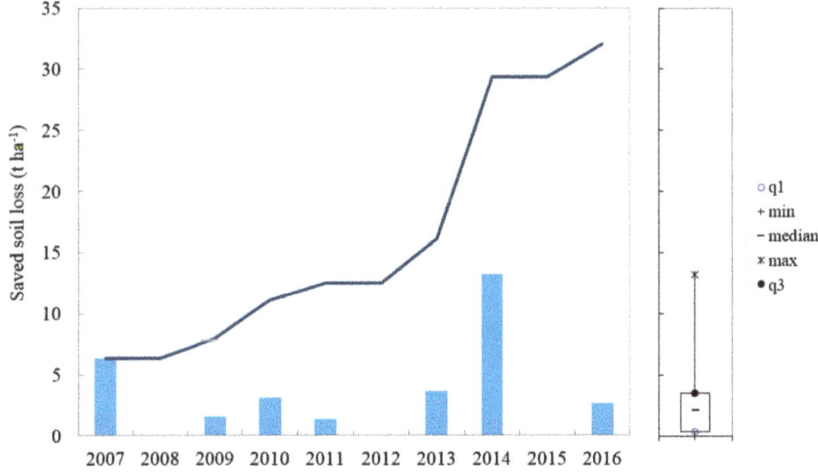

Figure 11. Impact of BMP scenario on saved soil loss. Solid line is the cumulative saved soil loss, whereas the histogram and the corresponding box-plot refer to the annual saved soil loss (q1 and q3 are the first and the third quartiles).

The same figure illustrates the cumulative saved soil loss. Results show that, usually, cover cropping management produced an important reduction of soil loss rate with respect to CP management with a total saved soil loss around 32 t ha^{-1} during the observed period (2007–2016) (Figure 11). The average saved soil loss over the 10-year period was equal to 3.2 t ha^{-1} with a maximum of 13.2 t ha^{-1} in 2014 (Figure 11). Interesting, the latter result reflects the seasonal R and C factor interactions discussed in Figure 6.

4. Conclusions

In the Mediterranean environment, hilly vineyards have higher soil losses compared to other rainfed cropping systems (cereals, olives) [27]. High erosion rates are mainly due to (i) bare soil conditions for most of the year and (ii) the vine rows orientation along highest slopes, which favor water runoff and sediment yields.

Several authors describe the effectiveness of Best Management Practice (BMP) as an important tool to prevent soil erosion rather than Conventional Practice (CP), but at the same time, they stress on the high soil loss variability due to the interaction of several factors characterized by large uncertainty [28,29].

In the near future, rainfall intensity will increase in mid-latitude areas [30,31]. One of the effects of higher rainfall intensity is increasing soil erosion, suggesting that cover crop management (BMP) can be used as an adaptation tool to this type of climate change [28].

This paper shows the importance of appropriate R and C factor assessment and their inter- and intra-annual interactions in determining soil erosion variability, by using climatic input data at a high temporal scale resolution for the R factor, and remotely sensed NDVI time-series for the C factor. It is known that the variability of C factor is managed by human activity and it is less subjected to the unpredictability of natural events as is the R factor. Indeed, the C factor is also subjected to uncertainty in relation to weather and to all related agronomic practices (species choice, seeding date, emergence, crop development, etc.). Moreover, the use of control measures (BMP) requires prudent knowledge of land management at the farm level [32] and a deep awareness of the driving forces determining soil erosion.

In this paper, starting from temporally detailed data-set, the RUSLE model was applied at different time-scales of aggregations for two different managed vineyards (BMP and CP farms). In particular, RUSLE application at yearly, seasonal, monthly and event scales of aggregation was performed. Results showed that at a yearly scale the soil loss mainly reflects the variability of the annual R factor, as a consequence of the slight annual C factor variability for the two selected vineyards, which appeared almost greater for the CP farm than the BMP farm. Moreover, at a yearly scale no interaction between R and C factors arose.

Differences in the C factor between the two plots, and their interactions with the R factor, were better evidenced at seasonal, monthly and event scale if referring to soil loss per unit LS and P factors, RKC. In particular, for the January–March period, the lower RKC values of the BMP farm provide lower soil loss values than the CP farm and a similar behaviour occurs in the November–December period. At these scales, high values of R factors generally occurred when the two plots were characterized by low C values, suggesting the need for cover crop soil management in the corresponding periods. Furthermore, the impact of BMP in preventing soil erosion was described in terms of average saved soil loss, which over the 10-year period of observation was equal to 3.2 t ha^{-1} with a maximum of 13.2 t ha^{-1} in 2014.

The evaluation of soil erosion at a different temporal scale and its implications can help stakeholders and scientists formulate better soil conservation practices and agricultural management, also considering that erosivity rates are expected to increase to account for the increase of rainfall intensity.

Finally, the introduction of a nature-based solution (large use of spontaneous cover cropping, permanent cover cropping, etc.), could be the new paradigm of soil conservation management, especially in semi-arid vineyards. All those management strategies increase resilience and reduce vineyard system vulnerability, representing a concrete adaptation strategy.

Author Contributions: Conceptualization, G.B., M.M., A.N. and L.G.; Methodology, G.B. and M.M.; Writing-Original Draft Preparation, G.B., A.N. and M.M.; Writing-Review & Editing, G.B., M.M., A.N. and L.G.

Funding: This research received no external funding.

Conflicts of Interest: The authors declare no conflict of interest.

References

1. Renard, K.G.; Foster, G.R.; Weesies, G.A.; McCool, D.K.; Yoder, D.C. *Predicting Soil Erosion by Water: A Guide to Conservation Planning with the Revised Universal Soil Loss Equation (RUSLE)*; Agriculture Handbook Number 703; US Government Printing Office: Washington, DC, USA, 1997.

2. De Santos Loureiro, N.; de Azevedo Coutinho, M. A new procedure to estimate the RUSLE EI_{30} Index, based on monthly rainfall data and applied to the Algarve region, Portugal. *J. Hydrol.* **2001**, *250*, 12–18. [CrossRef]

3. Nunes, A.N.; Lourenço, L.; Vieira, A.; Bento-Gonçalves, A. Precipitation and erosivity in Southern Portugal: Seasonal variability and trends (1950–2008). *Land Degrad. Dev.* **2016**, *27*, 211–222. [CrossRef]

4. Ferreira, V.; Panagopoulos, T. Seasonality of soil erosion under Mediterranean conditions at the Alqueva Dam watershed. *Environ. Manag.* **2014**, *54*, 67–83. [CrossRef]

5. Terranova, O.G.; Gariano, S.L. Rainstorms able to induce flash floods in a Mediterranean-climate region (Calabria, southern Italy). *Nat. Hazards Earth Syst. Sci.* **2014**, *14*, 2423–2434. [CrossRef]

6. D'Asaro, F.; D'Agostino, L.; Bagarello, V. Assessing changes in rainfall erosivity in Sicily during the twentieth century. *Hydrol. Process.* **2007**, *21*, 2862–2871. [CrossRef]

7. Land Processes Distributed Active Archive Center (LP DAAC). 1990. Available online: http://lpdaac.usgs.gov/ (accessed on 8 May 2019).

8. Karaburun, A.; Bhandari, A.K. Estimation of C factor for soil erosion modelling using NDVI in Buyukcekmece watershed. *Ozean J. Appl. Sci.* **2010**, *3*, 77–85.

9. Vicente, M.L.; Navas, A.; Machin, J. Identifying erosive periods by using RUSLE factors in mountain fields of the Central Spanish Pyrenees. *Hydrol. Earth Syst. Sci. Discuss.* **2007**, *4*, 2111–2142. [CrossRef]

10. Van der Knijff, J.M.; Jones, R.J.A.; Montanarella, L. *Soil Erosion Risk Assessment in Europe*; EUR 19044 EN; European Commission: Brussels, Belgium, 2000.

11. Panagos, P.; Borrelli, P.; Poesen, J.; Ballabio, C.; Lugato, E.; Meusburger, K.; Montanarella, L.; Alewell, C. The new assessment of soil loss by water erosion in Europe, *Environ. Sci. Policy* **2015**, *54*, 438–447. [CrossRef]

12. Benavidez, R.; Jackson, B.; Maxwell, D.; Norton, K. A review of the (Revised) Universal Soil Loss Equation ((R)USLE): With a view to increasing its global applicability and improving soil loss estimates. *Hydrol. Earth Syst. Sci.* **2018**, *22*, 6059–6086. [CrossRef]

13. Ferro, V.; Giordano, G.; Iovino, M. Isoerosivity and erosion risk map for Sicily. *Hydrol. Sci. J.* **1991**, *36*, 549–564. [CrossRef]

14. Ballabio, C.; Borrelli, P.; Spinoni, J.; Meusburger, K.; Michaelides, S.; Beguería, S.; Klik, A.; Petan, S.; Janeček, M.; Olsen, P.; et al. Mapping monthly rainfall erosivity in Europe. *Sci. Total Environ.* **2017**, *579*, 1298–1315. [CrossRef] [PubMed]

15. Baiamonte, G.; Mercalli, L.; Cat Berro, D.; Agnese, C.; Ferraris, S. Modelling the frequency distribution of interarrival times from daily precipitation time-series in North-West Italy. *Hydrol. Res.* **2018**, *50*, 339–357. [CrossRef]

16. Ma, B.L.; Dwyer, L.M.; Costa, C.; Cober, E.R.; Morrison, M.J. Early prediction of soybean yield from canopy reflectance measurements. *Agron. J.* **2001**, *93*, 1227–1234. [CrossRef]

17. De Asis, A.M.; Omasa, K. Estimation of vegetation parameter for modeling soil erosion using linear Spectral Mixture Analysis of Landsat ETM data. *ISPRS J. Photogramm. Remote. Sens.* **2007**, *62*, 309–324. [CrossRef]

18. Li, C.; Qi, J.; Yang, L.; Wang, S.; Yang, W.; Zhu, G.; Zou, S.; Zhang, F. Regional vegetation dynamics and its response to climate change—A case study in the Tao River Basin in Northwestern China. *Environ. Res. Lett.* **2014**, *9*, 12500. [CrossRef]

19. Christensen, S.; Goudriaan, J. Deriving light interception and biomass from spectral reflectance ratio. *Remote Sens. Environ.* **1993**, *43*, 87–95. [CrossRef]

20. Aparicio, N.; Villegas, D.; Casadesus, J.; Araus, J.L.; Royo, C. Spectral vegetation indices as nondestructive tools for determining durum wheat yield. *Agron. J.* **2000**, *92*, 83–91. [CrossRef]

21. Morgan, R.P.C. Soil Erosion and Conservation. In *Environmental Modelling: Finding Simplicity in Complexity*, 2nd ed.; Blackwell Publishing: Oxford, UK, 2005.
22. Panagos, P.; Meusburger, K.; Ballabio, C.; Borrelli, P.; Alewell, C. Soil erodibility in Europe: A high-resolution dataset based on LUCAS. *Sci. Total Environ.* **2014**, *479*, 189–200. [CrossRef]
23. Wischmeier, W.H.; Smith, D.D. *Predicting Rainfall Erosion Losses: Guide to Conservation Planning USDA*; Agriculture Handbook Number 537; U.S. Government Printing Office: Washington, DC, USA, 1978.
24. Desmet, P.J.; Govers, G. A GIS procedure for automatically calculating the USLE LS factor on topographically complex landscape units. *J. Soil Water Conserv.* **1996**, *51*, 427–433.
25. Adornado, H.A.; Yoshida, M.; Apolinares, H.A. Erosion vulnerability assessment in REINA, Quezon Province, Philippines with raster-based tool built within GIS environment. *Agric. Inf. Res.* **2009**, *18*, 24–31. [CrossRef]
26. Agnese, C.; Baiamonte, G.; Cammalleri, C. Modelling the occurrence of rainy days under a typical Mediterranean climate. *Adv. Water Res.* **2014**, *64*, 62–76. [CrossRef]
27. Kosmas, C.; Danalatos, N.; Cammeraat, L.H.; Chabart, M.; Diamantopoulos, J.; Farand, R.; Gutierreze, L.; Jacob, A.; Marques, H.; Martinez-Fernandezg, J.; et al. The effect of land use on runoff and soil erosion rates under Mediterranean conditions. *Catena* **1997**, *29*, 45–59. [CrossRef]
28. Novara, A.; Gristina, L.; Saladino, S.S.; Santoro, A.; Cerdà, A. Soil erosion assessment on tillage and alternative soil managements in a Sicilian vineyard. *Soil Tillage Res.* **2011**, *117*. [CrossRef]
29. Baiamonte, G.; D'Asaro, F. Discussion of "Analysis of extreme rainfall trends in sicily for the evaluation of depth-duration-frequency curves in climate change scenarios, by Lorena Liuzzo and Gabriele Freni". *J. Hydrol. E ASCE* **2016**, *21*. [CrossRef]
30. IPCC. *Climate Change 2013: The Physical Science Basis. Contribution of Working Group I to the Fifth Assessment Report of the Intergovernmental Panel on Climate Change*; Stock, T.F., Qin, D., Plattner, G.-K., Tignor, M., Allen, S.K., Boschung, J., Nauels, A., Xia, Y., Bex, V., Midgley, P.M., Eds.; Cambridge University Press: Cambridge, UK, 2013.
31. Trenberth, K.E. Changes in precipitation with climate change. *Clim. Res.* **2011**, *47*, 123–138. [CrossRef]
32. Kessler, A.; De Graaff, J.; Olsen, P. Farm-level adoption of soil and water conservation measures and policy implications in Europe. *Land Use Policy* **2010**, *27*, 1–3. [CrossRef]

Article

Recent and Future Changes in Rainfall Erosivity and Implications for the Soil Erosion Risk in Brandenburg, NE Germany

Andreas Gericke [1,*], Jens Kiesel [1,2], Detlef Deumlich [3] and Markus Venohr [1,4]

[1] Leibniz-Institute of Freshwater Ecology and Inland Fisheries, 12489 Berlin, Germany; kiesel@igb-berlin.de (J.K.); m.venohr@igb-berlin.de (M.V.)
[2] Department of Hydrology and Water Resources Management, Christian-Albrechts-University Kiel, 24118 Kiel, Germany
[3] Leibniz Centre for Agricultural Landscape Research, 15374 Müncheberg, Germany; ddeumlich@zalf.de
[4] Department of Geography, Humboldt-University of Berlin, 12489 Berlin, Germany
* Correspondence: gericke@igb-berlin.de

Received: 15 April 2019; Accepted: 26 April 2019; Published: 29 April 2019

Abstract: The universal soil loss equation (USLE) is widely used to identify areas of erosion risk at regional scales. In Brandenburg, USLE R factors are usually estimated from summer rainfall, based on a relationship from the 1990s. We compared estimated and calculated factors of 22 stations with 10-min rainfall data. To obtain more realistic estimations, we regressed the latter to three rainfall indices (total and heavy-rainfall sums). These models were applied to estimate future R factors of 188 climate stations. To assess uncertainties, we derived eight scenarios from 15 climate models and two representative concentration pathways (RCP), and compared the effects of index choice to the choices of climate model, RCP, and bias correction. The existing regression model underestimated the calculated R factors by 40%. Moreover, using heavy-rainfall sums instead of total sums explained the variability of current R factors better, increased their future changes, and reduced the model uncertainty. The impact of index choice on future R factors was similar to the other choices. Despite all uncertainties, the results indicate that average R factors will remain above past values. Instead, the extent of arable land experiencing excessive soil loss might double until the mid-century with RCP 8.5 and unchanged land management.

Keywords: climate change; EURO-CORDEX; Germany; model ensemble; R factor; rainfall erosivity; trend analysis; uncertainty; universal soil loss equation (USLE)

---

1. Introduction

Soils are a fundamental resource for life on Earth and provide numerous goods and services for the human society [1]. The degradation of soils poses a global threat to our well-being, mainly due to soil erosion by water [2,3] (henceforth soil erosion) exceeding the 'tolerable' natural formation rate of around 1 t ha^{-1} a^{-1} [4]. Among other consequences, soil erosion can hamper the sustainable agricultural production [5], impair water quality and habitats [6,7], and reduce the lifetime of reservoirs [8].

Unsustainable agriculture is a key driver of soil erosion, not only in Europe [3,9,10], especially when heavy rainfall meets inappropriate management. To mitigate agricultural impacts on soils, the Common Agricultural Policy (CAP) of the European Union introduced the Good Agricultural and Environmental Conditions (GAEC) in 2003 as a set of environmental standards and rules on cross compliance for financially supported farmers. Accordingly, German laws and regulations on both federal and state levels address soil erosion, not only to implement the GAEC, but also within the European Water Framework Directive.

To estimate potential soil erosion, derivatives of the universal soil loss equation (USLE) are commonly used. For instance, German federal states apply an adapted version of the USLE [11,12] to identify areas prone to soil loss and to impose countermeasures on farmers. The USLE estimates the soil erosion rate from five factors, namely rainfall erosivity (R factor), soil erodibility (K factor), slope length and steepness (LS factor), soil cover (C factor), and soil conservation (P factor).

The R factor of the USLE is the long-term average annual sum of the rainfall erosivity. The erosivity expresses the capacity of a rainfall to induce soil erosion and integrates its duration, amount, and intensity (cf. Appendix A, also for the units of the USLE factors). The calculation of R factors requires long time series of precipitation data at high temporal resolution which are often unavailable. Many studies thus rely on daily to annual data for extrapolation in space and time [13–15]. Accordingly, the German industrial norm (DIN) on soil erosion [11] lists linear regression models to estimate R factors from summer or annual rainfall. While they were derived at the state level, the explained variability of the calculated R was unsatisfying in Brandenburg and other federal states in the North German Plain, with Pearson's r being lower than in other parts of Germany [11]. The DIN models were established during the early 1990s, so they also reflect the climate from the 1960s to the 1980s (cf. [16]). The lack of accuracy and the age raise the question whether the DIN model can be recommended to estimate current and future R factors for land and water management in Brandenburg.

Climate change can exhibit direct effects (changes in the amount, intensity, and distribution of rainfall) and indirect effects on soil erosion (rainfall and temperature changes affect biomass production, soil moisture, and the growing season) [17]. Numerous trend studies have assessed past changes in precipitation, especially changes in extreme precipitation across Europe [18–21], in (Northeastern, NE) Germany [22–25] and neighboring countries [23,26,27]. These studies consistently found increasing winter rainfall in Central Europe, while the trends for summer precipitation are less coherent. The latter studies indicate less rainfall and longer dry spells during summer, although this holds not necessarily true for extreme rainfall [23] and when recent data is included [26]. The few regional to national studies on rainfall erosivity in Central/Western Europe show a multi-decadal variability with an increase in erosion risks since the end of the 20th century [28,29], although the strength and direction of regional trends are not necessarily valid everywhere [30].

Additionally, studies on future R factors show contradictory changes for NE Germany. While a European assessment proposed that R factors might approximately double from 2010 to 2050 even under moderate climate change scenarios [31], a German-wide analysis reported an average increase of 10% for 2011 to 2041 and a decline in the same order of magnitude for 2041 to 2071 compared to the reference period 1971–2000 [32]. Both studies did not assess how the choices of their respective climate model and rainfall indices to estimate the R factors affected their scenario results. According to more recent German ensemble studies, there might be a tendency of decreasing summer totals, but increasing summer extremes [33,34]. Thus, different choices of aggregated rainfall indices might change the direction of future R changes, with implications for discussions of a sustainable regional land and water management under climate change.

The main objective of this study is to assess the impact of climate change on rainfall erosivity and, subsequently, on the potential soil erosion risk in Brandenburg, without considering further impacts on land use and vegetation cover. Using an ensemble of climate scenarios, our study addresses the following main questions:

- Which aggregated rainfall index is best to estimate current R factors and their recent change in NE Germany?
- How does climate change affect regional R factors and the risk of soil erosion?
- How does the rainfall index affect future trends and how does the impact compare to other sources of uncertainty such as the choice of climate model, bias correction, and RCP scenario?

2. Materials and Methods

2.1. Study Area

The federal state of Brandenburg is situated in NE Germany. It shares the Eastern border with Poland and encompasses Berlin, Germany's capital city. Brandenburg covers an area of 30.000 km^2, of which one third is arable land. It completely belongs to the European ecoregion "Central Plains" [35]. Shaped during the Pleistocene, the landscape is characterized by flat to undulating terrain, a dense network of streams, many (often shallow) lakes and ponds, as well as sandy soils. The climate is temperate and fully humid (Cfb) according to the Köppen-Geiger classification with an annual mean air temperature of 9.6 °C and an annual precipitation of 560 mm (1996–2015, original data provided by Reference [36]).

2.2. Rainfall Indices and the Variability of Calculated R Factors

We calculated the rainfall erosivity and the R factors in kJ m^{-2} mm h^{-1} for 22 stations in and near Brandenburg (Table 1, squares in Figure 1) according to the German Norm DIN ([11,12], Appendix A). The required detailed precipitation data, sampled at 10 min intervals with PLUVIO OTT weighing gauges which partly replaced earlier NG-200 volumetric gauges, was provided by the German Meteorological Service (DWD) for the years 2000 to 2015. Implausible values were corrected (Appendix B). Table 2 provides an overview of the data used in this study.

Table 1. Stations with high-resolution data for which R factors were calculated (end year is 2015).

ID	Name	Longitude	Latitude	Start Year	Calculated R kJ m^{-2} mm h^{-1}	Data Gaps
400	Berlin Buch	13.500	52.633	2004	77.2	
410	Berlin Kaniswall	13.733	52.400	2004	68.7	
430	Berlin Tegel [3]	13.317	52.567	2000 [1]	61.9 [2]	
714	Neu Madlitz	14.250	52.367	2005	96.9	
880	Cottbus [3]	14.317	51.783	2000 [1]	83.9	
1052	Drewitz	12.167	52.217	2003	85.4	Jan–Mar 2003
1801	Groß Kreutz	12.800	52.400	2003	63.8	
2625	Kleßen	12.500	52.733	2003	80.2	Jan–Mar 2003 3 Nov–4 Apr
2733	Kremmen	13.017	52.733	2005	89.8	
2856	Langenlipsdorf	13.083	51.917	2004	71.2	
2997	Lieberose	14.300	51.983	2003	99.4	Jan–Apr 2003
3015	Lindenberg [3]	14.117	52.217	2000	63.9	
3376	Müncheberg	14.117	52.517	2004	99.1	Dec 2005
3881	Passow	14.100	53.150	2005	52.0	
3906	Perleberg	11.867	53.100	2004	67.7	
3967	Pohlitz	14.567	52.183	2005	91.2	Jan–Mar 2005
3987	Potsdam [3]	13.067	52.383	2000	76.6	Nov–Dec 2000
4555	Schollene	12.183	52.667	2007	73.2	
4637	Staaken	13.117	52.533	2009	51.8	
5614	Winterfeld-Sallenthin	11.250	52.750	2004	65.4	Jan–Mar 2004
5825	Berge	12.783	52.617	2003	64.7	
6170	Coschen	14.733	52.017	2003	95.9	Jan–Mar 2003

[1] Corrected values (Appendix B), [2] as used for the regression models without an extreme event (Appendix C), [3] change from volumetric to weighing gauge.

Three aggregated rainfall indices were derived from daily data and tested to explain the variability of the calculated R factors with linear regression models. We compared the total sum from May to September (P$_{sum}$) to the total sum of rainfall on heavy rainfall days. As no common criteria for "heavy rainfall day" exists, we considered exemplarily the 10 highest daily values (P$_{max10}$) and daily values above 11.8 mm (the 33th percentile of all calculated erosive rainfall events, which is above the threshold of 10 mm d^{-1} used for a "heavy rainfall day" e.g., by Reference [37]). In contrast to the DIN and similar empirical models, we excluded the October because rainfalls were less erosive than in summer months (cf. Figure 2 in the Results section). For each climate station, P_{sum}, P_{max10}, and $P_{11.8}$ were calculated as multi-annual means for the years in Table 1.

We applied the new regression models to the German-wide dataset of regionalized daily station data (REGNIE, [36]) to create new state-wide R maps for the years 2001–2015. These maps replaced the fixed R factor of 50 kJ m^{-2} mm h^{-1} of the current erosion map [38], e.g., used by the State Office of Environment to identify risk areas, i.e., arable land with erosion rates above 1 t ha^{-1} a^{-1}. Likewise, we obtained the current extent of risk areas. For this reason, we used the REGNIE data to establish the regression models. As REGNIE is derived from station data, differences between both only occur if stations are not considered for the regionalization.

Table 2. Overview of the climate data used in this study. The 188 data points comprise the 22 stations in Table 1. 2001–2015 was the reference period for scenario analyses.

Purpose	Data Source	Resolution	Data Set	Period	Chapter
R calculation	Station data	10 min	22	≥2000–2015	2.2
Regression analyses	REGNIE	Daily	22	≥2000–2015	2.2
Bias correction, ranking	REGNIE, climate models	Daily	188	1971–2015	2.3
R scenarios	Climate models	Daily	188	1971–2100	2.3
Erosion risk areas	REGNIE	Daily	(grid)	2001–2015	2.2
"	Climate models	Daily	188	2021–2100	2.4

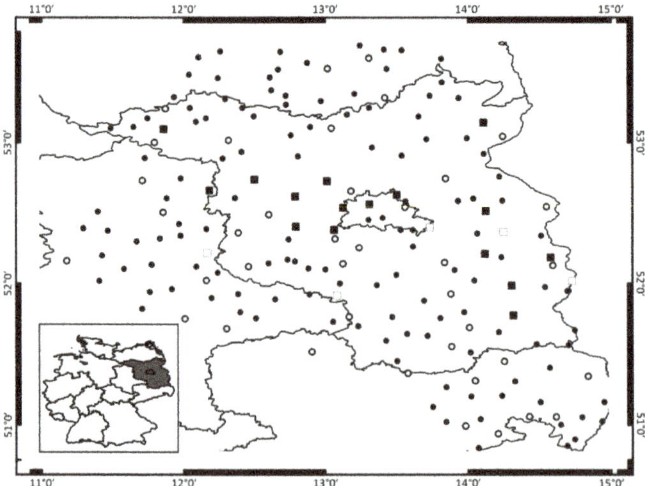

Figure 1. Climate stations used in this study and location of the study area in Germany (inset map), squares: stations with 10 min data for R calculation (Table 1), all stations with daily data to estimate R factors, bias correction of the climate models, and scenario analyses, black: stations with data for 1971–2015, hollow: stations with shorter time series. To harmonize the time series and to fill gaps we extracted the time series from regionalized station data.

2.3. Climate Scenarios

We used climate model data from the Coordinated Regional Downscaling Experiment which provides the most recent climate data for Europe (EURO-CORDEX domain, Reference [39]). Harmonized daily datasets of climate parameters are available in high spatial resolution from various GCM (general circulation models), RCM (regional circulation models), and RCP (representative concentration pathways). This combination makes the CORDEX data most suitable for assessing uncertainties related to the selection of climate models and RCP scenarios in regional studies.

The data availability in the CORDEX database differs among the RCP. To obtain a sufficiently large, yet homogeneous ensemble for the uncertainty assessment, we used 15 combinations of GCM and dynamical RCM in the highest possible spatial resolution of 0.11° (approx. 12 km, CORDEX domain EUR-11) available for RCP 4.5 (i.e., moderate climate change) and RCP 8.5 (i.e., extreme climate change) (Table 3).

Table 3. The output of climate models used in this study was available for RCP 4.5 and RCP 8.5.

ID	Institute	GCM	RCM	Ensemble	Version
1	CLMcom	CNRM-CERFACS-CNRM-CM5	CCLM4-8-17	r1i1p1	v1
2	SMHI	CNRM-CERFACS-CNRM-CM5	RCA4	r1i1p1	v1
3	CLMcom	ICHEC-EC-EARTH	CCLM4-8-17	r12i1p1	v1
4	DMI	ICHEC-EC-EARTH	HIRHAM5	r3i1p1	v1
5	KNMI	ICHEC-EC-EARTH	RACMO22E	r1i1p1	v1
6	SMHI	ICHEC-EC-EARTH	RCA4	r12i1p1	v1
7	IPSL-INERIS	IPSL-IPSL-CM5A-MR	WRF331F	r1i1p1	v1
8	SMHI	IPSL-IPSL-CM5A-MR	RCA4	r1i1p1	v1
9	CLMcom	MOHC-HadGEM2-ES	CCLM4-8-17	r1i1p1	v1
10	KNMI	MOHC-HadGEM2-ES	RACMO22E	r1i1p1	v2
11	SMHI	MOHC-HadGEM2-ES	RCA4	r1i1p1	v1
12	CLMcom	MPI-M-MPI-ESM-LR	CCLM4-8-17	r1i1p1	v1
13	MPI-CSC	MPI-M-MPI-ESM-LR	REMO2009	r1i1p1	v1
14	MPI-CSC	MPI-M-MPI-ESM-LR	REMO2009	r2i1p1	v1
15	SMHI	MPI-M-MPI-ESM-LR	RCA4	r1i1p1	v1

We downloaded the CORDEX data for the years 1971 to 2100 and extracted the time-series for 188 climate stations in and around Brandenburg (black circles in Figure 1). To reduce the deviations to station data, we applied a bias correction (distribution mapping with linear scaling) to adjust the monthly frequency distributions and means for the period 1971–2015. This period includes hindcasted data where both RCP are similar (1971–2005) and climate change projections where the RCP deviate (2006–2015). We used an adapted version of the software CMHyd (References [40,41], see Reference [42] for the conceptual and methodical background) to perform the bias correction. The software also transformed the 360-days calendar of the GCM MOHC-HadGEM2-ES to a 365-days calendar.

While most stations had sufficiently long time series for the bias correction (Figure 1), we considered additional stations with shorter time series until 2015 to increase the density for the spatial interpolation (Chapter 2.4). Thus, we used again the REGNIE data instead of the station data. P_{sum}, P_{max10}, and $P_{11.8}$ were calculated from the climate model output and the REGNIE data. We used a simple ranking scheme based on the average Kling-Gupta efficiency (KGE, [43]), the absolute percent bias (APB, in %), and the root-mean square error (RMSE) to assess the performance of the bias-corrected climate models for the period 1971–2015. The ranking considered the

- Spatial rank: For each station, the long-term average rainfall indices were calculated. KGE, APB and RMSE were obtained for each climate model.
- Trend rank: For each station, we calculated the annual indices from REGNIE and the climate models and determined KGE, APB and RMSE from the linear trend. For each climate model, we averaged KGE, APB, and RMSE for the ranking.

The rank of each climate models was determined with the function rank.avg in MS Excel which returns the average rank for equal values, e.g., eight if all 15 models would perform equally well for an index. To rank almost similar values equally, we rounded the KGE to the nearest 0.05 and APB and RMSE to the nearest integer. The five best-performing climate models were selected (subset) and compared to the whole ensemble of 15 models to evaluate the effect of model choice on R trends.

2.4. Impact of Climate Change on R Factors, Uncertainties, Consequences for the Extent of Erosion-Risk Areas

For all 188 climate stations and the 8 combination of climate models (whole ensemble or subset), bias correction (with or without), RCP (4.5 or 8.5), and index (heavy-rainfall or total sum), we derived the change in R relative to 2001–2015 for other periods of 15 years overlapping by five years (i.e., 2011–2025, 2021–2035, etc.) to assess how modelling choices influenced future trends. Each choice resulted in two different R factors (R and R′) for all the stations and periods. We considered this difference as uncertainty. We averaged the ratios of max(R, R′) and min(R, R′) over four periods in the first and the second half of this century (2011–2055, 2051–2095) to compare the impact of each choice.

Finally, the relative station values were interpolated using thin splines [44] (a) to assess the sub-regional pattern of R changes, and (b) to quantify the change in the extent of arable land [45] at risk of erosion rates exceeding the threshold of 1 t ha^{-1} a^{-1}. For the latter, we applied the interpolated data as correction factor to the revised state erosion map (Chapter 2.2).

3. Results

3.1. Rainfall Indices and the Spatial Variability of Current R Factors

The calculated R factors considerably exceeded the values estimated with the DIN regression model (Figure 2a), especially in the south-eastern part where the highest R factors were calculated. While the average calculated value was 76.8 kJ m^{-2} mm h^{-1}, the DIN model estimated only 42.7 kJ m^{-2} mm h^{-1}. The fixed value used for the existing erosion map corresponds to the maximum of the estimated values but to the minimum of the calculated values. In consequence, the current state-wide erosion map underestimates the potential erosion rate and the extent of risk areas.

About 90% of the annual rainfall erosivity occurred between May and September, 50% in July and August (Figure 2b). During the rest of the year, rainfall events were far less erosive. As aggregated rainfall indices cannot capture the lower peak intensities in these months and would overestimate their contribution to R factors, we did not consider these months in our rainfall indices for the regression analyses.

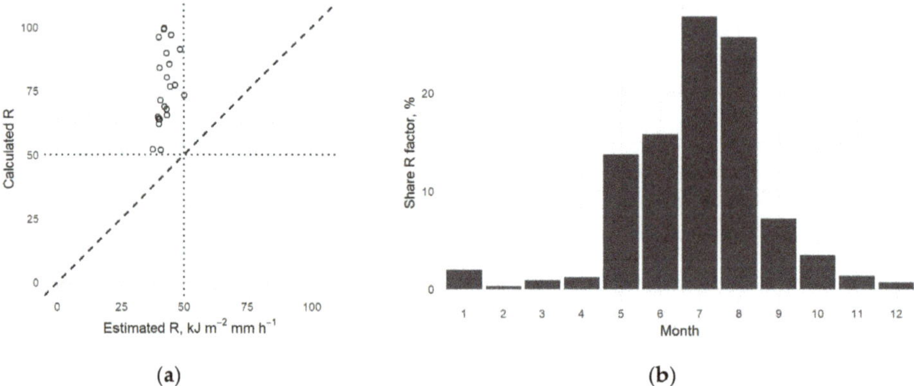

(**a**) (**b**)

Figure 2. Variability of rainfall erosivity: (**a**) R factors estimated with the existing regression model were lower and less variability than the calculated R factors (1:1 line = perfect model agreement). The fixed R value of 50 kJ m^{-2} mm h^{-1} used for the current state erosion map corresponds to the upper boundary of the estimated R factors (vertical line) and the lower boundary of the calculated R factors (horizontal line). Both, calculation and estimation of R factors according to the German norm DIN; (**b**) average intra-annual distribution of the rainfall erosivity. Rainfall from October to April was far less erosive than summer rainfall and therefore not considered in the revised regression models for R factors in Brandenburg.

The heavy-rainfall indices P_{max10} and $P_{11.8}$ explained the variability of the 22 calculated R factors better ($r^2 = 0.46$–0.50) than the total sum P_{sum} ($r^2 = 0.21$, and 0.08 if calculated for May to October as used in the DIN). However, the performance was strongly affected by the most extreme event, its erosivity (EI_{30} value) being more than twice the second highest value (494 kJ m^{-2} mm h^{-1}, Appendix C). Without this event, the corresponding R factor (of station 430, Table 1) decreased by 33% resulting in r^2 values of around 0.6 (Equations (2) and (3)) and 0.3 (Equation (1)).

Using heavy-rainfall indices instead of total sums affected the spatial pattern of the estimated R factor (Figure 3). While the outcomes of Equations (1)–(3) and even the DIN model were similar in Northern Brandenburg, the model deviations increased towards the South.

$$R = 0.541\ P_{sum} - 94.24 \tag{1}$$

$$R = 1.085\ P_{max10} - 105.92 \tag{2}$$

$$R = 0.695\ P_{11.8} - 22.58 \tag{3}$$

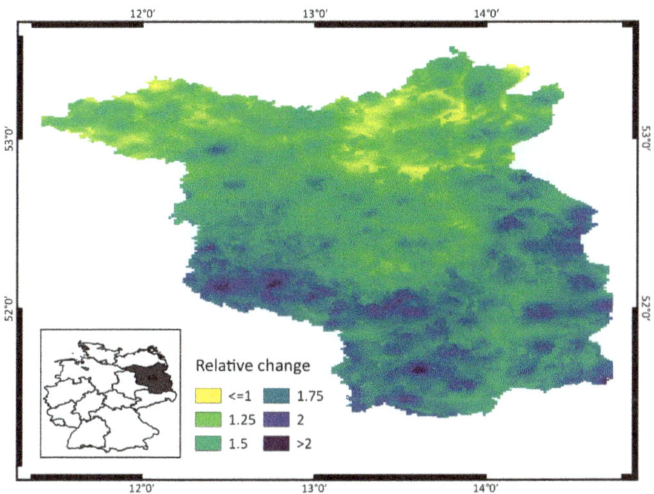

Figure 3. Ratio of R factors estimated with Equation (2) and the DIN regression model, based on REGNIE data (2001–2015); similar spatial pattern if Equation (2) was replaced by Equation (3) and the DIN model by Equation (1).

3.2. Ranking of Climate Models

No single GCM or RCM was unanimously ranked as "best matching" (Figure 4, Appendix D), since 4 combinations of GCM and RCM were equally best performing (ID 2, 5, 10, and 11, Table 3). The 5th rank was ambivalent with different results for both RCP. However, with the chosen model 7, the ensemble subset comprised 4 out of the 5 GCM and 3 out of the 6 RCM.

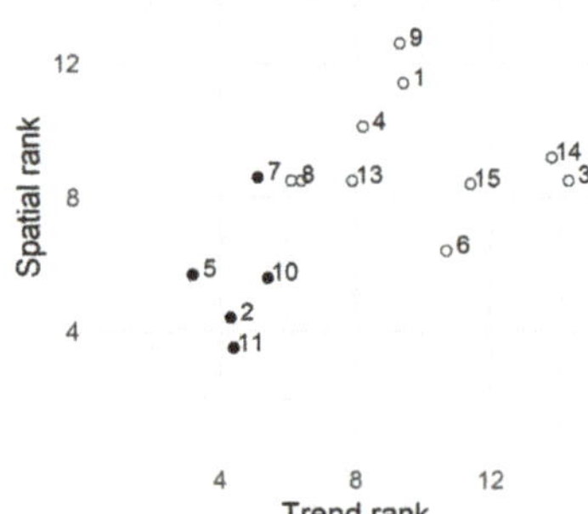

Figure 4. Average ranks for RCP 4.5 and 8.5 according to KGE, APB, and RMSE. Numbers correspond to the model ids in Table 3. The lower the rank, the better the model performance. Selected climate models for the ensemble subset in black.

3.3. Climate Scenarios and Soil Erosion Risk

Only the subset of the 5 top-ranked climate models captured, albeit underestimated, the general increase of the R factors during the past four decades (R values relative to 2001-15 below 1.0 in Figure 5). In contrast, the whole ensemble did not reveal this increase (past values $\approx$ 1.0). On average, the RCP 4.5 values (relative R of 0.81) were closer to the REGNIE data (relative R of 0.71) than the RCP 8.5 values (0.90, RCP not differentiated in Figure 5).

The variability was largest for the whole uncorrected ensemble, partly due to negative regional R factors obtained with the climate model ID 9 and RCP 4.5. For instance, the coefficient of variation (CV) was here above 22,500% for P_{sum} (Equation (1), but 22.9% without the negative values) and above 30% for P_{max10}, $P_{11.8}$, and RCP 8.5. The ranking helped to reduce the CV to 12–19%, which was close to the reference data (11%).

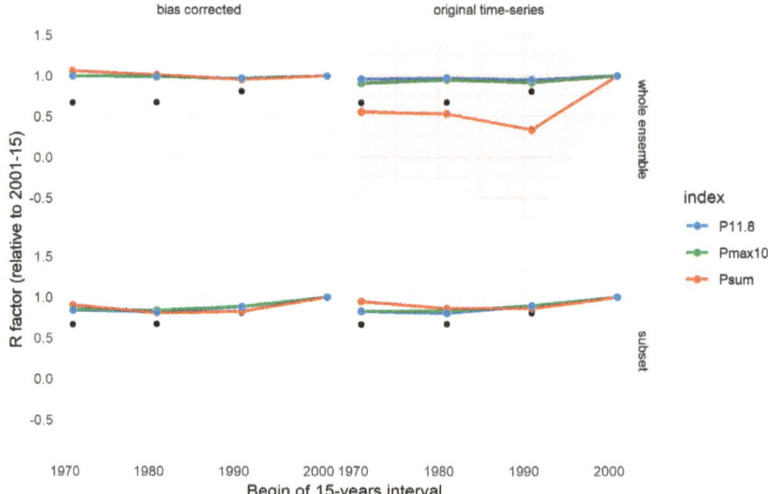

Figure 5. Historical R factors estimated with Equations (1)–(3), different ensembles of climate models (colored lines with 95%-confidence interval of the mean of climate models and stations) and the REGNIE data as reference (black); values relative to the reference period 2001–2015. Note: Both RCP and the three REGNIE results were averaged to improve readability.

Regarding future trends, all ensembles and regression models predicted increasing R factors towards the end of the 21st century (Figure 6), except for Equation (1) in combination with the whole uncorrected ensemble and RCP 4.5 (Figure 6a). However, this combination was again hampered by negative R factors.

As expected, the model ensemble and the RCP influenced the decadal variability. The ensemble subset amplified the differences between RCP 4.5 and 8.5. Until mid-century, the estimated regional R factors would generally remain above the reference period if the whole ensemble was used (solid lines in Figure 6). With the subset (broken line), the overall changes might be negligible and even negative for RCP 4.5, but increase more strongly for RCP 8.5.

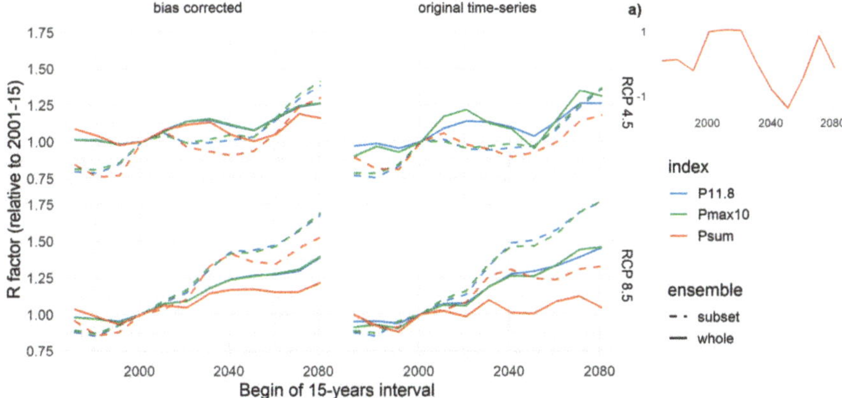

Figure 6. Variability of average R factors estimated with Equations (1)–(3), and the climate models, separated by bias correction (columns), RCP (rows), ensemble (line style), and rainfall index (color). (**a**) Changes were partly negative when Equation (1) (P_{sum}) was applied to the uncorrected model ensemble for RCP 4.5.

However, switching from P_{sum} to the heavy-rainfall indices also resulted in an increase of future R factors. The effect was larger for the uncorrected than for the bias-corrected climate models because the bias correction raised the ratio of heavy and total rainfall (Figure 7). The impacts of index choice on trends were comparable to the other modelling decisions (Figure 8).

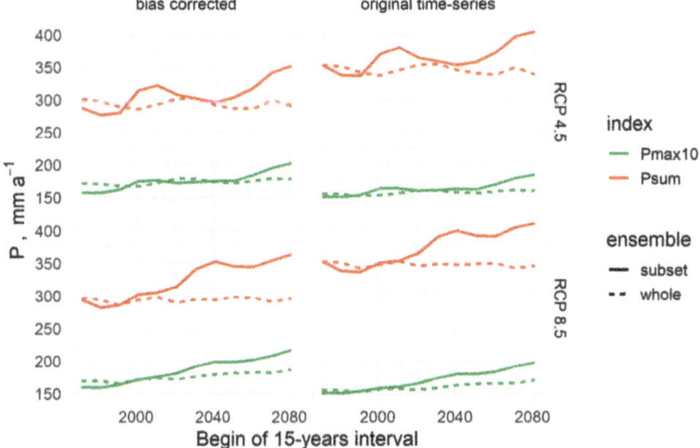

Figure 7. Trends of heavy and total rainfall sums averaged over 15 years and all stations. The bias correction resulted in lower total and slightly higher heavy-rainfall sums.

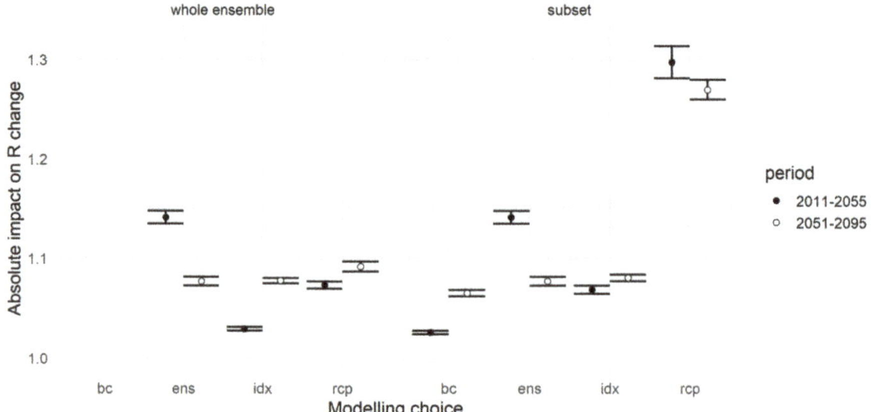

Figure 8. Effect of modelling decisions on future R changes relative to the R changes estimated with the bias-corrected subset for RCP 4.5 and Equation (1) (R change relative to 2001-15, cf. Figure 6), alternative choices of bias correction (bc), ensemble (ens), RCP (rcp), and index (idx, mean of Equations (2) and (3)). Mean and confidence interval (p = 95%) of climate stations and four overlapping 15-year periods. Values in left-most column outside value range (2011–2055: 1.98 ± 1.02, 2051–2095: 41 ± 75).

The combination of heavy-rainfall index, RCP 4.5, and ensemble subsetting suggested very low overall changes in R factors around mid-century (2041–2065) compared to 2001–2015 (Figure 6). However, the spatial aggregation masked distinctively opposite trends in different sub-regions. While R factors in some parts of Northern Brandenburg might increase by more than 15%, they can decrease by more than 10% in the South-Eastern part (Figure 9).

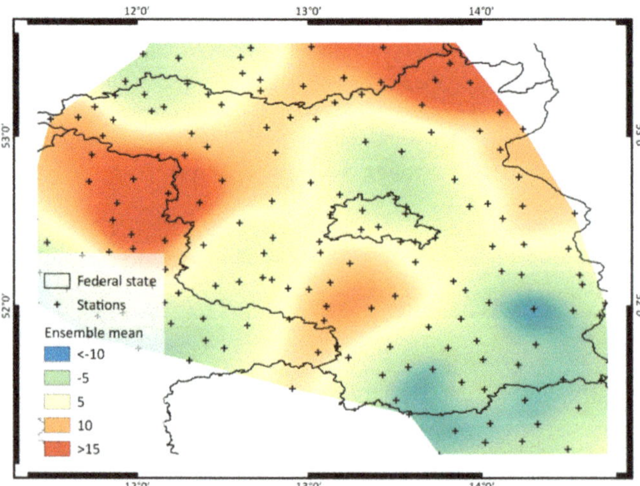

Figure 9. Percent change in the R factor 2041–2065 compared to 2001–2015 for a moderate climate change (RCP 4.5), using the bias-corrected ensemble subset and the average of Equations (2) and (3). Values interpolated with thin splines. The overall change for this combination is close to 0%.

Due to the spatial variability of R changes, of the other USLE factors, and of the distribution of arable land, changes of R and the extent of risk areas on arable land can differ. In Brandenburg, the risk areas might in fact increase more than the R factor, for instance from currently 3% to 6% by mid-century (2041–2065) with RCP 8.5 under status-quo land management, if the bias-corrected ensemble subset was applied (Figure 10), while the average R factor may raise by about 40% (Figure 6, left column). In accordance to the R factor, the heavy-rainfall indices favored a more pronounced change in risk areas than the total sum.

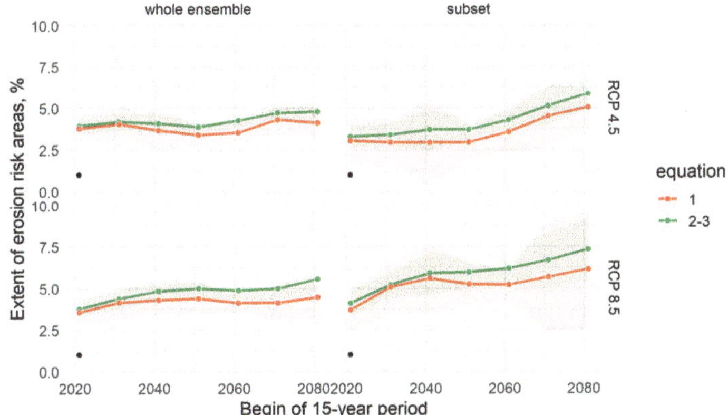

Figure 10. Future share of arable land with erosion risk, ensemble mean and confidence interval (p = 95%) of bias-corrected climate models for overlapping 15-year periods. Current value of 3.2% derived from the revised state erosion map in black.

4. Discussion

4.1. Indices to Estimate Current R Factors

The variability of total summer rainfall does not match the variability of R factors in Brandenburg. This explains the lack of accuracy of the existing DIN model, which cannot be resolved by a re-calibration. Instead, heavy-rainfall sums are more suitable for regional assessments of soil erosion risks. By changing the rainfall index, the estimated R factors and potential soil loss in the Southern part increase relatively to the northern part which corresponds to its higher calculated R factors (the 4 stations with $R > 90$ kJ m^{-2} mm h^{-1} are located in the South-Eastern part).

Given the low performance of the existing DIN models across the Central Plains ecoregion [11], this adjustment can help to improve the delineation of risk areas in other regions as well. However, the application of such simple empirical relationships is inherently impaired by the aggregated indices, and much of the unexplained variability of 40% has to be attributed to our use of daily data. Soil erosion often occurs during short peak intensities and the extreme erosivity of single events can considerably affect local R factors as multi-year averages. Likewise, interpolated station data cannot capture the high variability of rainfall erosivity even over short distances [46,47]. Better regional and national datasets are unavailable. Nonetheless, the data quality is continuously improving, including radar data [48,49].

4.2. Climate Change Impacts on R Factors and the Soil Erosion Risk

The significant underestimation of the calculated recent R factors by the existing DIN regression model indicates that the climate-based erosion risk increased since the 1990s in NE Germany. This corresponds well to rising total and heavy rainfalls in summer and is confirmed by an independent multi-decadal dataset with steadily increasing maximum and minimum annual values (Figure 11) and lower historical values for NE Germany [16]. Similar increases during the last decades were also reported for Western Germany [28] and recently deducted for the whole country [49].

Despite all model uncertainty, the climate scenarios consistently suggest that the rainfall erosivity and the erosion risk for the whole of Brandenburg might not return to levels of 2–3 decades ago—as observed during the 20th century in Western Germany [28] and suggested before to occur mid-century in NE Germany [32]. While the GCM HadGEM2 appeared twice in our ensemble subset, the rather low mid-century impact of a moderate climate change (RCP 4.5 in Figure 9) is nonetheless in contrast to the significant increase as derived in a European analysis [31]. However, the latter two studies relied only on a single and statistically downscaled GCM rather than an ensemble of dynamical RCM as used by CORDEX. The RCM type was recently shown to affect the direction of rainfall trends in Germany [33] (and also noted in Reference [32]).

A moderate climate change might, thus, open a window of opportunity to discuss and establish climate change adaptation, especially in southern Brandenburg. More soil conservation and better soil coverage on arable land are effective agricultural measures to mitigate the on-site and off-site impacts of intensified soil erosion under climate change.

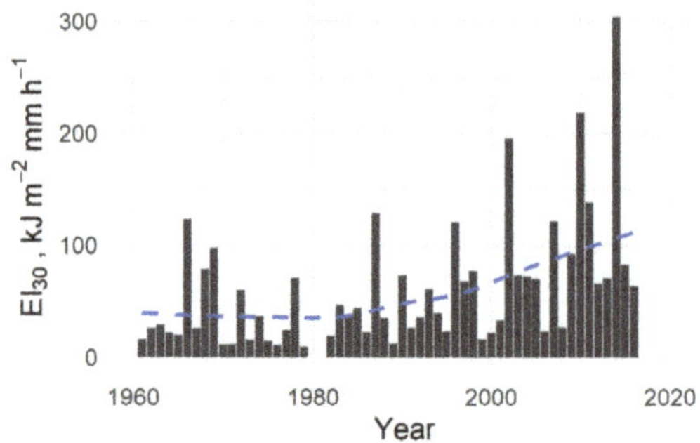

Figure 11. Annual variability of EI_{30} between April and October at Müncheberg (near station 3376), smoothed with a local weighted regression (blue line). Values for 1980-81 missing. Original data resolution: 10 min.

4.3. Impact of Index Choice on Future R Factors and other Sources of Uncertainty

The choice of the R estimator, i.e., the replacement of the currently used total sum by heavy-rainfall indices, can affect the trend similarly to other modelling decisions as sources of uncertainty in scenario analyses. The impact on trend strength—the increase of R and the erosion risk was stronger for heavy-rainfall than for total sums—supports findings, e.g., by References [33] and [34]: future summers might be characterized by more concentrated rainfall and longer dry spells, especially with RCP 8.5 (Figure 12).

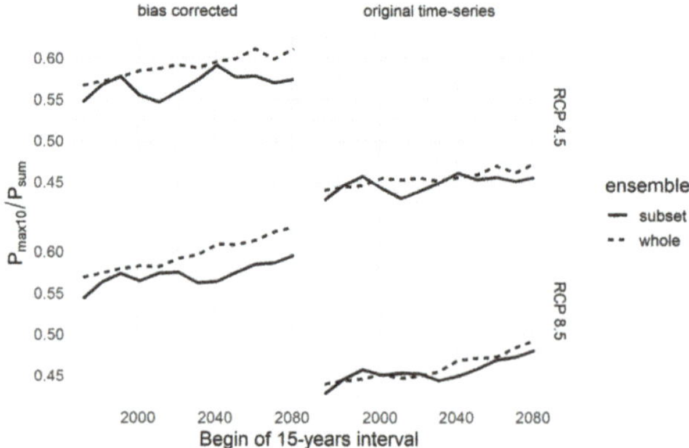

Figure 12. Decadal change in the ratio of P_{max10} and P_{sum} as derived from Figure 7.

Additionally, past and future R factors were more prone to model uncertainty if estimated from total instead of the heavy-rainfall sums. While a careful selection of the climate models and a bias correction can also help to reduce this uncertainty, both implicitly assume that model deviations are stationary which might not necessarily hold true under climate change conditions [50]. While certain combinations of GCM and RCM depicted the spatio-temporal variability of rainfall indices particularly well in the past, none can unanimously be recommended as "the best performing". Firstly, the CORDEX ensemble does not comprise all combinations of the available GCM and RCM. Secondly, the performance and ranking of climate models (and bias corrections) vary among ecoregions and depend on the hydro-climatic variables under consideration [41].

The similar results obtained with the two conceptually different heavy-rainfall indices show that the exact definition of "heavy rainfall days" is of minor relevance for the estimation of current and future R factors. The underestimation of the recent increase and the trend towards more and more heavy rainfall in summer point to the unknown effect of climate change on peak intensities of storm events. Should extreme events, such as the maximum value in our time-series (Appendix C) or the reported values by Reference [51], occur more frequently, even our revised empirical relationships would underestimate the future erosion risk and would thus become obsolete.

Author Contributions: Conceptualization, A.G.; Data curation, A.G. and M.V.; Formal analysis, A.G.; Funding acquisition, A.G. and M.V.; Investigation, A.G.; Methodology, A.G. and J.K.; Project administration, A.G.; Resources, J.K. (adapted CMHyd software) and D.D. (erosion map, data for Figure 12); Software, A.G. and J.K.; Supervision, M.V.; Validation, A.G. and D.D.; Visualization, A.G.; Writing—original draft, A.G.; Writing—review & editing, A.G., J.K., D.D. and M.V.

Funding: A.G. was funded by the German "Bundesministerium für Umwelt, Naturschutz und nukleare Sicherheit (Federal Ministry for the Environment, Nature Conservation and Nuclear Safety)", grant number 03DAS074. J.K. was funded by the German "Bundesministerium für Bildung und Forschung (Federal Ministry of Education and Research)" within the GLANCE project, grant number 01LN1320A. The publication of this article was funded by the Open Access Fund of the Leibniz Association.

Acknowledgments: We thank the German Meteorological Service (DWD) for providing the precipitation data and for the revision of suspicious values. We very much acknowledge the World Climate Research Programme's Working Group on Regional Climate and the Working Group on Coupled Modelling, former coordinating body of CORDEX and responsible panel for CMIP5 as well as the climate modelling groups for producing and making available their model outputs (cf. Table 3). We also appreciate the fast review process and the helpful comments of two anonymous reviewers. J.K. and A.G. express their gratitude to H. Rathjens for providing the CMHyd software and the support with the bias correction.

Conflicts of Interest: The authors declare no conflict of interest. The funders had no role in the design of the study; in the collection, analyses, or interpretation of data; in the writing of the manuscript, or in the decision to publish the results.

Appendix A

The R factor of the USLE is the long-term average annual sum of EI_{30}, the rainfall erosivity of a storm event. EI_{30} is the product of the total kinetic energy E (in kJ m^{-2} mm h^{-1}) and the peak 30-min intensity (I_{30}, in mm h^{-1}). Erosive storm events exceed the thresholds of 10 mm total rainfall or of 10 mm h^{-1} peak intensity within 30 min. Each event is separated by gaps of more than 6 h. Following [11,12], E was derived from the intensity I (mm h^{-1}) and precipitation P (mm) for each time step i (Equation (A1)). Note, that the unit kJ m^{-2} mm h^{-1} (or N h^{-1}) for our R factors equals 0.1 MJ mm ha^{-1} h^{-1}, e.g., used by [15]. Accordingly, the calculation of the soil loss in tons ha^{-1} requires USLE K factors in tons ha^{-1} h N^{-1} while the other USLE factors are dimensionless.

$$E = \sum_{i=1}^{n} E_i \text{ with}$$

$$E_i = \begin{cases} 0 & I_i < 0.05 \\ (11.89 + 8.73 \log I_i)P_i \times 10^{-3} & 5.05 \leq I_i \leq 76.2 \\ 28.33 P_i \times 10^{-3} & I_i > 76.2 \end{cases} \tag{A1}$$

Appendix B

The 10-min data was aggregated and compared to daily and hourly sums from [36]. As the control data is not a priori error-free, we asked the DWD to revise deviations (gaps and spurious values) with the strongest impact on the R factors ([52,53], Table A1). The missing data at station 880 was estimated from rainfall radar data (RADOLAN RW) [36] adjusted to the conventionally measured value of 85.7 mm/6 h by assuming linear changes during each hour (Figure A1). The EI_{30} value increased from 5.6 to 163.7 kJ m^{-2} mm h^{-1}.

Table A1. High-resolution data revised by DWD, station id according to Table 1.

Station ID	Date	Value (mm d^{-1})	Update
880	1999-08-03	134.7	0
880	2014-08-04	16.9	101.9 [1]
430	2003-01-08	76.1	0
430	2006-08-25	130	-
6170	2009-06-30	43.6	- [2]

[1] Missing data interpolated (see text for details), [2] daily data incorrect.

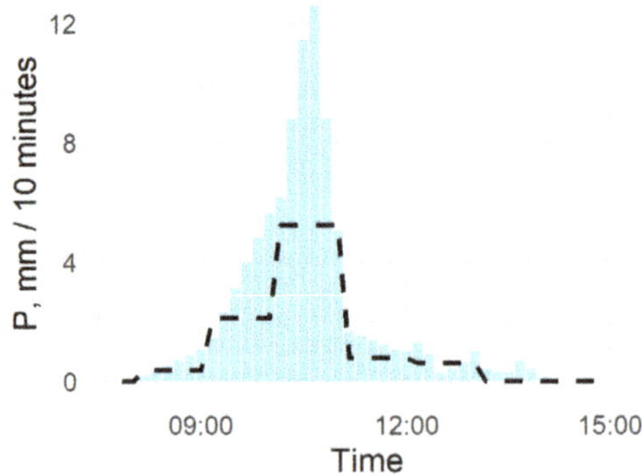

Figure A1. Interpolated rainfall (columns) from hourly radar data (average values as broken line).

Appendix C

The highest and most intense rainfall event in the dataset occurred on 25 August 2006 at Berlin Tegel (station ID 430 in Table 1). During 70 minutes, between 17:10 and 18:19, more than 122 mm was registered which corresponds to more than 20% of the total precipitation in this year (597 mm, Figure A2). The peak intensity of I_{30} = 135 mm/h resulted in an erosivity of 494 kJ m^{-2}, or 86% of the annual EI_{30}. The R factor for the years 2000-15 was 94 kJ m^{-2} mm h^{-1} with and 62 kJ m^{-2} mm h^{-1} without this event. Due to the much lower amount (cf. Figure A3) and intensity of rainfall, the daily erosivity at other climate stations was below 10 kJ m^{-2} mm h^{-1}. Information on the weather conditions in Central Europe at this time are available online [54] (in German).

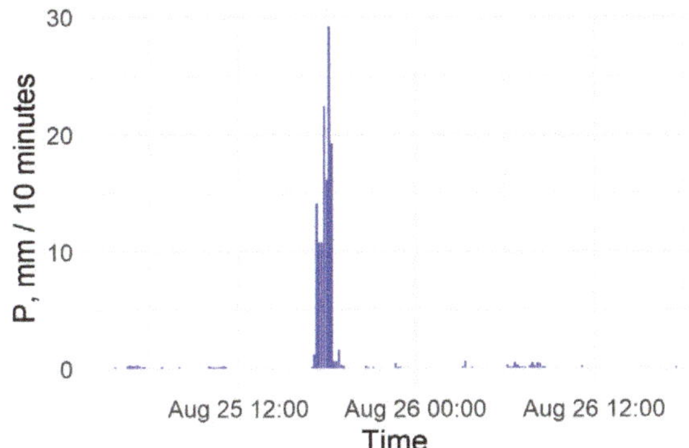

Figure A2. An extreme event: More than 20% of the annual precipitation occurred during 70 min.

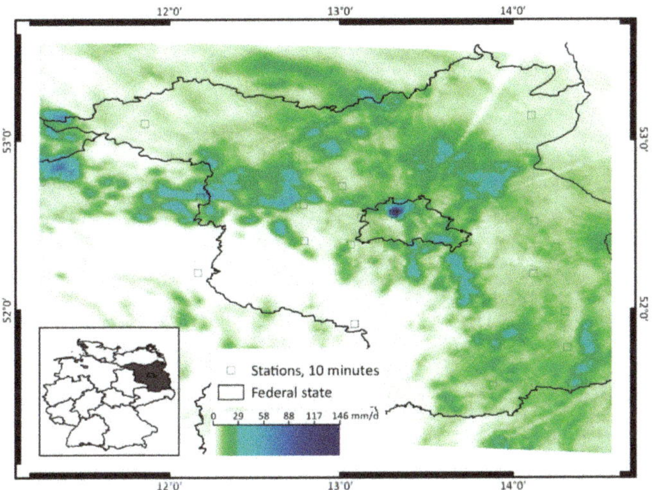

Figure A3. A singular event (dark blue): The rainfall amount on this day was significantly lower elsewhere, as well as the rainfall erosivity (not shown). Data source: RADOLAN RW [36]. The linear features are artifacts due to radar shadows.

Appendix D

Table A2 lists the aggregated rankings as shown in Figure 4. The spatial variability of P_{sum} was, as expected, almost identical between the bias-corrected (KGE $\approx$ 1) climate models and the REGNIE data, and in general very good for the two other variables (KGE > 0.7). Although the mostly negative KGE for the linear trends are not satisfactory, the bias correction could significantly reduce RMSE and APB (Table A3).

Table A2. Ranking of climate models (average of RCP 4.5 & RCP 8.5). The model ID corresponds to Table 3.

ID	Trend (KGE)	Trend (APB)	Trend (RMSE)	Trend (Mean)	Spatial (KGE)	Spatial (APB)	Spatial (RMSE)	Spatial (Mean)	Mean
1	7	10.6	10.6	9.4	10.8	11.8	11.7	11.4	10.4
2 [1]	4.1	4.5	4.4	4.3	5.1	4.1	4.1	4.4	4.35
3	14.3	14	14.4	14.3	8.1	8.8	8.6	8.5	11.4
4	7.1	9.1	8.5	8.2	8.1	11.1	11.3	10.1	9.15
5 [1]	5.7	1.8	2.1	3.2	6.2	5.8	5.3	5.7	4.45
6	11.2	10.3	10.7	10.7	7.2	5.8	6.3	6.4	8.55
7 [1]	4.2	5.5	5.6	5.1	9.9	8	7.8	8.6	6.85
8	5.4	6.6	6.3	6.1	8.1	8.8	8.8	8.5	7.3
9	7.7	10.4	9.8	9.3	12.4	12.7	12.7	12.6	10.95
10 [1]	8	4.2	4.2	5.4	6.2	5.3	5.4	5.6	5.5
11 [1]	5.9	3.7	3.7	4.4	3.8	3.3	3.3	3.5	3.95
12	6.4	6.5	6.3	6.4	9.2	8.3	8	8.5	7.45
13	7.3	8.2	8.3	7.9	8.1	8.3	9.3	8.5	8.2
14	14.2	13.4	13.9	13.8	8.1	9.8	9.8	9.2	11.5
15	11.7	11.3	11.3	11.4	8.8	8.3	8	8.4	9.9

[1] Selected for the ensemble subset.

Table A3. Average effect of bias correction on the performance of climate models.

Bias Correction	Variable	Trend (KGE)	Trend (APB)	Trend (RMSE)	Spatial (KGE)	Spatial (APB)	Spatial (RMSE)
Yes	P_{max10}	−1.3	10.0	17.7	0.9	8.1	13.0
	$P_{11.8}$	−0.8	18.5	24.7	0.8	12.8	16.0
	P_{sum}	−0.5	6.9	23.1	1.0	0.0	0.1
No	P_{max10}	−1.2	15.8	26.1	0.4	14.5	25.1
	$P_{11.8}$	−0.9	34.3	42.9	0.4	30.9	41.3
	P_{sum}	−0.6	29.0	87.5	0.4	27.7	86.0

References

1. FAO; ITPS. *Status of the World's Soil Resources*; Food and Agriculture Organization of the United Nations and Intergovernmental Technical Panel on Soils: Rome, Italy, 2015; p. 607.

2. Pimentel, D. Soil Erosion: A Food and Environmental Threat. *Environ. Dev. Sustain.* **2006**, *8*, 119–137. [CrossRef]

3. Panagos, P.; Imeson, A.; Meusburger, K.; Borrelli, P.; Poesen, J.; Alewell, C. Soil Conservation in Europe: Wish or Reality? *LDD* **2016**, *27*, 1547–1551. [CrossRef]

4. Verheijen, F.G.A.; Jones, R.J.A.; Rickson, R.J.; Smith, C.J. Tolerable versus actual soil erosion rates in Europe. *Earth-Sci. Rev.* **2009**, *94*, 23–38. [CrossRef]

5. Panagos, P.; Standardi, G.; Borrelli, P.; Lugato, E.; Montanarella, L.; Bosello, F. Cost of agricultural productivity loss due to soil erosion in the European Union: From direct cost evaluation approaches to the use of macroeconomic models. *LDD* **2018**, *29*, 471–484. [CrossRef]

6. Rickson, R.J. Can control of soil erosion mitigate water pollution by sediments? *Sci. Total Environ.* **2014**, *468–469*, 1187–1197. [CrossRef]

7. Zessner, M.; Zoboli, O.; Hepp, G.; Kuderna, M.; Weinberger, C.; Gabriel, O. Shedding Light on Increasing Trends of Phosphorus Concentration in Upper Austrian Rivers. *Water* **2016**, *8*, 404. [CrossRef]

8. Einsele, G.; Hinderer, M. Terrestrial sediment yield and the lifetimes of reservoirs, lakes, and larger basins. *Geol. Rundsch.* **1997**, *86*, 288–310. [CrossRef]

9. Vanmaercke, M.; Poesen, J.; Govers, G.; Verstraeten, G. Quantifying human impacts on catchment sediment yield: A continental approach. *Glob. Planet. Chang.* **2015**, *130*, 22–36. [CrossRef]

10. Borrelli, P.; Robinson, D.A.; Fleischer, L.R.; Lugato, E.; Ballabio, C.; Alewell, C.; Meusburger, K.; Modugno, S.; Schütt, B.; Ferro, V.; et al. An assessment of the global impact of 21st century land use change on soil erosion. *Nat. Commun.* **2017**, *8*, 2013. [CrossRef]

11. Deutsches Institut für Normung. *Bodenbeschaffenheit—Ermittlung der Erosionsgefährdung von Böden durch Wasser mit Hilfe der ABAG*; Norm DIN 19708:2005-02; Beuth: Berlin, Germany, 2005.

12. Schwertmann, U.; Vogl, W.; Kainz, M. *Bodenerosion durch Wasser: Vorhersage des Abtrags und Bewertung von Gegenmahahmen*; Ulmer: Stuttgart, Germany, 1987.

13. Diodato, N.; Borrelli, P.; Fiener, P.; Bellocchi, G.; Romano, N. Discovering historical rainfall erosivity with a parsimonious approach: A case study in Western Germany. *J. Hydrol.* **2017**, *544*, 1–9. [CrossRef]

14. Meusburger, K.; Steel, A.; Panagos, P.; Montanarella, L.; Alewell, C. Spatial and temporal variability of rainfall erosivity factor for Switzerland. *Hydrol. Earth Syst. Sci.* **2012**, *16*, 167–177. [CrossRef]

15. Panagos, P.; Ballabio, C.; Borrelli, P.; Meusburger, K.; Klik, A.; Rousseva, S.; Tadić, M.P.; Michaelides, S.; Hrabalíková, M.; Olsen, P.; et al. Rainfall erosivity in Europe. *Sci. Total Environ.* **2015**, *511*, 801–814. [CrossRef]

16. Deumlich, D. Beitrag zur Erarbeitung einer Isoerodentkarte Deutschlands. *Arch. Acker- Pfl. Boden.* **1993**, *37*, 17–24.

17. Li, Z.; Fang, H. Impacts of climate change on water erosion: A review. *Earth-Sci. Rev.* **2016**, *163*, 94–117. [CrossRef]

18. Zolina, O. Change in intense precipitation in Europe. In *Changes in Flood Risk in Europe*; Kundzewicz, Z.W., Ed.; IAHS Press: Wallingford, UK, 2012; pp. 97–120.

19. Zolina, O.; Simmer, C.; Belyaev, K.; Gulev, S.K.; Koltermann, P. Changes in the Duration of European Wet and Dry Spells during the Last 60 Years. *J. Clim.* **2013**, *26*, 2022–2047. [CrossRef]

20. Łupikasza, E.B. Seasonal patterns and consistency of extreme precipitation trends in Europe, December 1950 to February 2008. *Clim. Res.* **2017**, *72*, 217–237. [CrossRef]

21. van den Besselaar, E.J.M.; Klein Tank, A.M.G.; Buishand, T.A. Trends in European precipitation extremes over 1951–2010. *Int. J. Climatol.* **2013**, *33*, 2682–2689. [CrossRef]

22. Hänsel, S.; Petzold, S.; Matschullat, J. Precipitation Trend Analysis for Central Eastern Germany 1851–2006. In *Bioclimatology and Natural Hazards*; Střelcová, K., Mátyás, C., Kleidon, A., Lapin, M., Matejka, F., Blaženec, M., Škvarenina, J., Holécy, J., Eds.; Springer: Dordrecht, The Netherlands, 2009; pp. 29–38.

23. Łupikasza, E.B.; Hänsel, S.; Matschullat, J. Regional and seasonal variability of extreme precipitation trends in southern Poland and central-eastern Germany 1951–2006. *Int. J. Climatol.* **2011**, *31*, 2249–2271. [CrossRef]

24. Murawski, A.; Zimmer, J.; Merz, B. High spatial and temporal organization of changes in precipitation over Germany for 1951–2006. *Int. J. Climatol.* **2016**, *36*, 2582–2597. [CrossRef]

25. Zolina, O. Multidecadal trends in the duration of wet spells and associated intensity of precipitation as revealed by a very dense observational German network. *Environ. Res. Lett.* **2014**, *9*, 025003. [CrossRef]

26. Pińskwar, I.; Choryński, A.; Graczyk, D.; Kundzewicz, Z.W. Observed changes in extreme precipitation in Poland: 1991–2015 versus 1961–1990. *Theor. Appl. Climatol.* **2018**. [CrossRef]

27. Beranová, R.; Kyselý, J. Trends of precipitation characteristics in the Czech Republic over 1961–2012, their spatial patterns and links to temperature and the North Atlantic Oscillation. *Int. J. Climatol.* **2018**, *38*, e596–e606. [CrossRef]

28. Fiener, P.; Neuhaus, P.; Botschek, J. Long-term trends in rainfall erosivity-analysis of high resolution precipitation time series (1937–2007) from Western Germany. *Agric. For. Meteorol.* **2013**, *171–172*, 115–123. [CrossRef]

29. Verstraeten, G.; Poesen, J.; Demarée, G.; Salles, C. Long-term (105 years) variability in rain erosivity as derived from 10-min rainfall depth data for Ukkel (Brussels, Belgium): Implications for assessing soil erosion rates. *J. Geophys. Res. Atmos.* **2006**, *111*, 1–11. [CrossRef]

30. Hanel, M.; Pavlásková, A.; Kyselý, J. Trends in characteristics of sub-daily heavy precipitation and rainfall erosivity in the Czech Republic. *Int. J. Climatol.* **2016**, *36*, 1833–1845. [CrossRef]

31. Panagos, P.; Ballabio, C.; Meusburger, K.; Spinoni, J.; Alewell, C.; Borrelli, P. Towards estimates of future rainfall erosivity in Europe based on REDES and WorldClim datasets. *J. Hydrol.* **2017**, *548*, 251–262. [CrossRef]

32. Wurbs, D.; Steininger, M. *Wirkungen der Klimaänderungen auf die Böden—Untersuchungen zu Auswirkungen des Klimawandels auf die Bodenerosion durch Wasser*; UBA: Dessau-Roßlau, Germany, March 2011; p. 225.

33. Hübener, H.; Bülow, K.; Fooken, C.; Früh, B.; Hoffmann, P.; Höpp, S.; Keuler, K.; Menz, C.; Mohr, V.; Radtke, K.; et al. *ReKliEs-De Ergebnisbericht*; World Data Center for Climate (WDCC) at DKRZ: Hamburg, Germany, December 2017; p. 56.

34. Schwarzak, S.; Hänsel, S.; Matschullat, J. Projected changes in extreme precipitation characteristics for Central Eastern Germany (21st century, model-based analysis). *Int. J. Climatol.* **2015**, *35*, 2724–2734. [CrossRef]

35. European Environment Agency. Ecoregions for Rivers and Lakes. Available online: https://www.eea.europa.eu/ds_resolveuid/2C5D9354-5FC8-4EB3-BD7A-299D95EBAF (accessed on 13 March 2019).

36. Deutscher Wetterdienst. Available online: ftp://ftp-cdc.dwd.de (accessed on 12 May 2016).

37. ECA&D. Website of the European Climate Assessment & Dataset Project. Available online: https://eca.knmi.nl/ (accessed on 22 February 2019).

38. Deumlich, D.; Völker, L. *Bodenabtragskarte für Brandenburg in 25m-Auflösung*; Leibniz Centre for Agricultural Landscape Research: Müncheberg, Germany, 2012.

39. Jacob, D.; Petersen, J.; Eggert, B.; Alias, A.; Christensen, O.B.; Bouwer, L.M.; Braun, A.; Colette, A.; Déqué, M.; Georgievski, G.; et al. EURO-CORDEX: New high-resolution climate change projections for European impact research. *Reg. Environ. Chang.* **2014**, *14*, 563–578. [CrossRef]

40. Rathjens, H.; Bieger, K.; Srinivasan, R.; Chaubey, I.; Arnold, J.G. *CMhyd User Manual—Documentation for Preparing Simulated Climate Change Data for Hydrologic Impact Studies*. 2016, p. 16. Available online: https://swat.tamu.edu/media/115265/bias_cor_man.pdf (accessed on 21 December 2018).

41. Kiesel, J.; Gericke, A.; Rathjens, H.; Wetzig, A.; Kakouei, K.; Jähnig, S.C.; Fohrer, N. Climate change impacts on ecologically relevant hydrological indicators in three catchments in three European ecoregions. *Ecol. Eng.* **2019**, *127*, 404–416. [CrossRef]

42. Teutschbein, C.; Seibert, J. Bias correction of regional climate model simulations for hydrological climate-change impact studies: Review and evaluation of different methods. *J. Hydrol.* **2012**, *456–457*, 12–29. [CrossRef]

43. Kling, H.; Fuchs, M.; Paulin, M. Runoff conditions in the upper Danube basin under an ensemble of climate change scenarios. *J. Hydrol.* **2012**, *424–425*, 264–277. [CrossRef]

44. Nychka, D.; Furrer, R.; Paige, J.; Sain, S. *Fields: Tools for Spatial Data*, R package version 9.6; University Corporation for Atmospheric Research: Boulder, CO, USA, 2017.

45. Ministry of Rural Development, Environment and Agriculture of the Federal State of Brandenburg. *Digitales Feldblockkataster des Landes Brandenburg 2019*. Available online: http://geobroker.geobasis-bb.de/index.php (accessed on 22 October 2018).

46. Fiener, P.; Auerswald, K. Spatial variability of rainfall on a sub-kilometre scale. *Earth Surf. Process. Landf.* **2009**, *34*, 848–859. [CrossRef]

47. Fischer, F.K.; Winterrath, T.; Auerswald, K. Temporal- and spatial-scale and positional effects on rain erosivity derived from point-scale and contiguous rain data. *Hydrol. Earth Syst. Sci.* **2018**, *22*, 6505–6518. [CrossRef]

48. Fischer, F.K.; Kistler, M.; Brandhuber, R.; Maier, H.; Treisch, M.; Auerswald, K. Validation of official erosion modelling based on high-resolution radar rain data by aerial photo erosion classification. *Earth Surf. Process. Landf.* **2018**, *43*, 187–194. [CrossRef]

49. Auerswald, K.; Fischer, F.K.; Winterrath, T.; Brandhuber, R. Rain erosivity map for Germany derived from contiguous radar rain data. *Hydrol. Earth Syst. Sci.* **2019**, *32*, 1819–1832. [CrossRef]

50. Ehret, U.; Zehe, E.; Wulfmeyer, V.; Warrach-Sagi, K.; Liebert, J. HESS Opinions "Should we apply bias correction to global and regional climate model data?". *Hydrol. Earth Syst. Sci.* **2012**, *16*, 3391–3404. [CrossRef]

51. Vogel, E.; Deumlich, D.; Kaupenjohann, M. Bioenergy maize and soil erosion—Risk assessment and erosion control concepts. *Geoderma* **2016**, *261*, 80–92. [CrossRef]

52. Köcher, A. (DWD, Offenbach, Hessen, Germany). Personal communication, 2016.

53. Heil, U. (DWD, Offenbach, Hessen, Germany). Personal communication, 2016.

54. Karlsruhe Institute of Technology. Wettergefahren-Frühwarnung. Available online: http://www.wettergefahren-fruehwarnung.de/Ereignis/20060830_e.html (accessed on 20 March 2019).

Article

Numerical Analysis of Recharge Rates and Contaminant Travel Time in Layered Unsaturated Soils

Adam Szymkiewicz [1,*], Julien Savard [2,3] and Beata Jaworska-Szulc [1]

[1] Faculty of Civil and Environmental Engineering, Gdańsk University of Technology, ul. Narutowicza 11/12, 80-233 Gdańsk, Poland; bejaw@pg.edu.pl

[2] Conservatoire National des Arts et Métiers ChampagneArdenne, CEDEX 03, F-75141 Paris, France; jsavard@laposte.net

[3] Nancy Agency, SociétéAuxiliaire de Distribution de l'Eau-CompagnieGénérale des TravauxHydrauliques (SADE-CGTH), 54220 Malzéville, France

* Correspondence: adams@pg.edu.pl; Tel.: +48-58-347-1085

Received: 12 February 2019; Accepted: 12 March 2019; Published: 16 March 2019

Abstract: This study focused on the estimation of groundwater recharge rates and travel time of conservative contaminants between ground surface and aquifer. Numerical simulations of transient water flow and solute transport were performed using the SWAP computer program for 10 layered soil profiles, composed of materials ranging from gravel to clay. In particular, sensitivity of the results to the thickness and position of weakly permeable soil layers was carried out. Daily weather data set from Gdańsk (northern Poland) was used as the boundary condition. Two types of cover were considered, bare soil and grass, simulated with dynamic growth model. The results obtained with unsteady flow and transport model were compared with simpler methods for travel time estimation, based on the assumptions of steady flow and purely advective transport. The simplified methods were in reasonably good agreement with the transient modelling approach for coarse textured soils but tended to overestimate the travel time if a layer of fine textured soil was present near the surface. Thus, care should be taken when using the simplified methods to estimate vadose zone travel time and vulnerability of the underlying aquifers.

Keywords: unsaturated zone; Richards equation; contaminant transport; time lag; heterogeneous soils

---

1. Introduction

In view of the increasing use of groundwater resources worldwide, there is a need to develop efficient methods to quantify the recharge rate (i.e., the amount of water from precipitation reaching the groundwater table) and the time of migration of contaminants from the ground surface to the groundwater table. These two tasks are closely related, since they both require knowledge of water velocity in the vadose zone, which is generally variable in space and time. In order to achieve this goal, numerical models of unsaturated flow and transport are often used [1–12]. They can be considered as an approach alternative or complementary to other, more costly and time-consuming methods (e.g., lysimeters, tracer experiments). Several computer codes can be used for this purpose, some of them freely available, for example, HYDRUS-1D [13], SWAP [14], UNSAT-H [15], HELP [16]. There is also a growing body of literature and online resources from which the input data for numerical simulations can be obtained, including parameters of soil water retention, permeability, root water uptake and detailed weather data. Nevertheless, the widespread use of numerical modelling is still hampered by the limited availability of realistic data for specific scenarios, need for the user expertise, relatively long times of computation and possible convergence problems in transient flow simulations. For this reason, simplified approaches to quantify recharge and contaminant travel time remain a useful tool in hydrogeological practice [17–20].

Simplified methods of travel time estimation are usually based on the assumption of uniform steady vertical water flow through the vadose zone, corresponding to the average recharge rate. Only transport by advection is considered and the advective velocity is obtained by dividing the recharge rate by the mobile water content (often assumed equal to the total water content). In the simplest approach, the water content is assumed constant within each soil layer constituting the vertical profile. More detailed methods consider variability of the water content in each material layer. The purely advective transport model is considered to provide a safe estimate of the arrival time of contaminant at the groundwater table, due to neglecting adsorption and other processed causing degradation and retardation of the dissolved substance. However, the arrival time may be substantially shorter if the hydrodynamic dispersion processes are taken into account [21].

Sousa et al. [19] investigated the performance of three methods based on steady flow assumption for 8 layered soil profiles on a site near Woodstock, Ontario, Canada. They solved numerically the equation describing steady unsaturated water flow, in order to obtain a detailed distribution of the water content in each profile. This method was considered as the most accurate in the context of their study. Additionally, Sousa et al. [19] estimated the travel time with two other approaches: using hydrostatic profile of the water content above the groundwater table and using constant average values of the mobile water content in each soil layer. The recharge rates were estimated from earlier tracer experiments. Significant differences were observed in the values of travel time obtained with various methods. However, no attempt was undertaken to compare them with simulations of unsteady flow and transport driven by time-variable fluxes at the soil surface.

More recently, Szymkiewicz et al. [21] carried out a comparison between transient and steady-state based methods for estimating time lag for soil profiles with and without root zone. They showed that even if the average recharge rate obtained from transient simulations is used as the input in the steady state methods, the resulting travel times vary considerably. None of the simple methods was able to match the transient simulation results for all scenarios, although the agreement was better for a coarse textured soil (sand) than for a fine textured soil (clay loam). However, in Reference [21] only homogeneous soil profiles were considered. Moreover, the root water uptake was modelled in a simplified manner. In the scenarios with vegetation it was assumed that there is no evaporation from the soil surface and the total amount of potential evapotranspiration was assigned as a sink term to the root zone, regardless of the season of the year. For a more realistic result, the potential evapotranspiration should be split between the evaporation from the soil surface and the transpiration by roots, depending on the stage of plant growth. In the current study we extended our previous analysis by taking into account: (i) layered structure of soil profiles and (ii) a more detailed model of vegetation cover, including variable-in-time split between evaporation and transpiration. For this purpose, we used SWAP numerical code, which contains a detailed module for grass growth and transpiration [14]. We considered 2 homogeneous and 8 layered soil profiles. The recharge rate was mostly affected by the texture of the top soil layer. The travel time was more sensitive to the thickness of soil layers than to their position in the profile (except for the top layer). The presence of root zone significantly decreased recharge and increased travel time, although to a less extent than in our earlier study. The methods based on steady flow assumption gave results relatively close to the transient simulations for coarse textured soils but tended to overestimate the travel time in fine textured soils, in agreement with our previous findings [21].

2. Materials and Methods

2.1. General Assumptions

Following the state-of-the-art approach, for example, [7], we assume that the water flow is described by the Richards Equation (1) and the solute transport by an advection-dispersion Equation (2):

$$\frac{\partial \theta(h)}{\partial t} + \frac{\partial q}{\partial z} + S(h) = 0, \quad q = -k(\theta(h))\left(\frac{\partial h}{\partial z} + 1\right),$$ (1)

$$\frac{\partial(\theta c)}{\partial t} = \frac{\partial}{\partial z}\left(\theta D \frac{\partial c}{\partial z}\right) - \frac{\partial(qc)}{\partial z},$$ (2)

where: t—time, z—vertical coordinate (oriented upwards), h—soil water pressure head (negative in the unsaturated zone), θ—volumetric water content, q—volumetric water flux (Darcy velocity), $S(h)$—function representing water uptake by plant roots, k—hydraulic conductivity, c—solute concentration, D—hydrodynamic dispersion coefficient.

The root water uptake function $S(h)$ was computed according to the model of Feddes et al. [22], with additional compensation term, as described in Reference [14]. Reduction factors due to oxygen stress and drought stress were included.

The dispersion coefficient depends only on the mechanical dispersion (molecular diffusion is neglected): $D = a|q| + D_m{}^*$, where a is the longitudinal dispersivity. A constant value of $a = 6$cm was used in all simulations. This value corresponds to 0.01 of the length scale of profiles A–F described below. We chose it in line with our previous simulations described in Reference [21], in order to minimize the influence of dispersion on the solute travel time and facilitate comparison with simplified methods, which take into account only advection. As shown in Reference [21], larger values of dispersivity (e.g., equal to 0.1 profile depth [5]) lead to significantly earlier occurrence of the solute at the groundwater table. The dependence of dispersivity on the length scale is often associated with local scale variability of soil hydraulic parameters and the presence of heterogeneity patterns more complex than the simple structure of horizontal layers considered here. Similarly, preferential flow and transport phenomena and dual-porosity/dual-permeability behaviour, described for example in References [23,24], are not considered in this work. The present analysis is limited to one-dimensional setting and thus does not take into account the possibility of horizontal flow within soil layers and along layer interfaces, which may affect the recharge rate and contaminant travel time.

2.2. Structure and Hydraulic Properties of Soils

Calculations were performed for 10 soil profiles, as shown in Figure 1. In the first group of profiles (Figure 1A–G) the depth to groundwater table was set to 6m. The homogeneous sand and clay profiles (A and B) were considered as baseline cases. Profiles C and D represent simple layered structures with half of the profile occupied by each soil type. Profiles E–G contain a thin clay layer in sand material, positioned at different depths. This kind of structure has been encountered on glacial outwash plain in the region of Tuchola (northern Poland). Profile H can be considered typical for glacial moraine uplands in the region of Puck (northern Poland), where a confined aquifer is overlaid by multiple layers of glacial till deposits. Finally, profiles I and J are taken from [19] (respectively profiles 1 and 6 in the original paper). All soil materials are characterized by the van Genuchte-Mualem hydraulic functions [25]:

$$S_e = \frac{\theta - \theta_r}{\theta_s - \theta_r} = \left[1 + (\alpha|h|)^{n_g}\right]^{-m_g}, \quad k = k_s k_r = k_s \sqrt{S_e}\left[1 - \left(1 - S_e^{1/m_g}\right)^{m_g}\right]^2$$ (3)

where S_e—effective (normalized) water saturation, θ_r—residual water content, θ_s—water content at fully saturated (or field saturated) conditions, α—parameter related to the air entry pressure (depending on the size of pores), n_g, m_g parameters related to the pore size distribution ($m_g = 1 - 1/n_g$),

k_s—hydraulic conductivity at saturation, k_r—relative hydraulic conductivity. Parameters of each soil material are listed in Table 1. Soils in profiles A–H are characterized by average parameters for their textural classes, as reported in Reference [26]. For profiles I and J we used the parameters from the original study [19].

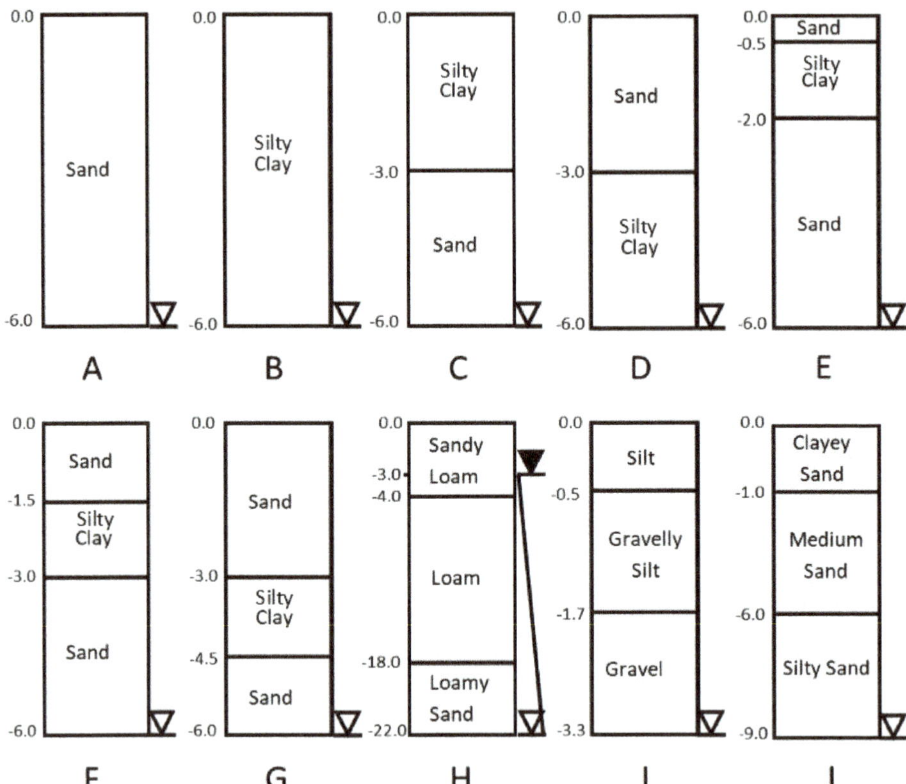

Figure 1. Soil profiles used in numerical simulations.

Table 1. Parameters of the soil materials.

Soil Type	θ_r (-)	θ_s (-)	α (m^{-1})	n_g (-)	k_s (m s^{-1})	θ_{field} Range (-)
Sand [25]	0.045	0.430	14.50	2.68	8.25×10^{-5}	0.07–0.10
Silty clay [25]	0.07	0.36	0.50	1.09	5.56×10^{-8}	0.24–0.38
Sandy loam [25]	0.065	0.41	7.50	1.89	1.22×10^{-5}	0.18–0.26
Loam [25]	0.078	0.43	3.60	1.56	2.89×10^{-6}	0.24–0.38
Loamy sand [25]	0.057	0.41	12.40	2.28	4.05×10^{-5}	0.18–0.26
Silt [19]	0.021	0.43	0.66	1.68	8.00×10^{-8}	0.30–0.36
Gravelly silt [19]	0.016	0.41	2.67	1.45	1.00×10^{-6}	0.18–0.36
Gravel [19]	0.001	0.28	49.30	2.19	5.00×10^{-2}	0.05–0.10
Clayey sand [19]	0.020	0.40	3.48	1.75	5.00×10^{-5}	0.18–0.26
Medium sand [19]	0.019	0.36	3.52	3.18	5.00×10^{-3}	0.07–0.10
Silty sand [19]	0.018	0.37	3.48	1.75	5.00×10^{-4}	0.18–0.26

2.3. Initial and Boundary Conditions

Water flow in vadose zone is driven by precipitation and evaporation fluxes and detailed weather data are necessary to carry out transient analysis. For all profiles we used daily data recorded at the weather station of the Gdańsk University of Technology (Poland) in the period 2011–2015. The average annual precipitation was 550 mm/year. For bare soil scenarios the potential evaporation was calculated by the SWAP code using the Penman-Monteith method, based on the provided values of temperatures, wind speed, relative humidity and solar radiation and the bare soil resistance factor. The average annual potential evaporation for bare soil scenarios was 617 mm/year. The atmospheric boundary condition implemented at the soil surface in SWAP and other similar codes, for example, [13] alternates between flux type and pressure type condition, depending on the state of the soil surface. The infiltration flux is limited by the condition that the maximum pressure at the surface cannot exceed the prescribed ponding depth (in our case equal to zero). Similarly, a limit on the evaporation flux is imposed. In our case we used the limiting condition of minimum allowable water pressure corresponding to the relative humidity of the atmosphere.

In scenarios with grass cover, the detailed model for grass growth and root water uptake was used, as provided in SWAP package [14] and the potential evaporation and transpiration, were calculated by the direct application of the Penman-Monteith equation for specific crop (grass) data. The maximum depth of root zone was set to 50 cm and the root density function decreased nonlinearly with depth. The interception was described by the Hoyningen and Braden model [27,28].

At the bottom of all soil profiles except H a constant value of the water pressure head $h = 0$ was specified, which corresponds to ground water table. In profile H, $h = 19$ m was assigned as the bottom boundary condition, representing the piezometric level of a confined aquifer.

Each simulation started with a 5-year warm-up period, in order to establish a realistic initial distribution of the water content in soil profile. In the warm-up period no solute was added to the soil and the concentration was kept equal to zero. Then the same set of weather data was used for the simulation of solute transport, with the solute concentration in precipitation water set to 1 mg/cm^3. The bottom boundary condition for solute transport was set to zero concentration gradient. Since the transport equation includes dispersion and the solute concentration in soil water increases due to evaporation and transpiration, the concentration at the bottom increased gradually from 0 to values larger than 1 mg/cm^3. Here we report the arrival time of concentration equal to 0.01 mg/cm^3 and 0.99 mg/cm^3, as indicators of the travel time from surface to the aquifer. In weakly permeable soils the 5-year period of weather data was repeated multiple times before the solute breakthrough occurred.

2.4. Numerical Discretization

Transient simulations of water flow and solute transport were performed using the SWAP code (version 4.01, Wageningen University & Research, Wageningen, The Netherlands) [14]. SWAP solves Equation (1) by means of cell-centred finite difference scheme, with first-order implicit time discretization. The solution accuracy depends on the size of grid cells and the method of calculating the average hydraulic conductivity between adjacent cells, for example, [29]. In our case the soil profiles were discretized using 0.5 cm grid cells in the uppermost 5cm of the profile and 1cm grid cells in the remaining part. The average hydraulic conductivity was calculated as the arithmetic mean. Preliminary calculations showed that further refining of the grid does not lead to appreciable changes in the solution—the recharge and evaporation fluxes obtained using grid sizes of 0.25 cm and 0.5 cm differed by less than 1%. The time step varied in the range 10^{-7} to 0.2 days and was adjusted automatically by the SWAP algorithm [14].

2.5. Steady-State Methods for Contaminant Travel Time

Results of transient simulations were compared to simplified estimations of contaminant travel time. We considered 4 methods, which showed the best performance in our previous study focused on homogeneous soils [21]. All methods use the same general formula to calculate the advective velocity v and the corresponding increment of the travel time Δt over a soil compartment of thickness Δz:

$$v(z) = \frac{R}{\theta(z)},$$ (4)

$$\Delta t = \frac{\Delta z \theta(z)}{R}$$ (5)

where R is the recharge rate, which must be known a priori. The total travel time is obtained by summing Δt for all compartments. The methods differ in the way the values of water content θ in each compartment are estimated. Some methods assume that it is constant for each material layer, while other consider a more detailed distribution of θ in the soil profile. All four simplified methods described below were used to calculate the solute travel time using the average values of recharge obtained for each profile from the SWAP simulations of transient flow. They are summarized in Table 2.

Table 2. Summary of simplified methods for travel time calculation.

Name	Reference	Method to Calculate $\theta(z)$
hydrostatic	[19]	θ variable in each soil layer, $\theta(z) = \theta(h(z))$, $h(z)$ corresponds to hydrostatic equilibrium above the groundwater table
steady flow	[19]	θ variable in each soil layer, $\theta(z) = \theta(h(z))$, $h(z)$ obtained from the solution of steady flow equation with uniform flux equal to the average groundwater recharge
Charbeneau and Daniel	[17,30]	θ uniform in each soil layer, calculated from Equation (6)
Witczak and Żurek	[31,32]	θ uniform in each soil layer, chosen from a range of typical field values θ_{field} provided in Reference [29,30]

The first method applied in this study is referred to as hydrostatic, since the assumption of hydrostatic equilibrium is used to obtain the distribution of the water content above the groundwater table. While the hydraulic equilibrium condition is not consistent with the occurrence of recharge (downward water flux), it can be considered a reasonable approximation in some situations [19].

The second method, called here steady flow, was suggested in Reference [19]. It makes use of the numerical solution to the steady flow equation, with the lower boundary condition corresponding to the groundwater table position and the condition at the surface representing steady water flux equal to the assumed recharge rate. The solution results in a $\theta(z)$ profile, which can be used in conjunction with Equation (4) to estimate the travel time. Sousa et al. [19] pointed out to significant differences between the hydrostatic and steady flow method for a number of soil profiles. Obviously, the hydrostatic profile results in lower values of the water content than steady flow profile, leading to larger velocities and shorter time lags for the same recharge rate. However, our earlier study [21] showed that the hydrostatic assumption may not necessarily be less accurate than steady state assumption, if compared to transient flow and transport simulations. This point is further elaborated in the discussion section.

The third method is taken from Charbeneau and Daniel [30] and Charbeneau [17]. In this case the assumption of gravity-dominated flow is invoked, which means that the hydraulic gradient is equal to unity and thus $k(\theta) = R$. If the relationship between k and θ is known, the corresponding value of the water content can be calculated for each material layer. Specifically, Charbeneau and Daniel [30] used Brooks-Corey type conductivity function, which led to the following expression for the travel time in a homogeneous soil layer:

$$\Delta t = \frac{L}{R}\left[\theta_r + (\theta_s - \theta_r)(R/k_s)^{(3\lambda+2)/\lambda}\right],$$ (6)

Following [17] the values of λ were assumed equal to $\lambda = n_g - 1$ for each soil in Table 1. In profile H, which is mostly in the saturated zone, Equation (6) was used only for the upper 3 m of the profile, while in the remaining part Equation (5) was used with $\theta(z)$ set to the saturated water content θ_s for each soil layer.

Finally, we also consider the approach proposed by Witczak and Żurek [31] and later modified by Duda et al. [32]. They suggested to use Equations (4) and (5), and provided a range of θ values for different soil types occurring in Poland in typical field conditions. In Table 1 we provide the range of values for each soil, based on the recommendations in Reference [31,32], denoted as θ_{field}. As a result, for each profile we obtained the lower and upper estimations of the travel time, corresponding to the choice of minimum and maximum values of the water content for the soil materials. In contrast to the previous methods, this one does not require the knowledge of the retention or conductivity functions but its accuracy depend on the choice of θ_{field}.

3. Results

Table 3 shows the average annual values of recharge, recharge-precipitation ratio (R/P) and solute arrival time for the case of bare soil. Results for profiles A to G show that the type of soil in the uppermost layer has the largest influence on the value of recharge. Profiles D–G, where the top layer consists of sand, have very similar recharge rates, close to the value calculated for homogeneous sand profile A. The R/P ratio for profiles A and D to G is between 0.5 and 0.6, which is in agreement with previous studies based on numerical modelling [2,21] and can be considered as the upper limit of groundwater recharge under specific climate conditions. It is larger than the maximum values reported in Reference [32] for highly permeable soils (0.36). Smaller R/P ratios were obtained for profiles H (0.45) and J (0.35), where the top layer consists of sandy loam or clayey sand, respectively. In profiles B-C and I the presence of weakly permeable clay or silt at the surface reduced the R/P ratio to 0.11 and 0.18 respectively. We also note that the estimated value of recharge for profiles I and J are considerably smaller than those reported by Sousa et al. [19], based on tracer experiments (469 mm/year and 412 mm/year, respectively), however, this can be probably explained by higher precipitation at the location of the study [19].

Table 3. Recharge rates and travel times obtained from numerical simulations of transient flow for non-vegetated soil.

Quantity	Mean Annual Recharge (mm/year)	Recharge/Precipitation Ratio (-)	Arrival Time $c = 0.01$ mg/cm^3 (days)	Arrival Time $c = 0.99$ mg/cm^3 (days)
Profile A	312	0.57	424	661
Profile B	62	0.11	7962	10010
Profile C	61	0.11	4460	6198
Profile D	325	0.59	895	1535
Profile E	316	0.57	746	1030
Profile F	319	0.58	736	1022
Profile G	320	0.58	738	1020
Profile H	247	0.45	10512	12642
Profile I	98	0.18	808	1524
Profile J	195	0.35	794	1308

In contrast to the value of recharge, the solute travel time seems to be significantly affected by the presence of clay layers embedded deeper in sand (profiles D–G). The travel time in profile D, with the lower half composed of clay is more than twice as long as in homogeneous sand (A). In profiles E–G, where the clay layer is 1.5 m thick, the travel times do not depend on the position of the clay layer within the profiles. All three values are very similar to each other and are between the results for profile A (no clay layer) and profile D (3.0 m clay layer). Due to dispersion, there are visible differences in the arrival times of 1% and 99% of the boundary concentration. The largest relative difference is observed for profile I and the smallest for profile H.

The obtained range of values for the arrival time can be compared with travel time estimates based on steady state approaches, given in Table 4. For soil profiles with significant proportion of highly permeable materials (A, E–G) the assumption of hydrostatic distribution leads to considerably shorter travel time than the assumption of steady downward flow. These two estimations provide a range of values that is comparable to the range of travel times calculated from transient simulations. As can be seen in Figure 1, for profile E the hydrostatic distribution of θ approximately corresponds to the summer season and the steady flow distribution—to the winter season, with some water ponding on the top of the clay layer. A very similar range of values is obtained using the approach by Witczak and Żurek, with average values of θ_{ield} reported in Table 1. Considering application of this method to profile E, one has to note that the values of field water content θ_{field} in Table 1 are in good agreement with the results shown in Figure 1 for the lower part of the profile. However, in the upper sand layer the values are generally larger, especially in winter and the clay layer remain almost fully saturated, with $\theta = 0.36$ corresponding to the upper limit of θ_{field} in Table 1. The lower estimates from Witczak and Żurek, which are in agreement with the early arrival of solute at the groundwater table, were obtained for smaller values of θ_{field} from literature, not consistent with the results of simulations.

Table 4. Estimates of travel time (days)obtained with steady-state methods for non-vegetated soil.

Method	Hydrostatic	Steady Flow	Charbeneau & Daniel [28]	Witczak & Żurek [29]
Profile A	383	655	629	491–702
Profile B	12,011	12,650	11,445	8477–11,303
Profile C	7106	7948	7164	5565–7539
Profile D	1325	1545	1453	1044–1415
Profile E	849	1124	1030	758–1044
Profile F	879	1155	1057	780–1074
Profile G	880	1141	1045	770–1163
Profile H	12,512	12,747	12,590	12,728–13,078
Profile I	1334	2121	1419	1661–2875
Profile J	1111	1310	830	2003–2883

In profiles B, C and D the discrepancies between estimations based on transient and steady state methods are larger. First, we note that hydrostatic, steady flow and Charbeneau and Daniel methods lead to similar estimates of the travel time and all three are overestimating the solute arrival time, compared to the transient simulations. Also, the results from Witczak and Żurek approach differed from the transient simulations more significantly than in the previous case, for the range of considered water contents. However, the estimate obtained for lower values of the field water content was closer to the breakthrough time obtained from SWAP.

For profile H, all methods based on steady state solutions were close to each other and predicted travel time values close to the arrival time of 99% of input concentration in transient simulation. A good agreement between all methods for this profile can be explained by the fact, that most of its depth is in saturated conditions, with uniquely defined water content value.

In contrast, for profiles I and J the differences between various methods were significant. In profile I the steady flow method overestimated the travel time, while both hydrostatic and Charbeneau and Daniel methods gave results comparable to the values obtained from transient simulations. In profile J all three methods were consistent with transient simulations. In our case the differences between the hydrostatic and steady flow approach were much smaller than reported in Reference [19] for the same profiles but larger recharge values. For profiles I and J the method of Witczak and Żurek appeared to be less accurate than the other simplified methods, substantially overestimating the travel time. This can be explained by very low water content occurring in coarse soil materials (gravel and medium sand), which according to transient flow, steady flow and hydrostatic results (Figure 2) is below the lower limit of field water content given in Table 1. Thus, the resulting advective velocity is larger than predicted by the method of Witczak and Żurek.

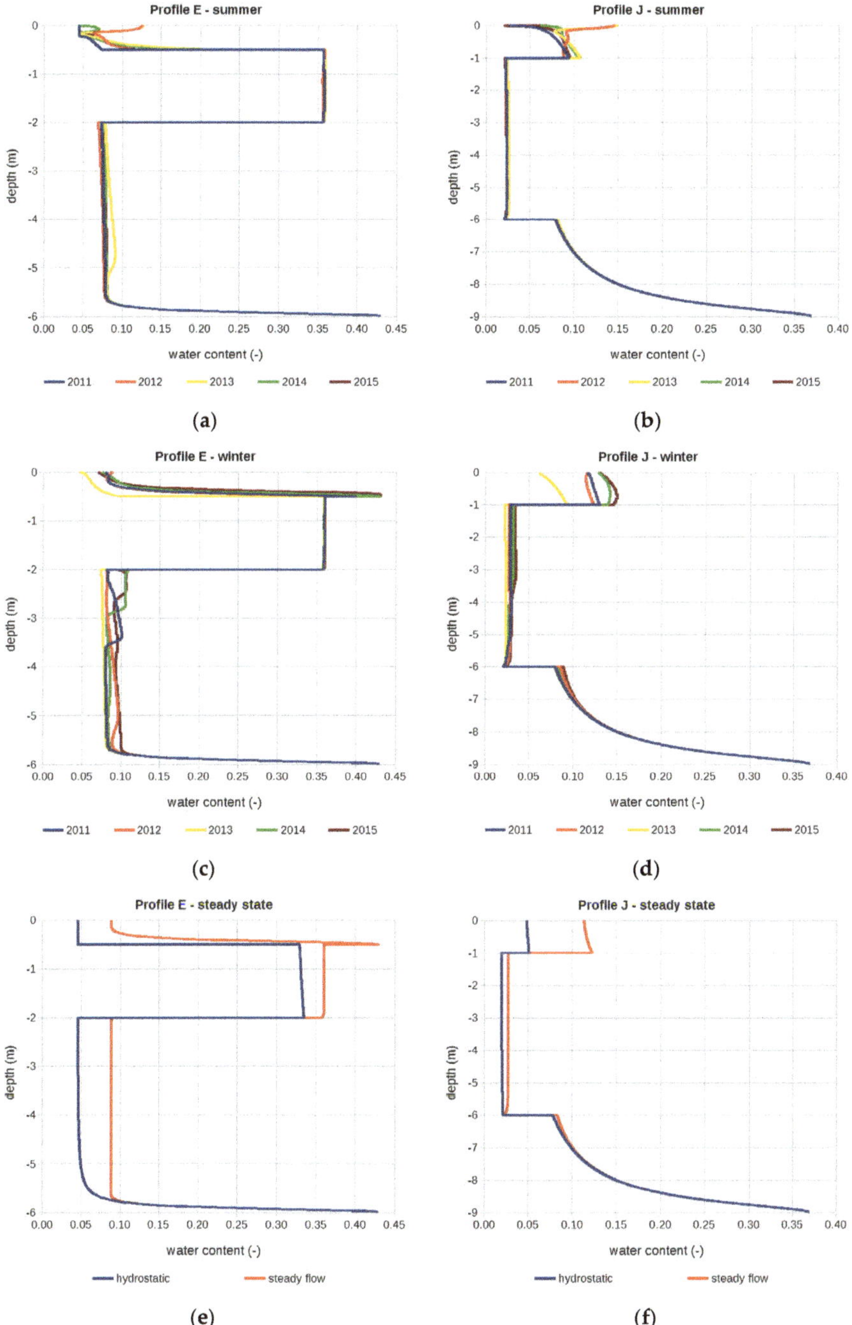

Figure 2. Volumetric water content distributions obtained from transient simulations and steady state approximations for soil profiles E and J, non-vegetated soil, (**a**) profile E, summer (30 Jun.), (**b**) profile J, summer; (**c**) profile E, winter (31 Dec.); (**d**) profile J, winter; (**e**) profile E, steady state, (**f**) profile J, steady state.

In Table 5 we report recharge and travel time values obtained with transient simulations for soil with grass cover. There is a significant reduction of recharge, ranging from 22% for profile H to 50% for profile I. In our previous study [21], we observed even larger decrease in homogeneous profiles of sand and silty clay. However, in that work, we assumed that the reference evapotranspiration corresponds entirely to transpiration and the root density decreases linearly with depth. Here, a more detailed description is used, where only a part of the reference evapotranspiration is assigned to transpiration depending on the growth of plants and the root density distribution is such that great majority of roots is very close to the soil surface. Consequently, there is less difference between the bare soil and grass scenarios. Keese at al. [2] also reported more significant reduction of recharge due to vegetation than obtained here, however their studies focused on locations with much higher potential evapotranspiration. The influence of various models of root water uptake and their parameters should be further investigated in conjunction with soil textural variability and climate.

Table 5. Recharge rates and travel times obtained from numerical simulations of transient flow for soil with grass cover.

Quantity	Mean Annual Recharge (mm/year)	Recharge/Precipitation Ratio (-)	Arrival Time $c = 0.01$ mg/cm^3 (days)	Arrival Time $c = 0.99$ mg/cm^3 (days)
Profile A	220	0.40	531	806
Profile B	38	0.07	11,782	13,555
Profile C	38	0.07	6336	8706
Profile D	223	0.42	1480	1876
Profile E	215	0.39	851	1526
Profile F	224	0.41	834	1503
Profile G	224	0.41	830	1501
Profile H	195	0.35	12,956	15,477
Profile I	52	0.09	963	1852
Profile J	156	0.28	1314	1660

The travel time estimates obtained with steady state approximations for vegetated soils are summarized in Table 6. Again, the performance of simplified methods is case-dependent. In particular, for profiles E–G the hydrostatic and steady flow methods cannot capture the early breakthrough time of the solute and for profiles B–D they considerably overestimate the travel time even for $c = 0.99$ mg/cm^3. In contrast, both methods still perform well in the case of homogeneous sand. For profiles B–D the method of Witczak and Żurek provides a range of time lag values more comparable to the transient simulations but again, this is due to the use of smaller, literature-based water content, not corresponding to the simulation results. In profile I all steady state approximations significantly overpredict the solute travel time. On the other hand, in profile J the hydrostatic and steady flow methods produce results close to the transient flow simulations, while the method of Charbeneau and Daniel leads to shorter arrival time than the one obtained from transient flow simulations. Similarly to the non-vegetated case, the method of Witczak and Żurek predicts much longer travel times than SWAP calculations.

Table 6. Estimates of travel time (days) obtained with steady-state methods for soil with grass cover.

Method	Hydrostatic	Steady Flow	Charbeneau & Daniel [28]	Witczak & Żurek [29]
Profile A	543	892	857	697–995
Profile B	19,598	20,550	18,392	13,832–21,900
Profile C	11,407	12,630	11,271	8933–13,832
Profile D	1880	2170	2027	1482–2295
Profile E	1283	1661	1514	1146–1732
Profile F	1241	1598	1457	1100–1662
Profile G	1257	1598	1457	1100–1662
Profile H	15,849	16,120	15,926	16,116–16,565
Profile I	2513	3905	2454	3131–5419
Profile J	1388	1623	1011	2504–3603

4. Summary and Conclusions

Numerical simulations of transient flow and transport in layered soil profiles were compared to simplified methods of estimating solute travel time, based on the assumption of steady flow and purely advective transport. Even with the relatively small value of dispersion constant used in this study (6 cm), the arrival of the solute at ground water table was spread in time. Thus, the unsaturated zone time lag should be considered as a time interval, rather than a single value. The simplified methods reproduced the results of transient simulations with mixed success. They performed best for profiles dominated by sand materials without vegetation. In the case of vegetated profiles or profiles with fine-textured material at the surface, the simplified method were generally not able to capture the earlier arrival time of concentration $c = 0.01$ mg/cm^3, the results were closer to the travel time for $c = 0.99$ mg/cm^3 but in some cases the time was greatly overestimated. It is difficult to distinguish any of the simple methods as superior to the other ones. The methods based on hydrostatic and steady-state distribution of the water content can be viewed as complementary in the case of coarse-textured soils but for fine textured soils they both predict time lags significantly longer than the transient simulations. The method based on the use of average values of the water content [29] performed quite well despite its simplicity but the agreement with transient simulations was often obtained for values of the water content taken from the literature and not corresponding to the particular setting under consideration. Thus, from the point of view of practical application of this method, the use of water content values measured in the field may not necessarily lead to improved results.

This study confirmed the utility of numerical modelling as a tool to estimate groundwater recharge, especially due to the ease of implementing different scenarios related to lithology, vegetation and weather data. The obtained results are in agreement with earlier reports, showing the importance of climate, soil texture in the surface layer and vegetation as the main controls of groundwater recharge [2–5].

Author Contributions: Conceptualization, A.S.; Methodology, A.S. and B.J.-S.; Investigation, J.S. and A.S.; Writing-Original Draft Preparation, A.S. and J.S.

Funding: The work of A.S. and B.J.-S. was funded by National Centre for Research and Development (NCBR), Poland, in the framework of the project BIOSTRATEG3/343927/3/NCBR/2017 "Modelling of the impact of the agricultural holdings and land-use structure on the quality of inland and coastal waters of the Baltic Sea set up on the example of the Municipality of Puck region—Integrated info-prediction Web Service Water PUCK"—BIOSTRATEG Programme and by National Science Centre (NCN), Poland, in the framework of the project 2015/17/B/ST10/03233 "Groundwater recharge on outwash plain. "The work of J.S. was carried out during his stay at the Gdańsk University of Technology in the framework of ERASMUS student exchange programme, funded by EU.

Acknowledgments: The weather data was obtained from meteorological station of Gdańsk University of Technology, Faculty of Civil and Environmental Engineering, Department of Hydraulic Engineering, in Sopot (54.2 N, 18.3 E). The authors would like to thank two anonymous reviewers for their suggestions.

Conflicts of Interest: The authors declare no conflict of interest.

References

1. Scanlon, B.R.; Christman, M.; Reedy, R.C.; Porro, I.; Simunek, J.; Flerchinger, G.N. Intercode comparisons for simulating water balance of surficial sediments in semiarid regions. *Water Resour. Res.* **2002**, *38*. [CrossRef]
2. Keese, K.E.; Scanlon, B.R.; Reedy, R.C. Assessing controls on diffuse groundwater recharge using unsaturated flow modeling. *Water Resour. Res.* **2005**, *41*. [CrossRef]
3. Ordens, C.M.; Post, V.E.; Werner, A.D.; Hutson, J.L. Influence of model conceptualisation on one-dimensional recharge quantification: Uley South, South Australia. *Hydrogeol. J.* **2014**, *22*, 795–805. [CrossRef]
4. Vero, S.E.; Ibrahim, T.G.; Creamer, R.E.; Grant, J.; Healy, M.G.; Henry, T.; Kramers, G.; Richards, K.G.; Fenton, O. Consequences of varied soil hydraulic and meteorological complexity on unsaturated zone time lag estimates. *J. Contam. Hydrol.* **2014**, *170*, 53–67. [CrossRef] [PubMed]
5. Wang, T.; Franz, T.E.; Zlotnik, V.A. Controls of soil hydraulic characteristics on modeling groundwater recharge under different climatic conditions. *J. Hydrol.* **2015**, *521*, 470–481. [CrossRef]
6. Fenton, O.; Vero, S.; Ibrahim, T.G.; Murphy, P.N.C.; Sherriff, S.C.; Ó hUallacháin, D. Consequences of using different soil texture determination methodologies for soil physical quality and unsaturated zone time lag estimates. *J. Contam. Hydrol.* **2015**, *182*, 16–24. [CrossRef] [PubMed]
7. Vero, S.E.; Healy, M.G.; Henry, T.; Creamer, R.E.; Ibrahim, T.G.; Richards, K.G.; Mellander, P.-E.; McDonald, N.T.; Fenton, O. A framework for determining unsaturated zone water quality time lags at catchment scale. *Agric. Ecosyst. Environ.* **2017**, *236*, 234–242. [CrossRef]
8. Batalha, M.S.; Barbosa, M.C.; Faybishenko, B.; van Genuchten, M.T. Effect of temporal averaging of meteorological data on predictions of groundwater recharge. *J. Hydrol. Hydromech.* **2018**, *66*, 143–152. [CrossRef]
9. Szymkiewicz, A.; Gumuła-Kawęcka, A.; Šimůnek, J.; Leterme, B.; Beegum, S.; Jaworska-Szulc, B.; Pruszkowska-Cacers, M.; Gorczewska-Langner, W.; Jacques, D. Simulations of freshwater lens recharge and salt/freshwater interfaces using the HYDRUS and SWI2 packages for MODFLOW. *J. Hydrol. Hydromech.* **2018**, *66*, 246–256. [CrossRef]
10. Bashir, R.; Pastora Chevez, E. Spatial and seasonal variations of water and salt movement in the vadose zone at salt-impacted sites. *Water* **2018**, *10*, 1833. [CrossRef]
11. Zhou, Y.; Wang, X.S.; Han, P.F. Depth-dependent seasonal variation of soil water in a thick vadose zone in the Badain Jaran Desert, China. *Water* **2018**, *10*, 1719. [CrossRef]
12. Beegum, S.; Šimůnek, J.; Szymkiewicz, A.; Sudheer, K.P.; Nambi, I.M. Implementation of solute transport in the vadose zone into the 'HYDRUS package for MODFLOW'. *Groundwater* **2018**. [CrossRef]
13. Šimůnek, J.; Šejna, M.; Saito, H.; Sakai, M.; van Genuchten, M.T. *The HYDRUS-1D Software Package for Simulating the One-Dimensional Movement of Water, Heat, and Multiple Solutes in Variably-Saturated Media, Version 4.0*; HYDRUS Software Series 3; University of California: Riverside, CA, USA, 2008.
14. Kroes, J.G.; van Dam, J.C.; Bartholomeus, R.P.; Groenendijk, P.; Heinen, M.; Hendriks, R.F.A.; Mulder, H.M.; Supit, I.; van Walsum, P.E.V. *SWAP Version 4. Theory Description and User Manual*; Report 2780; Wageningen University & Research: Wageningen, The Netherlands, 2017.
15. Fayer, M.J.; Jones, T.L. *UNSAT-H Version 2.0: Unsaturated Soil Water and Heat Flow Model (No. PNL-6779)*; Pacific Northwest Lab.: Richland, WA, USA, 1990.
16. Schroeder, P.R.; Dozier, T.S.; Zappi, P.A.; McEnroe, B.M.; Sjostrom, J.W.; Peton, R.L. *The Hydrologic Evaluation of Landfill Performance (HELP) Model: Engineering Documentation for Version 3*; EPA/600/R-94/168b; US. Environmental Protection Agency, Risk Reduction Engineering Laboratory: Cincinnati, OH, USA, 1994.
17. Charbeneau, R.J. *Groundwater Hydraulics and Pollutant Transport*; Waveland Press: Long Grove, IL, USA, 2006; ISBN 1478608315.
18. Fenton, O.; Schulte, R.P.; Jordan, P.; Lalor, S.T.; Richards, K.G. Time lag: A methodology for the estimation of vertical and horizontal travel and flushing timescales to nitrate threshold concentrations in Irish aquifers. *Environ. Sci. Policy* **2011**, *14*, 419–431. [CrossRef]
19. Sousa, M.R.; Jones, J.P.; Frind, E.O.; Rudolph, D.L. A simple method to assess unsaturated zone time lag in the travel time from ground surface to receptor. *J. Contam. Hydrol.* **2013**, *144*, 138–151. [CrossRef]
20. Potrykus, D.; Gumuła-Kawęcka, A.; Jaworska-Szulc, B.; Pruszkowska-Caceres, M.; Szymkiewicz, A. Assessing groundwater vulnerability to pollution in the Puck region (denudation moraine upland) using vertical seepage method. *E3S Web Conf.* **2018**, *44*, 00147. [CrossRef]

21. Szymkiewicz, A.; Gumuła-Kawęcka, A.; Potrykus, D.; Jaworska-Szulc, B.; Pruszkowska-Caceres, M.; Gorczewska-Langner, W. Estimation of conservative contaminant travel time through vadose zone based on transient and steady flow approaches. *Water* **2018**, *10*, 1417. [CrossRef]
22. Feddes, R.A.; Kowalik, P.J.; Zaradny, H. *Simulation of Field Water Use and Crop Yield*; Simulation Monographs; Pudoc: Wageningen, The Netherlands, 1978.
23. Šimůnek, J.; van Genuchten, M.T. Modeling nonequilibrium flow and transport processes using HYDRUS. *Vadose Zone J.* **2008**, *7*, 782–797. [CrossRef]
24. Szymkiewicz, A.; Lewandowska, J. Unified macroscopic model for unsaturated water flow in soils of bimodal porosity. *Hydrol. Sci. J.* **2006**, *51*, 1106–1124. [CrossRef]
25. Van Genuchten, M.T. A closed-form equation for predicting the hydraulic conductivity of unsaturated soils. *Soil Sci. Soc. Am. J.* **1980**, *44*, 892–898. [CrossRef]
26. Carsel, R.F.; Parrish, R.S. Developing joint probability distributions of soil water retention characteristics. *Water Resour. Res.* **1988**, *24*, 755–769. [CrossRef]
27. Von Hoyningen-Hüne, J. Die Interception des Niederschlags in landwirtschaftlichen Beständen. *Schr. DVWK* **1983**, *57*, 1–53.
28. Braden, H. Ein Energiehaushalts- und Verdunstungsmodell für Wasser und Stoffhaushaltuntersuchungen landwirtschaftlich genutzer Einzugsgebiete. *Mitt. Deutsch.Bodenkd. Geselschaft* **1985**, *42*, 294–299. (In German)
29. Szymkiewicz, A. Approximation of internodal conductivities in numerical simulation of one-dimensional infiltration, drainage, and capillary rise in unsaturated soils. *Water Resour. Res.* **2009**, *45*, W10403. [CrossRef]
30. Charbeneau, R.J.; Daniel, D.E. Contaminant transport in unsaturated flow. In *Handbook of Hydrology*; Maidment, D.R., Ed.; McGraw-Hill: New York, NY, USA, 1993.
31. Witczak, S.; Żurek, A. Wykorzystanie map glebowo-rolniczych w ocenie ochronnej roli gleb dla wód podziemnych (Use of soil-agricultural maps in the evolution of protective role of soil for groundwater). In *Metodyczne Podstawy Ochrony Wód Podziemnych*; Methodical Principles of Groundwater Protection; Kleczkowski, A., Ed.; AkademiaGórniczo-Hutnicza: Kraków, Poland, 1994. (In Polish)
32. Duda, R.; Winid, B.; Zdechlik, R.; Stępień, M. *Metodyka Wyboru Optymalnej Metody Wyznaczania Zasięgu Stref Ochronnych Ujęć Zwykłych Wód Podziemnych z Uwzględnieniem Warunków Hydrogeologicznych Obszaru RZGW w Krakowie*; Methodology of Selecting the Optimal Method of the Wellhead Protection Area Delineation Taking into Account the Hydrogeological Conditions in Areas Administered by the Regional Water Management Board in Cracow; Akademia Górniczo-Hutnicza: Kraków, Poland, 2013; ISBN 9788388927294. (In Polish)

 *water*

Article

Influence of Abandoning Agricultural Land Use on Hydrophysical Properties of Sandy Soil

Edyta Hewelke

Faculty of Civil and Environmental Engineering, Warsaw University of Life Sciences-SGGW, Nowoursynowska 166, 02-787 Warsaw, Poland; edyta_hewelke@sggw.pl

Received: 23 January 2019; Accepted: 9 March 2019; Published: 13 March 2019

Abstract: Soil water repellency can significantly degrade its agricultural utility and bring about negative environmental consequences (i.e., reduced infiltration capacity, enhanced overland flow, increased erosion rates, and water infiltration occurred in irregular patterns). The presented study aimed to establish whether excluding albic Podzols from agricultural production and their spontaneous inhabitation by a pine tree stand affected their hydrophysical properties. Studies with the application of the water drop penetration time (WDPT) test showed that a change in the land use increased the potential water repellency of the surface layer (horizon A) and caused its changeover from strongly repellent class (Class 2) to extremely repellent (Class 5). The relationship between soil moisture content and wettability made it possible to determine the critical soil moisture content (CSMC) for the occurrence of the phenomenon of water repellency. It was confirmed that the CSMC value increased along with a change in use. For the site under arable use, it was 9–10 vol.%, whereas for the site formerly under arable use and currently covered predominantly by a pine tree stand, a value in the range of 14–16 vol.% was reached. A laboratory experiment on surface runoff of the soil formerly under arable use showed that over half of the rainfall may be transformed into surface runoff as a result of occurring water repellency. This means that exceeding the critical soil moisture content makes the recharge of soil retention difficult and may significantly influence the water balance of soil, as well as increasing its susceptibility to drought.

Keywords: soil water repellency; land use change; agrohydrology; water retention; surface runoff

---

1. Introduction

The amount of water which sandy soil can retain is low and results from the distribution of soil pores, which is dominated by large pores that do not contribute to water storage. This phenomenon is described by the soil water desorption curve. In soils characterized by a high contact angle, the London dispersion forces make the wetting of soil difficult to different degrees. As a result, full use may not be made of the potential retention ability of soil, seeing as how rainwater, instead of infiltrating, will gather on the surface of the soil and run off. In connection with the above, some researchers [1,2] treat soil water repellency (SWR) as one of the most important properties of soil, which determines its physical and chemical properties, and is decisive to its production and regulation functions. Water repellency may have significant agrohydrological consequences by increasing the susceptibility of soils to drought. The reasons behind water repellency are being identified to an increasingly wider extent and are related to the occurrence of organic carbon, especially humic and fulvic acids, as well as waxes and lipids of different origins [2–6]. Water repellency also occurs following forest fires and the burning of grasses [7–11] as well as soil contamination with crude oil derivatives [12–15]. Water repellency is of a seasonal nature and is strictly connected with soil moisture content [16–21]. Many authors [22–27] draw attention to the connection of water repellency with the type of soil use. A deciding factor when it comes to the ability of soil to absorb and retain rainwater is the wettability of the soil material, which depends on the level of humification of soil organic matter [28]. In Poland, a common phenomenon connected with political transformation is abandoning agricultural production on sandy soils characterized by low productivity [29]. These areas are intentionally afforested or, most often, become spontaneously covered by forest plants with a large share of pine trees.

The dimension of the agricultural areas abandoned or converted into production forests in Europe varies widely between scenarios [30]. According to intermediate scenarios in Verburg and Overmars [31], between 10 and 29 million ha of land will be released from agriculture between 2000 and 2030.

Water infrastructure (drainage system) for purposes of agricultural production undergoes fast degradation in these areas. As a result of changing agrohydrological conditions and the balance of organic substances, the areas that once had been an agricultural ecosystem undergo fast transformations, and soil properties may change the ecosystem services [32]. The complexity of the planning process is increasing, especially in the context of the sustainable use of forest resources and its adaptation to climate change [33,34]. To avoid further land degradation and promote land restoration, multifunctional use of land is needed within the boundaries of the soil–water system [35]. A robust soil–water system is essential for achieving most of the UN sustainable development goals [36,37], as interlinked goals. Moreover, sustainable solutions need to embed short-term management in long-term landscape planning in the direction of long-term sustainability. The aim of this study was to assess whether resigning from agricultural production on sandy (nutrient-poor) soil and the uncontrolled succession of a pine stand can significantly influence the shaping of hydrophysical properties of soil. Identification of the main parameters driving system dynamics is essential to solve land and water-management related problems [38].

2. Materials and Methods

2.1. Description of the Study Site

The study site is located in central Poland in the Mazowieckie Province, Stanisławów Commune. Sandy soils, mainly albic-Podzols [39] with a low production potential, are found here and are the reason behind the significant limitation or abandonment of agricultural production. This is, at the same time, influenced by the proximity of Warsaw, which is an attractive job market. Prior to ceasing plant production at the beginning of the 90s, mainly rye and potatoes had been grown here. Currently, part of the site is covered by self-sown Scots pine (*Pinus sylvestris* L.) 80%, silver birch (*Betula pendula* Roth) 15%, and aspen poplar (*Populus tremula* L.). Grasses, blackberry (*Rubus* L.), wood club-rush (*Scirpus sylvaticus* L.), and European goldenrod (*Solidago virgaurea* L.) are found in the poorly developed undergrowth. The drainage system has been overgrown and has undergone partial degradation. The land is flat, with local denivelations of approx. 5–10 cm. This leads to long-term high soil moisture content, usually in Spring, with the level of the water table at 5–10 cm below the surface in a formerly arable area. In the Summer, upon the falling of the water table, the soil becomes more susceptible to drought. On part of the site of the same soil unit, the land remains under arable use. The study was carried out at two points. The first was located in a formerly arable area (Site 1, N 52°28′28.21″ and E 21°21′72.78″), while the latter remains under active extensive agricultural use (Site 2, N 52°28′39.52″ and E 21°52′29.16″). All samples were collected from the top layer (horizon A, 0–10 cm) during a wet period at the beginning of April 2018. At this term, soil moisture content was approximately at the level of field capacity.

2.2. Evaluation of Soil Water Repellency

The soil water repellency (SWR) was determined using the water drop penetration time (WDPT) test. This test is the most widespread [5,16] and the most suitable [40] method, as it is relatively simple and cheap. In order to determine the potential SWR value, the soil samples were dried at room temperature (20 °C) to a constant weight. Triplicate samples of about 20 g of soil were placed in Petri dishes and 5 drops (the volume of water in a droplet was equal to 60 µL), using a standard medicine dropper, of distilled water were deposited onto smoothed soil samples. The sample surfaces were gently smoothed by hand for these tests. The median values of the WDPT test were used to assess the SWR class. The classification of SWR presented in Table 1 was proposed by Dekker and Jungerius [41] and comprises up to 5 classes, further subdividing the extremely repellent class into 2 classes [16]. In order to establish the relationship between soil moisture content and SWR, the WDPT test was performed for different moisture contents that had been adjusted by equilibrating the material at 7 pF levels (i.e., 2.0, 2.3, 2.7, 3.0, 3.3, 3.7, 4.2) in triplicate on undisturbed soil samples (100 cm^3).

Table 1. Classification of soil water repellency using the water drop penetration time (WDPT) test., Dekker et al. [16,41].

Classification	Threshold WDPT Test	Class
Hydrophilic	≤5 s	0
Slightly water repellent	5–60 s	1
Strongly water repellent	60–600 s	2
Severely water repellent	600 s–1 h	3
Extremely water repellent	1–3 h	4
Extremely water repellent	3–6 h	5
Extremely water repellent	>6 h	6

2.3. Determination of Basic Physical and Chemical Properties of Soil

The particle size distribution was assessed using the Bouyoucos method with modifications by Casagrande and Prószyński (the aerometric method) for particles smaller than 0.1 mm, and the sieve method for particles larger than 0.1 mm [42]. The bulk density was assessed by dividing the core samples at 105 °C. Measurement of this parameter was conducted in five replicates. Total porosity (p) was calculated as $= 1 - \frac{\rho_b}{\rho_s}$, where ρ_b is the dry bulk density of the soil (kg m^{-3}) and ρ_s is the particle density assumed to be 2650 kg m^{-3}. Soil pH was measured in a 1:5 soil:water suspension using a standard potentiometric method. Organic carbon content was determined using Tiurin's method [43], with total carbon measured with the Kjeldahl method (Kjeltec–Tecator analyser). Measurements of pH, organic carbon, and total nitrogen were done in triplicates.

2.4. Determination of Soil Hydraulic Properties

Soil moisture retention characteristics were measured in a laboratory in triplicate on undisturbed soil samples (100 cm^3) using a reference method [44]. The saturation of soil to its full water-holding capacity was carried out in laboratory for three days by gradually increasing the water table upwards from the bottom of each sample. The moisture content values of pF between 0.4 and 2.0 were determined in a standard sand box, whereas the amounts of water at pF 2.3, 2.7, 3.4, and 4.2 were measured in pressure chambers. Laboratory measured saturated hydraulic conductivity (K_s) was determined by the constant head method. Metal cores (7.3 cm diameter, 6 cm height) were used to collect undisturbed samples of soil. In the laboratory, the samples were saturated with water from bottom up (capillary rise) for 3 days prior to measurements. The amount of surface runoff was tested on disturbed samples in the laboratory, maintaining a bulk density (ρ_s) similar to the natural one. A rainfall intensity of 2 mm lasting 420 min., which corresponds to a total dose of 14 mm, at a terrain slope of 5‰, was simulated. Surface runoff was captured by an open drain located on the border of the tested microplot, perpendicularly to the slope. Next, water was directed to a measurement tank, where registration of the volume of surface runoff was carried out every 30 min.

3. Results

3.1. Basic Soil Properties

Basic physical and chemical properties of soil are presented in Table 2. According to the USDA classification [45], the analyzed soil was classified as fine sand. Soil bulk density and total porosity were practically the same at the post-arable site and the site remaining under extensive arable use. However, formerly arable land was characterized by a lower (pH = 4.7) than soil of the same complex still under cultivation (pH = 5.3). Soil organic carbon content also varied, amounting to 1.25% in the surface layer of formerly arable land (0–10 cm), as compared to cultivated soil (0.89%).

Table 2. Basic properties of genetic horizon A (0–10 cm) of soil of the two study sites, ($\pm$ = standard deviation).

Characteristic	Site 1 Forest (after Arable Usage)	Site 2 Extensive Arable Usage
Sand (%)	94	94
Silt (%)	4	4
Clay (%)	2	2
Soil bulk density, n = 5 (kg m^{-3})	151040.1	1490 $\pm$ 45.2
Total porosity, n = 5 (%)	43.01 $\pm$ 0.16	43.8 $\pm$ 1.7
Soil organic carbon n = 3 (%)	1.25 $\pm$ 0.19	0.89 $\pm$ 0.13
Nitrogen total n = 3 (%)	0.0939 $\pm$ 0.0007	0.0676 $\pm$ 0.0003
C:N	13.3	13.2
pH (H$_2$0) n = 3 (−)	4. 7 $\pm$ 0.1	5.3 $\pm$ 0.1

3.2. Saturated Hydraulic Conductivity and Water Retention

Samples taken from the surface layer of soil (genetic horizon A) were characterized by soil moisture content similar to that of field capacity at the time of sampling, at which the analyzed soil was wettable (Class 0). The obtained soil saturation was 0.98 $\pm$ 0.01 of total porosity. For Site 1, the average value of saturated hydraulic conductivity for n = 6 was K_s = 2.66 $\times$ 10^{-5} $\pm$ 0.45 $\times$ 10^{-5} ms^{-1} at a coefficient of variation v = 16.9%, and for site 2: K_s = 3.44 $\times$ 10^{-5} $\pm$ 0.66 $\times$ 10^{-5} ms^{-1}, v = 19.2%. The obtained conductivity results were similar to those provided in literature for albic Podzols [46,47], and the coefficients of variation indicate low variability of data. Extensive research on the saturated soil conductivity under conditions of abandonment of agricultural use was conducted by Di Prima et al. [48] and methods, with a characterization based exclusively on a stabilized infiltration process, yielded also an appreciably low variability of the conductivity results. The pF curves measured for both sites (Figure 1), as well as saturated water conductivity K_s, did not vary by the manner of soil use. The total water content available to plants indicated from the retention curve was 0.13 cm^3 cm^{-3}.

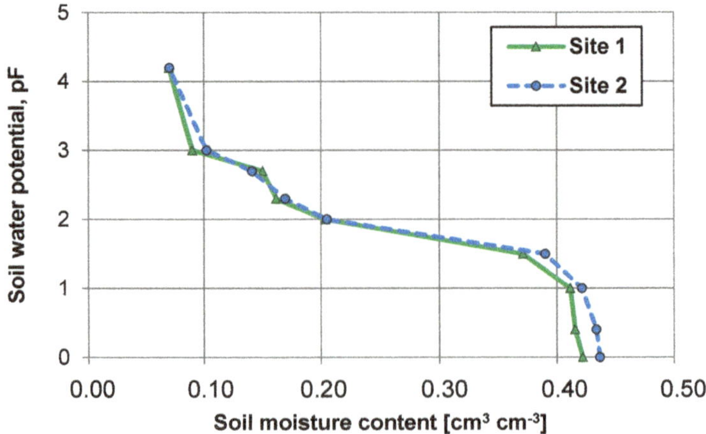

Figure 1. pF curves for post-arable land site (Site 1) and site under arable use (Site 2).

3.3. Assessment of Soil Water Repellency

The basic statistic measures of potential water repellency for the n = 15 number of replicates are presented in Table 3. The median of measured WDPT values for Site 1 was 17,700 s, which classifies Site 1 as an extremely repellent class (Class 5). The measured extreme WDPT values were max = 19,200 s and WDPT min = 16,080 s, which also belong to Class 5. The median of measured WDPT values for Site 2 was 90 s, which classifies the soil as being strongly repellent (Class 2). The maximum value was 284 s which still was Class 2, while the minimum value of 38 s belongs to Class 1 (slightly water repellent).

Table 3. Values of potential soil water repellency as derived from water drop penetration time (WDPT) test under forest following arable use (Site 1) and under arable use (Site 2), n = 15.

WDPT Characteristic	Site 1	Site 2
Median (s)	17,700	90
Average (s)	17,760	123
Max (s)	19,200	284
Min (s)	16,080	38
Range (s)	3120	246

The relationship between soil water potential (in terms of pF) and the share of WDPT classes found for the respective soil water potential is presented in Figure 2. With rising pressure heads, SWR increased significantly in both A horizons. In the case of low soil water tension up to field capacity on Site 1 and up to pF = 3.0 on Site 2, the soils were wettable. At higher pF the soil became increasingly repellent with decreasing soil moisture content.

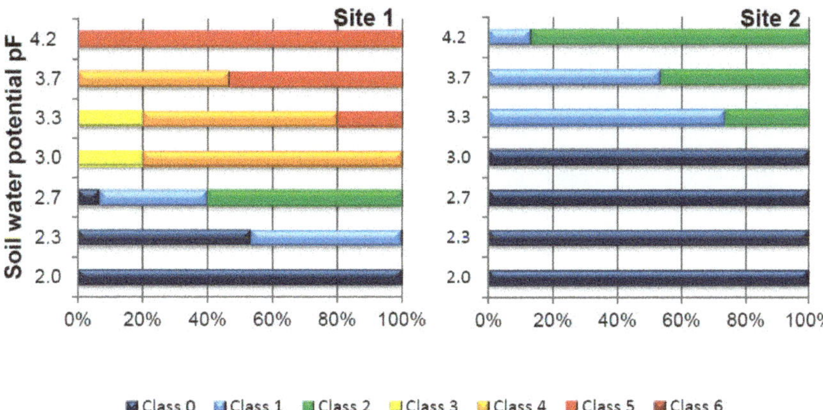

Figure 2. Soil water repellency, in terms of WDPT classes [16], of the A horizons of Site 1 and Site 2, as a function of soil water potential in terms of pF.

The critical moisture content for repellency (CSMC), delivered from the relationship between WDPT and soil water potential, on the site formerly under arable use was 0.16–0.14 cm^3 cm^{-3} which corresponds to pF = 2.3–2.7. At the same time, the value of CSMC for extensive arable use was 0.10–0.09 cm^3 cm^{-3}, which corresponds to a pF = 3.0–3.3.

3.4. Surface Runoff in Soil Formerly under Arable Use

Taking into account the extreme potential SWR of the Site 1 A horizon, surface runoff was analyzed with this material. A visualization of the wetting of the soil surface is presented in Figure 3. What is characteristic is the uneven wetting of the surface. Soil surfaces of high moisture content, from which surface runoff takes place, as well as completely dry surfaces are noticeable. The thickness of the wetted layer after completion of the experiment was approximately 5 mm, while the soil below was completely dry. The course of runoff during the experiment is presented in Figure 4. The measured total runoff was 6.72 mm, meaning that, of the total rainfall, the soil retained merely 48%. The obtained results of surface runoff confirm that high water repellency can significantly affect the agrohydrological regime. It can significantly decrease the amount of water available to plants, causing increased susceptibility to drought, accelerated mineralization of organic substances, and additional CO_2 emissions [21,49].

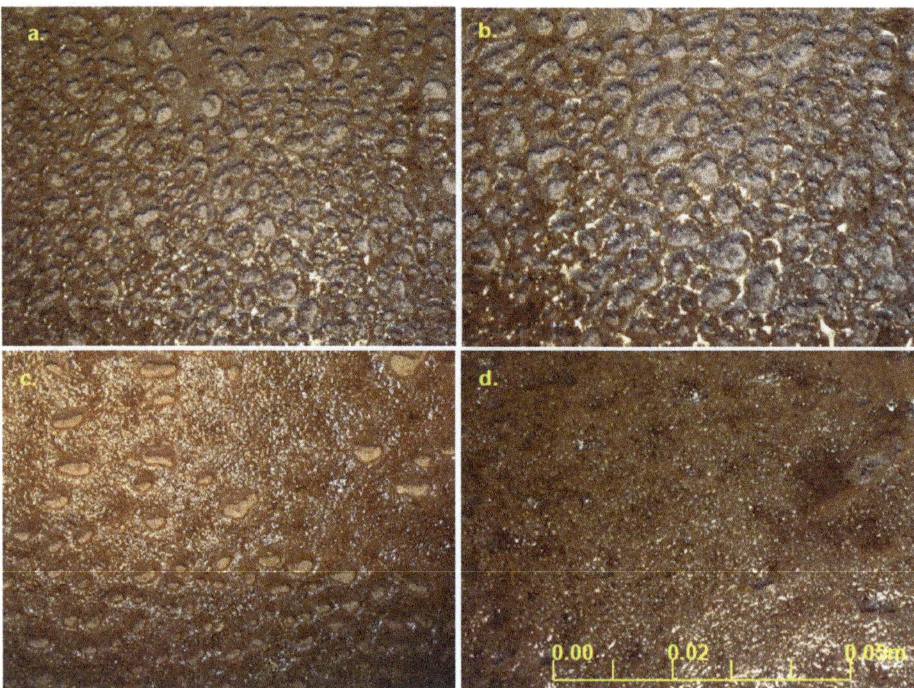

Figure 3. Visualization of wetted surface area over time during simulated rainfall, Site 1: (**a**) after 1 h; (**b**) after 2 h; (**c**) after 4 h; and (**d**) after 7 h.

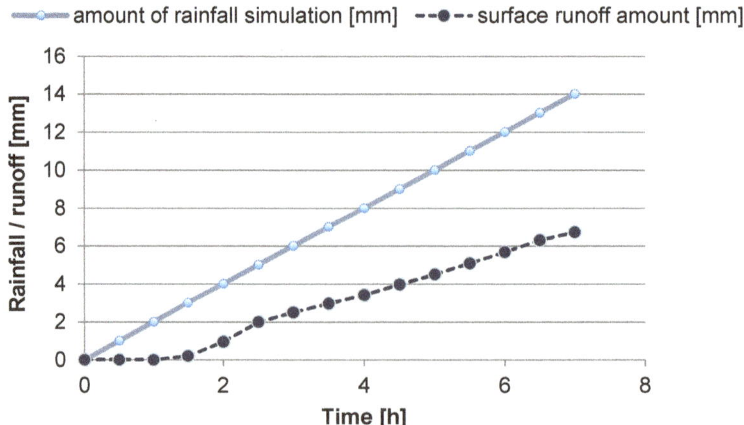

Figure 4. Course of rainfall and surface runoff caused by soil water repellency (Site 1).

4. Discussion

Global warming is causing severe soil droughts to occur more and more often in the continental climate [50–52]. Thus, risk connected with the occurrence of water repellency increases, especially in sandy soils. This contributes, especially, to increasing surface runoff and decreasing periodical soil water retention [53,54]. The degree of retention capabilities of forest soils are much less known than the retention of land used for agriculture [55,56]. On the other hand, pedotransfer functions have a local nature, as a result of which differences in precision of retention assessment may be significant [57]. Additionally, the development of a new methodology to compare the connectivity processes at the catchment with the pedon scale indicates the possibility of allowing inclusion of the absent micro-topographical information (e.g., [58]). Abandoning agricultural production on soils characterized by low productivity potential and their transformation into forests is economically and environmentally ratified, but should be preceded by an individual analysis of each case. The study site was characterized by a low nitrogen content, typical of albic Podzols. The soil organic content (SOC) content (1.25%) was significantly higher on the post-arable site, spontaneously afforested with a dominance of pine, in relation to cultivated soil (0.89%). Sibielec et al. [59], on the base of many-year studies, states that sandy soils in Poland contain SOC in a range from 1.01% to 2% in 63% of their data set. The C:N ratio for both sites was the same and amounted to 13.3–13.2, similar to average for sandy soils in Poland. Abandoning arable use along with changes in the air–water ratios lead resulted in a decrease in pH, from acidic to highly acidic. Afforestation with *Pinus sylvestris* (80%) also affected the SOC composition, enriching the soil in waxes [26,28]. Many authors (e.g., [17,18,60]) indicate waxes as one of the reasons for the water repellency of forest soils. The strongest SWR under thicker layers of litter was reported by Buczko et al. [17,61], and the authors [61] suggested that it is caused by the changing chemistry of the soil organic matter, along with depth, and/or varied bonding of this organic matter to the soil particles.

The deposit of organic matter in the soil on Site 1 is dominated by *Pinus silvestris* trees, its decomposition and penetration into A horizon is the most likely cause for the observed distinct water repellency, as indicated from the WDPT test. A very distinct difference in water repellency was; however, observed in the surface layer of soil between the two sites. The reduction in organic matter content due to soil tillage promotes the reduction in repellency by reducing the CSMC, beyond which hydrophobic soils become hydrophilic as well as persistence of water repellency [62]. Repellency on the post-arable site occurred during drought, that is at higher moisture content than on the site under arable use. Critical soil moisture content (CSMC) for the occurrence of water repellency in post-arable, afforested land was 0.14–0.16 cm cm^{-3}, as compared to 0.09–0.10 cm cm^{-3} in arable soil. Here, increased CSMC and distinct increase in water repellency on the afforested post-arable site was classified in Class 5 (i.e., extremely repellent), while arable soil was found to fall into Class 2 (i.e., strongly water repellent).

In literature on the subject, variability in CSMC values can be found in relation to the type of soil (i.e., 2% for dune sand [63], 9.3%–15% for sand [64], 14%–27% for loamy sand [19], 3%4–38.5% for clayey peat [65], 41%–49% for moorsh formations, 64%–69% for alder peats, and 83%–86% in reed peats [20]). CSMC results obtained in this study confirm the reports presented by Ziogas et al. [64] for sandy soils.

On the other hand, in the analyzed case, increasing the values of CSMC and the high increase in water repellency occurred as a result of abandoning agricultural production and the succession of forest vegetation with a dominance of pine. The hydrological consequences of SWR have also been indicated by other authors [66–72]. According to studies of Butzen et al. [73] on coniferous forest sites in Germany, water repellency effects were an important factor triggering overland flow generation. For the post-agricultural site, from the experiment with simulated rainfall, a surface runoff of 50% was indicated. Despite simplifications in reflecting field conditions (e.g., the lack of vegetation), the obtained results indicate the direction of changes in the rainfall–runoff relationship when the phenomenon of water repellency occurs. Water repellency was characterized by seasonality, which was also observed by Buczko et al. [17,18], Leighton-Boyce et al. [26], and Hewelke et al. [11,15]. On the analyzed site, the soil was wettable in the period of early spring, whereas the phenomenon of water repellency occurred after a longer period without rainfall. The persistency and the severity of water repellency is decisive to the shaping of the dynamics of soil moisture content and requires further research.

5. Conclusions

The present study confirmed that abandoning arable use and allowing for spontaneous afforestation with the succession of pine had a negative influence on soil hydraulic properties. Changes in use led to a decrease in the CSMC and a significant increase in water repellency. The basic strategy of preventing water repellency, and its consequences, is maintaining an adequately high moisture content of soil. In the case of excluding land from agricultural production, its afforestation with a dominance of pine should not be allowed, and introducing, rather, a mixed stand of trees, appropriate for the soil type and climate conditions, should be considered. It should be kept in mind that progressing climate change and the increased frequency of the occurrence of soil droughts may lead to an increased significance of water repellency in the water management of soils. The overview of studies on water repellency, caused by both natural as well as anthropogenic factors, indicates that it ought to be treated as one of the indicators of soil quality, with the present work indicating the linkages of soil properties to ecosystem services and to UN sustainable goals for development.

Funding: This research received no external funding.

Acknowledgments: I thank the anonymous reviewers and the Academic Editor for their insightful suggestions and comments that have helped to improve the quality of the manuscript.

Conflicts of Interest: The author declares no conflict of interest.

References

1. Czachor, H.; Flis-Bujak, M.; Ksiezpolska, A.; Niewczas, J.; Falski, M. Analysis of factors affecting the wettability of mineral soils. *Acta Agrophys.* **2009**, *2*, 84.
2. Goebel, M.O.; Bachmann, J.; Reichstein, M.; Janssens, I.A.; Guggenberger, G. Soil water repellency and its implications for organic matter decomposition—Is there a link to extreme climatic events? *Glob. Chang. Biol.* **2011**, *17*, 2640–2656. [CrossRef]
3. Franco, C.M.M.; Michelsen, P.P.; Oades, J.M. Amelioration of water repellency: Application of slow-release fertilisers to stimulate microbial breakdown of waxes. *J. Hydrol.* **2000**, *231*, 342–351. [CrossRef]
4. Franco, C.M.M.; Tate, M.E.; Oades, J.M. Studies on non-wetting sands. 1. The role of intrinsic particulate organic-matter in the development of water-repellency in non-wetting sands. *Soil Res.* **1995**, *33*, 253–263. [CrossRef]
5. Doerr, S.H.; Shakesby, R.A.; Walsh, R. Soil water repellency: Its causes, characteristics and hydro-geomorphological significance. *Earth-Sci. Rev.* **2000**, *51*, 33–65. [CrossRef]
6. Lachacz, A.; Nitkiewicz, M.; Kalisz, B. Water repellency of post-boggy soils with a various content of organic matter. *Biologia* **2009**, *64*, 634–638. [CrossRef]
7. Moody, J.A.; Kinner, D.A.; Úbeda, X. Linking hydraulic properties of fire-affected soils to infiltration and water repellency. *J. Hydrol.* **2009**, *379*, 291–303. [CrossRef]
8. Certini, G. Effects of fire on properties of forest soils: A review. *Oecologia* **2005**, *143*, 1–10. [CrossRef]
9. Granged, A.J.; Jordán, A.; Zavala, L.M.; Bárcenas, G. Fire-induced changes in soil water repellency increased fingered flow and runoff rates following the 2004 Huelva wildfire. *Hydrol. Process.* **2011**, *25*, 1614–1629. [CrossRef]
10. Bogacz, A.; Łabaz, B.; Woźniczka, P. Impact of fire on values of organic material transformation Indicators. *Rocz. Glebozn. Soil Sci. Annu.* **2013**, *64*, 88–92. [CrossRef]
11. Hewelke, E.; Oktaba, L.; Gozdowski, D.; Kondras, M.; Olejniczak, I.; Górska, E.B. Intensity and persistence of soil water repellency in pine forest soil in a temperate continental climate under drought conditions. *Water* **2018**, *10*, 1121. [CrossRef]

12. Adams, R.H.; Osorio, F.G.; Cruz, J.Z. Water repellency in oil contaminated sandy and clayey soils. *Int. J. Environ. Sci. Technol.* **2008**, *5*, 445–454. [CrossRef]

13. Takawira, A.; Gwenzi, W.; Nyamugafata, P. Does hydrocarbon contamination induce water repellency and changes in hydraulic properties in inherently wettable tropical sandy soils? *Geoderma* **2014**, *235*, 279–289. [CrossRef]

14. Klamerus-Iwan, A.; Błońska, E.; Lasota, J.; Kalandyk, A.; Waligórski, P. Influence of oil contamination on physical and biological properties of forest soil after chainsaw use. *Water Air Soil Pollut.* **2015**, *226*, 389. [CrossRef]

15. Hewelke, E.; Szatyłowicz, J.; Hewelke, P.; Gnatowski, T.; Aghalarov, R. The Impact of Diesel Oil Pollution on the Hydrophobicity and CO_2 Efflux of Forest Soils. *Water Air Soil Pollut.* **2018**, *229*, 51. [CrossRef]

16. Dekker, L.W.; Ritsema, C.J.; Oostindie, K.; Moore, D.; Wesseling, J.G. Methods for determining soil water repellency on field-moist samples. *Water Resour. Res.* **2009**, *45*. [CrossRef]

17. Buczko, U.; Bens, O.; Hüttl, R.F. Variability of soil water repellency in sandy forest soils with different stand structure under Scots pine (*Pinus sylvestris*) and beech (*Fagus sylvatica*). *Geoderma* **2005**, *126*, 317–336. [CrossRef]

18. Buczko, U.; Bens, O.; Hüttl, R.F. Changes in soil water repellency in a pine–beech forest transformation chronosequence: Influence of antecedent rainfall and air temperatures. *Ecol. Eng.* **2007**, *31*, 154–164. [CrossRef]

19. Leighton-Boyce, G.; Doerr, S.H.; Shakesby, R.A.; Walsh, R.P.D.; Ferreira, A.J.D.; Boulet, A.K.; Coelho, C.O.A. Temporal dynamics of water repellency and soil moisture in eucalypt plantations, Portugal. *Aust. J. Soil Res.* **2005**, *43*, 269–280. [CrossRef]

20. Hewelke, E.; Szatyłowicz, J.; Gnatowski, T.; Oleszczuk, R. Effects of soil water repellency on moisture patterns in a degraded Sapric Histosol. *Land Degrad. Dev.* **2016**, *27*, 955–964. [CrossRef]

21. Urbanek, E.; Doerr, S.H. CO_2 efflux from soils with seasonal water repellency. *Biogeosciences* **2017**, *14*, 4781–4794. [CrossRef]

22. Imeson, A.C.; Verstraten, J.M.; Van Mulligen, E.J.; Sevink, J. The effects of fire and water repellency on infiltration and runoff under Mediterranean type forest. *Catena* **1992**, *19*, 345–361. [CrossRef]

23. Harper, R.J.; McKissock, I.; Gilkes, R.J.; Carter, D.J.; Blackwell, P.S. A multivariate framework for interpreting the effects of soil properties, soil management and landuse on water repellency. *J. Hydrol.* **2000**, *231*, 371–383. [CrossRef]

24. Mataix-Solera, J.; Arcenegui, V.; Guerrero, C.; Mayoral, A.M.; Morales, J.; González, J.; García-Orenes, F.; Gómez, I. Water repellency under different plant species in a calcareous forest soil in a semiarid Mediterranean environment. *Hydrol. Process.* **2007**, *21*, 2300–2309. [CrossRef]

25. Zavala, L.M.; González, F.A.; Jordán, A. Intensity and persistence of water repellency in relation to vegetation types and soil parameters in Mediterranean SW Spain. *Geoderma* **2009**, *152*, 361–374. [CrossRef]

26. Lichner, L.; Holko, L.; Zhukova, N.; Schacht, K.; Rajkai, K.; Fodor, N.; Sándor, R. Plants and biological soil crust influence the hydrophysical parameters and water flow in an aeolian sandy soil. *J. Hydrol. Hydromech.* **2012**, *60*, 309–318. [CrossRef]

27. Orfánus, T.; Dlapa, P.; Fodor, N.; Rajkai, K.; Sándor, R.; Nováková, K. How severe and subcritical water repellency determines the seasonal infiltration in natural and cultivated sandy soils. *Soil Tillage Res.* **2014**, *135*, 49–59. [CrossRef]

28. Prusinkiewicz, Z.; Kosakowski, A. The wettability of soil organic matter as the forming factor of the water properties of forest soils. *Rocz. Glebozn.-Soil Sci. Annu.* **1986**, *37*, 3–23.

29. Pudełko, R.; Kozak, M.; Jędrejek, A.; Gałczyńska, M.; Pomianek, B. Regionalisation of unutilised agricultural area in Poland. *Polish J. Soil Sci.* **2018**, *51*, 119. [CrossRef]

30. Navarro, L.M.; Pereira, H.M. Rewilding abandoned landscapes in Europe. In *Rewilding European Landscapes*; Springer: Cham, Switzerland, 2015; pp. 3–23.

31. Verburg, P.H.; Overmars, K.P. Combining top-down and bottom-up dynamics in land use modeling: Exploring the future of abandoned farmlands in Europe with the Dyna-CLUE model. *Landsc. Ecol.* **2009**, *24*, 1167–1181. [CrossRef]

32. Schwilch, G.; Lemann, T.; Berglund, Ö.; Camarotto, C.; Cerdà, A.; Daliakopoulos, I.N.; Kohnová, S.; Krzeminska, D.; Marañón, T.; Rietra, R.; et al. Assessing impacts of soil management measures on Ecosystem Services. *Sustainability* **2018**, *10*, 4416. [CrossRef]

33. Borecki, T.; Łopiński, Ł.; Kędziora, W.; Orzechowski, M.; Wójcik, R.; Stępień, E. The Concept of Regulating Forest Management in a Region Subject to High Environmental Pressure. *Forests* **2018**, *9*, 539. [CrossRef]

34. Borecki, T.; Orzechowski, M.; Stępień, E.; Wójcik, R. Expected impact of climate change on forest ecosystems and its consequences in forest management planning. *Sylwan* **2017**, *161*, 531–538.

35. Keesstra, S.; Mol, G.; de Leeuw, J.; Okx, J.; de Cleen, M.; Visser, S. Soil-related sustainable development goals: Four concepts to make land degradation neutrality and restoration work. *Land* **2018**, *7*, 133. [CrossRef]

36. Griggs, D.; Stafford-Smith, M.; Gaffney, O.; Rockström, J.; Öhman, M.C.; Shyamsundar, P.; Steffen, W.; Glaser, G.; Kanie, N.; Noble, I. Policy: Sustainable development goals for people and planet. *Nature* **2013**, *495*, 305–307. [CrossRef]

37. Steffen, W.; Richardson, K.; Rockström, J.; Cornell, S.E.; Fetzer, I.; Bennett, E.M.; Biggs, R.; Carpenter, S.R.; de Vries, W.; de Wit, C.A.; et al. Planetary boundaries: Guiding human development on a changing planet. *Science* **2015**, *347*, 1259855. [CrossRef]

38. Keesstra, S.; Nunes, J.P.; Saco, P.; Parsons, T.; Poeppl, R.; Masselink, R.; Cerdà, A. The way forward: Can connectivity be useful to design better measuring and modelling schemes for water and sediment dynamics? *Sci. Total Environ.* **2018**, *644*, 1557–1572. [CrossRef]

39. IUSS Working Group WRB. *World Reference Base for Soil Resources 2014, Update 2015 International Soil Classification System for Naming Soils and Creating Legends for Soil Maps*; World Soil Resources Reports No. 106; IUSS: Austria Vienna, 2015; p. 192.

40. Papierowska, E.; Matysiak, W.; Szatyłowicz, J.; Debaene, G.; Urbanek, E.; Kalisz, B.; Łachacz, A. Compatibility of methods used for soil water repellency determination for organic and organo-mineral soils. *Geoderma* **2018**, *314*, 221–231. [CrossRef]

41. Dekker, L.W.; Jungerius, P.D. Water repellency in the dunes with special reference to The Netherlands. *Catena* **1990**, *18*, 173–183.

42. Ryżak, M.; Bartminski, P.; Bieganowski, A. Method for determination of particle size distribution of mineral soils. *Acta Agrophys.* **2009**, *175*, 1–84.

43. Lityński, T.; Jurkowska, H.; Gorlach, E. *Chemical and Agriculture Analysis*; PWN: Warszawa, Poland, 1976; pp. 129–132.

44. Klute, A. *Methods of Soil Analysis. Part 1. Physical and Mineralogical Methods. Agronomy Monographs*; ASA and SSA, A. Klute: Madison, WI, USA, 1986; Volume 9.

45. Soil Survey Division Staff. *Soil Survey Manual*; United States Department of Agriculture: Washington, DC, USA, 1993; p. 315.

46. Buczko, U.; Bens, O.; Huttl, R.F. Water infiltration and hydrophobicity in forest soils of a pine–beech transformation chronosequence. *J. Hydrol.* **2006**, *331*, 383–395. [CrossRef]

47. Lichner, L'.; Capuliak, J.; Zhukova, N.; Holko, L.; Czachor, H.; Kollár, J. Pines influence hydrophysical parameters and water flow in a sandy soil. *Biologia* **2013**, *68*, 1104–1108. [CrossRef]

48. Di Prima, S.; Lassabatere, L.; Rodrigo-Comino, J.; Marrosu, R.; Pulido, M.; Angulo-Jaramillo, R.; Úbeda, X.; Keesstra, S.; Cerdà, A.; Pirastru, M. Comparing transient and steady-state analysis of single-ring infiltrometer data for an abandoned field affected by fire in Eastern Spain. *Water* **2018**, *10*, 514. [CrossRef]

49. Tezza, L.; Vendrame, N.; Pitacco, A. Disentangling the carbon budget of a vineyard: The role of soil management. *Agric. Ecosyst. Environ.* **2019**, *272*, 52–62. [CrossRef]

50. Boczoń, A.; Kowalska, A.; Dudzińska, M.; Wróbel, M. Drought in Polish Forests in 2015. *Polish J. Environ. Stud.* **2016**, *25*, 1857–1862. [CrossRef]

51. Stojanovic, M.; Drumond, A.; Nieto, R.; Gimeno, L. Anomalies in Moisture Supply during the 2003 Drought Event in Europe: A Lagrangian Analysis. *Water* **2018**, *10*, 467. [CrossRef]

52. Koutroulis, A.G.; Papadimitriou, L.V.; Grillakis, M.G.; Tsanis, I.K.; Wyser, K.; Betts, R.A. Freshwater vulnerability under high end climate change. A pan-European assessment. *Sci. Total Environ.* **2018**, *613*, 271–286. [CrossRef]

53. Ferreira, C.S.S.; Walsh, R.P.D.; Shakesby, R.A.; Keizer, J.J.; Soares, D.; González-Pelayo, O.; Ferreira, A.J.D. Differences in overland flow, hydrophobicity and soil moisture dynamics between Mediterranean woodland types in a peri-urban catchment in Portugal. *J. Hydrol.* **2016**, *533*, 473–485. [CrossRef]

54. Rye, C.F.; Smettem, K.R.J. The effect of water repellent soil surface layers on preferential flow and bare soil evaporation. *Geoderma* **2017**, *289*, 142–149. [CrossRef]

55. Hewelke, P.; Gnatowski, T.; Hewelke, E.; Tyszka, J.; Zakowicz, S. Analysis of Water Retention Capacity for Select Forest Soils in Poland. *Polish J. Environ. Stud.* **2015**, *24*, 1013–1019. [CrossRef]

56. Hewelke, P.; Hewelke, E.; Chołast, S.; Żakowicz, S.; Lesak, M. Assessment of the possibility of applying selected pedotransfer functions for indicating the retention of forest soils in Poland. *Sci. Rev. Eng. Environ. Sci.* **2017**, *26*, 336–345. [CrossRef]

57. Hewelke, P.; Hewelke, E.; Oleszczuk, R.; Kwas, M. The application of pedotransfer functions in the estimation of water retention in alluvial soils in Żuławy Wiślane, northern Poland. *Soil Sci. Annu.* **2018**, *69*, 3–10. [CrossRef]

58. Rodrigo Comino, J.; Keesstra, S.D.; Cerdà, A. Connectivity assessment in Mediterranean vineyards using improved stock unearthing method, LiDAR and soil erosion field surveys. *Earth Surface Process. Landf.* **2018**, *43*, 2193–2206. [CrossRef]

59. Siebielec, G.; Smreczak, B.; Klimkowicz-Pawlas, A.; Kowalik, M.; Kaczyński, R.; Koza, P.; Ukalska-Jaruga, A.; Łysiak, M.; Wójtowicz, U.; Poręba, L.; et al. *Report on the Third Phase of the Contract "Monitoring of Arable Soil Chemistry in Poland in 2015–2017"*; IUNG-PIB: Puławy, Poland, 2017; p. 190.

60. Orfánus, T.; Bedrna, Z.; Lichner, L.; Hallet, P.D.; Kňava, K.; Sebíň, M. Spatial variability of water repellency in pine forest soil. *Soil Water Res.* **2008**, *3*, 123–129. [CrossRef]

61. Buczko, U.; Bens, O.; Fischer, H.; Hüttl, R.F. Water repellency in sandy luvisols under different forest transformation stages in northeast Germany. *Geoderma* **2002**, *109*, 1–18. [CrossRef]

62. Vogelmann, E.S.; Reichert, J.M.; Prevedello, J.; Consensa, C.O.B.; Oliveira, A.É.; Awe, G.O.; Mataix-Solera, J. Threshold water content beyond which hydrophobic soils become hydrophilic: The role of soil texture and organic matter content. *Geoderma* **2013**, *209*, 177–187. [CrossRef]

63. Dekker, L.W.; Doerr, S.H.; Oostindie, K.; Ziogas, A.K.; Ritsema, C.J. Water repellency and critical soil water content in a dune sand. *Soil Sci. Soc. Am. J.* **2001**, *65*, 1667–1674. [CrossRef]

64. Ziogas, A.K.; Dekker, L.W.; Oostindie, K.; Ritsema, C.J. Soil water repellency in north-eastern Greece with adverse effects of drying on the persistence. *Soil Res.* **2005**, *43*, 281–289. [CrossRef]

65. Dekker, L.W.; Ritsema, C.J. Variation in water content and wetting patterns in Dutch water repellent peaty clay and clayey peat soils. *Catena* **1996**, *28*, 89–105. [CrossRef]

66. Cerdà, A.; Doerr, S.H. Soil wettability, runoff and erodibility of major dry-Mediterranean land use types on calcareous soils. *Hydrol. Process.* **2007**, *21*, 2325–2336. [CrossRef]

67. Miyata, S.; Kosugi, K.I.; Gomi, T.; Onda, Y.; Mizuyama, T. Surface runoff as affected by soil water repellency in a Japanese cypress forest. *Hydrol. Process. Int. J.* **2007**, *21*, 2365–2376. [CrossRef]

68. Neris, J.; Tejedor, M.; Rodríguez, M.; Fuentes, J.; Jiménez, C. Effect of forest floor characteristics on water repellency, infiltration, runoff and soil loss in Andisols of Tenerife (Canary Islands, Spain). *Catena* **2013**, *108*, 50–57. [CrossRef]

69. Olorunfemi, I.E.; Fasinmirin, J.T. Land use management effects on soil hydrophobicity and hydraulic properties in Ekiti State, forest vegetative zone of Nigeria. *Catena* **2017**, *155*, 170–182. [CrossRef]

70. Hejduk, L.; Hejduk, A.; Baryła, A.; Hewelke, E. Influence of selected factors on erodibility in catchment scale on the basis of field investigation. *J. Ecol. Eng.* **2017**, *18*, 256–267. [CrossRef]

71. Cerdà, A.; Rodrigo-Comino, J.; Novara, A.; Brevik, E.C.; Vaezi, A.R.; Pulido, M.; Giménez-Morera, A.; Keesstra, S.D. Long-term impact of rainfed agricultural land abandonment on soil erosion in the Western Mediterranean basin. *Prog. Phys. Geogr. Earth Environ.* **2018**, *42*, 202–219. [CrossRef]

72. Mao, J.; Nierop, K.G.; Dekker, S.C.; Dekker, L.W.; Chen, B. Understanding the mechanisms of soil water repellency from nanoscale to ecosystem scale: A review. *J. Soils Sediments* **2019**, *19*, 1–15. [CrossRef]

73. Butzen, V.; Seeger, M.; Marruedo, A.; de Jonge, L.; Wengel, R.; Ries, J.B.; Casper, M.C. Water repellency under coniferous and deciduous forest—Experimental assessment and impact on overland flow. *Catena* **2015**, *133*, 255–265. [CrossRef]

Article

Effects of No-Tillage and Conventional Tillage on Physical and Hydraulic Properties of Fine Textured Soils under Winter Wheat

Mirko Castellini [1,*], Francesco Fornaro [1], Pasquale Garofalo [1], Luisa Giglio [1], Michele Rinaldi [2], Domenico Ventrella [1], Carolina Vitti [1] and Alessandro Vittorio Vonella [1]

[1] Council for Agricultural Research and Economics-Research Center for Agriculture and Environment (CREA-AA), Via C. Ulpiani 5, 70125 Bari, Italy; francesco.fornaro@crea.gov.it (F.F.); pasquale.garofalo@crea.gov.it (P.G.); luisa.giglio@crea.gov.it (L.G.); domenico.ventrella@crea.gov.it (D.V.); carolina.vitti@crea.gov.it (C.V.); vittorio.vonella@crea.gov.it (A.V.V.)

[2] Council for Agricultural Research and Economics - Research Centre for Cereal and Industrial Crops (CREA-CI), S.S. 673 km 25,200, 71121 Foggia, Italy; michele.rinaldi@crea.gov.it

* Correspondence: mirko.castellini@crea.gov.it; Tel.: +39-080-5475039

Received: 29 January 2019; Accepted: 1 March 2019; Published: 7 March 2019

Abstract: The conversion from conventional tillage (CT) to no-tillage (NT) of the soil is often suggested for positive long-term effects on several physical and hydraulic soil properties. In fact, although shortly after the conversion a worsening of the soil may occur, this transition should evolve in a progressive improvement of soil properties. Therefore, investigations aiming at evaluating the effects of NT on porous media are advisable, since such information may be relevant to better address the farmers' choices to this specific soil conservation management strategy. In this investigation, innovative and standard methods were applied to compare CT and NT on two farms where the conversion took place 6 or 24 years ago, respectively. Regardless of the investigated farm, results showed negligible differences in cumulative infiltration or infiltration rate, soil sorptivity, saturated hydraulic conductivity, conductive pores size, or hydraulic conductivity functions. Since relatively small discrepancies were also highlighted in terms of bulk density or soil organic carbon, it was possible to conclude that NT did not have a negative impact on the main physical and hydraulic properties of investigated clay soils. However, a significantly higher number of small pores was detected under long-term NT compared to CT, so we concluded that the former soil was a more conductive pore system, i.e., consisting of numerous relatively smaller pores but continuous and better interconnected. Based on measured capacity-based indicators (macroporosity, air capacity, relative field capacity, plant available water capacity), NT always showed a more appropriate proportion of water and air in the soil.

Keywords: BEST-procedure; soil hydraulic conductivity; capacity-based soil indicators; conventional tillage; no-tillage; durum wheat

---

1. Introduction

No-tillage, zero tillage, and direct drilling or sod-seeding are terms to define a soil management system in which field crops are sown without any main soil tillage, determining a very limited disturbance of the soil (i.e., lower than 5 cm), which may arise by the passage of the drill coulters during sowing [1]. In this farming system, at least a third of the soil surface may remain covered with plant residues, but this soil surface can even reach 100% [2], thus promoting soil protection from water erosion [3,4] and potentially increasing both the organic matter content and the presence of microorganisms into the soil [5]. Moreover, conversion to no-tillage systems may improve soil physical properties [6] and increase the soil water retention in rainfed environments [7]. In addition, savings on operating costs and reductions in machinery emissions are expected [8].

Overall, no-tillage and sod-seeding is used nowadays, especially for cereal cultivation. Compared to the different areas where such techniques are quite common, i.e., America and Australia [2], in Europe (EU 27), minimum tillage and sod-seeding is only applied on 3.5 million hectares (Eurostat SAPM and FSS, 2010) which represent 3.5% of total arable land area. Italy follows this trend since, in mid-2013, the proportion between conventional tillage (CT) and no-tillage (NT) was respectively of about 5.2 and 0.6 million hectares [9]. Several reasons can be assumed for the reduced application of these agronomic techniques in Mediterranean environments, including: (i) the lack of policies encouraging their adoption, (ii) a prejudice by farmers, since detectable positive effects are often not immediately apparent as they can only be observed after that a new equilibrium in soil properties has been established, (iii) appropriate drilling equipment and great skill and expertise are needed [10]. Regarding the second point, literature references suggest that the success of this option basically depends on the water availability for the crop; for example, comparable wheat yields between soil management strategies (CT≈NT) may be obtained under dry climates, while greater differences (CT»NT) can occur under humid climates [11]. In fact, exceptions have been identified with reference to specific soils, i.e., Vertisols, that is to say fertile structured soils and with a good water holding capacity, for which the shift from CT to NT can be less problematic and some level of economically acceptable yields even during dry periods can be reached [12,13]. Consequently, as fine textured soils are quite common in several cereal-growing areas (i.e., Southern Italy), conversion to NT may be considered a viable solution as compared to less suited soils [14]. The mentioned trend could be reversed as, starting from 2014, more and more farmers in Italy are making a conversion from CT to NT, taking advantage of the benefits of public funding for rural development (PSR 2014–2020), such that an increase of more than twice the NT area is expected over the next few years [15,16]. Consequently, as durum wheat is the main cereal crop in Italy, with a cultivated area of about 1.2–1.3 Mha (ISTAT, 2017), an increased focus on the effects of conversion from CT to NT on soil physical and hydraulic properties is desirable [1,2].

A relevant topic discussed in the literature concerns the transition time from CT to NT after which soil properties and crop yields begin to benefit from an expected higher complexity of the soil system [1,17,18]. A summary of long-term NT impact on physical and hydraulic soil properties was reported by Strudley et al. [19], and a recent review by Chandrasekhar et al. [20] discussed the modeling approaches to study the soil porosity in the transition from CT to NT. In particular, shortly after the conversion to NT and up to about 4–5 years, a worsening in soil properties, i.e., an increase in dry bulk density and a corresponding decrease in soil porosity or in hydraulic conductivity, can occur [2,19] regardless on the soil type [2]. However, this relatively short transition should evolve in a progressive improvement of soil structure, stabilization of the organic carbon content and/or of its improvement, i.e., in more complex fractions, also thanks to the increase of the microflora and the microfauna into the soil [1,2]. References also suggest that a period of about a decade may be considered the reasonable minimum time-frame for reaching a steady-state, namely characterized by an enhanced soil physical quality and stabilized production levels [1,11,19,21]. Reichert et al. [22], investigating the impact of conventional tillage on soils previously under no-tillage, suggested a temporal schematization of soil reconsolidation and soil aggregate creation processes, which can be divided into four phases: initial (1.5 years), intermediary (3.5 years), transitional (5 years), and stabilized (14 years) conditions. Regarding the impact on crop production, a global meta-analysis by Pittelkow et al. [11] has suggested that yields in the first 1–2 years following NT implementation declined for all crops except oilseeds and cotton, but matched CT yields after 3–10 years, except for maize and wheat in humid climates. Johnston and Poulton [23] suggested that research on this topic should be conducted investigating long-term experiments in order to assess the environmental sustainability and profitability when it works at full capacity. Long-term experiments, although expensive, can represent ideal laboratories to elucidate, for a given agro-environment, the impact of NT on soil physical and hydraulic properties [24–26]. Therefore, as the time elapsed in the conversion from CT to NT may be a crucial factor to evaluate the impact of alternative soil management strategies on soil properties, it is desirable to establish comparisons selecting farms under transition or farms where it is plausible to hypothesize that steady conditions of soil properties were reached, with farms in which the conventional tillage of the soil is adopted.

The assessment of NT impact on the soil obviously also depends on the specific evaluated soil property, soil texture, agro-environment, and their interaction [1,19,26]. Chang and Lindwall [27], for example, suggest that after 8 years on a loam soil in Canada, soil bulk density did not differ among crop rotations, and after 10 years of NT with continuous winter wheat, soil bulk density was higher than under CT. Conversely, an initial increase in bulk density near the surface of NT soils was showed by Vogeler et al. [21], but after 6 years a decline to values equal to or even less than that after ploughing were detected. Contrasting results were reported also in terms of hydrodynamic soil properties. Liepic et al. [28], for example, reported a reduction of about 60% of the cumulative infiltration under long-term NT as compared to CT, while Azooz and Arshad [29] showed that long-term NT generally increased ponded infiltration rates under initial dry, near field capacity and field capacity, but not under near saturated soil conditions.

The general conclusions drawn from the examined literature suggest that, if a trend exists, NT increases the aggregate stability and the pores' connectivity, generating inconsistent responses in total porosity and soil bulk density compared to CT; this corresponds to a very likely increase in the saturated or near-saturated hydraulic conductivity of NT soil management [19]. Therefore, suggested advances in this research area should be addressed to dispel existing doubts in order to: (i) fill the gap of knowledge for specific agro-environments, (ii) provide information of NT under long-term experiments to quantify pros and cons compared to CT, (iii) obtain improvements in the soil hydraulic characterization accuracy, by applying parametric estimation methods that allow both to evaluate soil properties changes and to obtain input data for agro-environmental simulation purposes. However, to our knowledge, a lack of information exists for Mediterranean agro-environments of Italy, as only few investigations were carried out in orchards [30] or for irrigated vegetable crops [31], and no reference has assessed the impact of NT on hydrostatic and hydrodynamic soil properties under rainfed wheat cultivation.

The Beerkan Estimation of Soil Transfer (BEST) parameters procedure [32] is an attractive, easy, robust, and inexpensive way for a parametric soil hydraulic characterization [33,34]; namely, it allows for the simultaneous determination of both the water retention curve, i.e. the relationship between volumetric water content, θ, and pressure head, h, and the hydraulic conductivity function, i.e. the relationship between the soil hydraulic conductivity, K, and θ (or h) [32]. Because of its simplicity, accuracy, and versatility of application (i.e., a standard or simplified application of the experimental procedure can be chosen), in fact, BEST was adopted in various agronomic-forest or -environmental investigations to compare, for example, different land uses [35,36], to assess the physical quality of Brazilian [37], Sicilian [38], or Burundian [33] soils, to investigate the effect of forest restoration on soil hydraulic conductivity [39] or to establish the role of soil sealing in water infiltration [40,41]. To date, the experimental procedure has received increasing attention and about eighty theoretical or applicative manuscripts were published, as reported on the Scopus database. Application of the BEST-procedure, for example, recently allowed the direct measurement of basic hydrodynamic soil properties, i.e., saturated hydraulic conductivity and soil sorptivity, and to estimate water dynamic based indicators, i.e., flow-weighted mean pore size and the corresponding number of hydraulically active pores of agricultural [42,43] and marginal [36] soils. Consequently, as the methodology allows one to obtain a high number of accurate measurements in a relatively short time, it allows one to minimize the spatial and temporal variability of the investigated soil properties. Moreover, since it is expected that alternative soil management strategies, such as CT or NT, can impact sorptivity [44], structure [45], and hydrostatic and hydrodynamic soil properties [46], BEST has proven to be a technique to accurately evaluate the impact of agronomic treatments on the physical and hydraulic properties of cultivated soils. However, since the considered soil management strategy could also affect the optimal balance between water and air in the soil, the water retention curve has been experimentally determined in the lab, and some capacity-based indicators (e.g., macroporosity, air capacity, relative field capacity, plant available water capacity) were estimated to evaluate the impact of MT and NT on soil water conservation [36].

The general objective of this investigation was to assess the impact of alternative soil management strategies on physical and hydraulic properties of fine textured soils, applying field and lab procedures, to help fill a gap of information on this topic for Mediterranean environments. In particular, two private farms located in Puglia (southern Italy), in which the long-term NT for durum wheat was established from a different time frame, six or twenty-four years, were selected and compared to CT. To achieve these goals, basic physical and hydraulic properties of the soil, i.e., texture, bulk density, total organic carbon, cumulative infiltration or infiltration rate, saturated hydraulic conductivity and sorptivity, and soil water retention curve were directly measured; two indicators of soil porosity were estimated to account for the hydrodynamic soil properties (flow-weighted mean pore size and number of hydraulically active pores), and four capacity-based indicators (macroporosity, air capacity, relative field capacity, and plant available water capacity) were considered as hydrostatic soil properties.

2. Materials and Methods

2.1. Experimental Sites

Selected experimental sites were two private farms located in central (Giglio farm, Gravina in Puglia, 40°52′57.1″N, 16°22′31.4″E) and north (Casone farm, Candela, 41°08′04.8″N, 15°31′56.0″E) Apulia, southern Italy (Figure 1). Hereafter, the two farms investigated will be identified as Gravina and Candela. Two alternative soil management strategies for durum wheat cultivation were considered in this investigation, namely conventional tillage, CT and no-tillage, NT. Two surface areas of approximately 2.6–5.3 and 5.5–8.2 ha, were respectively selected in Gravina and Candela farms (Figure 1). In Gravina, the NT management plot was repeated for approximately 24 years while in CT plot the monoculture of wheat had been carried out for 28 years. Protein pea fallow has preceded durum wheat cultivation on NT. Therefore, NT run under conservative agriculture, as it respects the principles of direct sowing, residuals on the surface and crop rotation. As an example, Figure 1 shows some views of fields at the Gravina site. At Candela farm, the transition was relatively more recent as NT practice had been implemented from 6 years, and a biennial crop rotation, "cereals (barley or durum wheat)—field bean" was established. Additional information on the main cultivation practices carried out in the investigated farms were reported in Table S1.

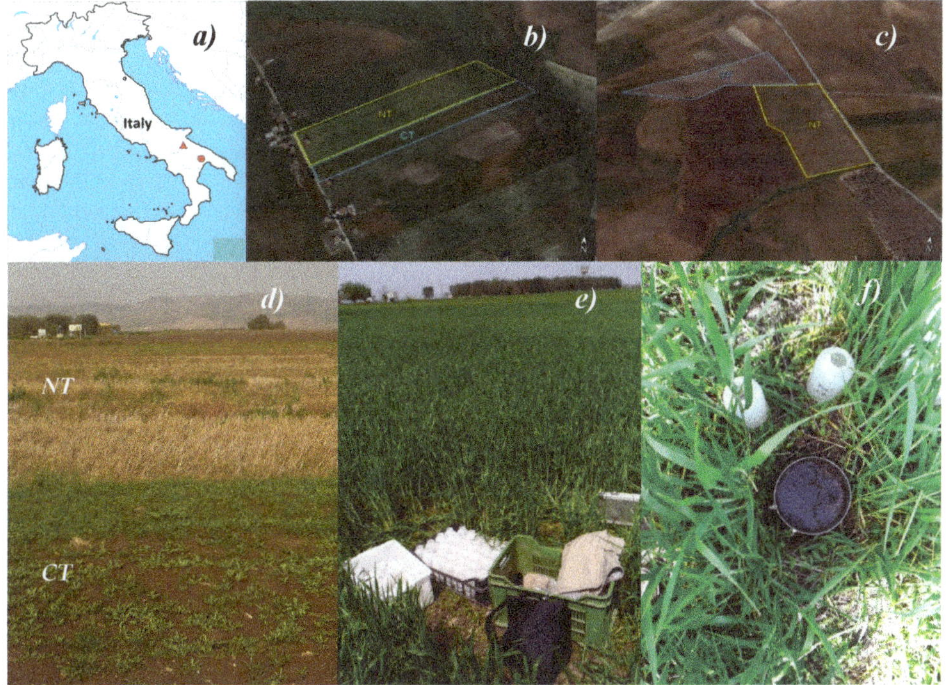

Figure 1. Geographical location of the studied sites (**a**) of Gravina (**b**) and Candela (**c**) (point and triangle respectively on the map); view of Gravina site (early November) under no-tillage, NT and conventional tillage, CT (**d**); view of the field site in April (**e**) and detail of the cylinder used for Beerkan tests (**f**) under NT in Gravina.

2.2. Soil Sampling and Measurements

Eleven to thirteen BEST experiments were randomly carried out in the spring season (between the end of April and the beginning of May, at heading-flowering stages) of 2017 to obtain soil hydraulic characterization (i.e., saturated hydraulic conductivity, K_s and the hydraulic conductivity function) of each experimental area (i.e., CT and NT of Gravina and Candela sites). In particular, as is common for BEST application, a falling head infiltration experiment of Beerkan type was carried out at each sampling point using a metal ring with a cutting edge. In detail, Beerkan tests were performed using a 15-cm-inner diameter cylinder, inserted to a depth of about 1 cm to avoid lateral loss of ponded water [32]; a known volume of water (200 mL) was repeatedly poured into the cylinder, establishing a height of water of 1.1 cm, and the time needed for complete infiltration was logged. The procedure was repeated for a total of 15 water volumes, and experimental cumulative infiltration, I(t), and infiltration rate q(t), were thus deduced.

For each infiltration point, two undisturbed soil cores (0.05 m in height by 0.05 m in diameter) were collected at the 0 to 0.05 m and 0.05 to 0.10 m depths near the cylinder (e.g., <10 cm) to determine the soil water content at the beginning of an infiltration experiment, θ_i, and the soil dry bulk density, BD. Following a procedure commonly used for BEST application, a unique value of both θ_i and BD was determined for each sampling point by averaging the values measured at the two depths [36,40,47]. A disturbed soil sample (0–0.10 m depth) was collected at each sampling point to determine both the soil particle size distribution (PSD) and the total organic carbon (TOC) content. The clay, silt, and sand percentages were determined according to the USDA standards [48]. The TOC content was quantified through dry combustion using a TOC Vario Select analyzer (Elementar, Germany), which conducts a catalytic combustion by high temperatures in an air environment [49].

For each soil management strategy (CT and NT) and farm (Gravina and Candela) considered in this investigation, from seven to eleven undisturbed soil cores (8 cm-inner-diameter by 5 cm-height) were randomly collected at the 0 to 0.10 m depth to determine some soil water retention values at high pressure heads ($h \geq -100$ cm); a corresponding number of disturbed soil samples were also collected to determine other water retention curve values at low pressure heads ($h \leq -330$ cm). In detail, volumetric water retention, θ, data were determined on each undisturbed soil core by a hanging water column apparatus for pressure head, h, values ranging from -5 to -100 cm, and on repacked soil cores by pressure plate method for h values ranging from -330 to $-15{,}300$ cm [50]. Finally, the soil water retention function was obtained fitting the experimental data with the van Genuchten [51] model, as it is common in the parameterization procedures [50].

At the end of the crop season, each farmer provided the total grain yields (q ha^{-1}) obtained in the fields under CT and NT. Additional measurements on the crop growth, i.e., leaf area index (*LAI*) and aboveground biomass, were also carried out once at Gravina site at the end of stem elongation stage. LAI was directly measured by LAI-2000 Plant Canopy Analyzer (Li-Cor).

2.3. Application of BEST Procedure and Estimation of Soil Porosity Indicators

The BEST procedure [32] was applied in order to estimate the hydraulic conductivity function, i.e., the relationship between the hydraulic conductivity, K, and the volumetric soil water content, θ. according to the Brooks and Corey [52] model:

$$\frac{K(\theta)}{K_s} = \left(\frac{\theta - \theta_r}{\theta_s - \theta_r} \right)^{\eta} \tag{1a}$$

$$\text{with } \eta = \frac{2}{mn} + 2 + p \tag{1b}$$

where θ [L^3·L^{-3}] is the volumetric soil water content; K [L T^{-1}] is the soil hydraulic conductivity; n (>2), m, and η are shape parameters; p is a tortuosity parameter set equal to 1 following Burdine's condition [53]; field saturated soil water content, θ_s [L^3·L^{-3}], residual soil water content, θ_r [L^3·L^{-3}], and field saturated hydraulic conductivity, K_s [L T^{-1}], are scale parameters. In BEST, θ_r is assumed to be zero. Both shape and scale parameters must be estimated.

The texture-dependent shape parameters of the model were estimated from the particle-size analysis and *BD* assuming a shape similarity between the *PSD* and the water retention curve [54], while structure dependent scale parameters are estimated by a three-dimensional (3D) field infiltration experiment at zero pressure head, using the two-term transient infiltration equation by Haverkamp et al. [55]. A simplified approach was adopted in this investigation in order to apply the BEST-procedure in the simplest way possible (as the choices made below about the infiltration constants). In detail, basic soil texture fractions, i.e., sand, sa (%, USDA classification) and clay, cl (%) content were used to estimate n parameter, using a specifically developed pedotransfer function, PTF, by Minasny and McBratney [56]:

$$n = 2.18 + 0.11 [48.087 - 44.954 S(x_1) - 1.023 S(x_2) - 3.896 S(x_3)] \tag{2a}$$

where:

$$x_1 = 24.547 - 0.238 sa - 0.082 cl \tag{2b}$$

$$x_2 = -3.569 + 0.081 sa \tag{2c}$$

$$x_3 = 0.694 - 0.024 sa + 0.048 cl \tag{2d}$$

$$S(x) = \frac{1}{1 + \exp(-x)} \tag{2e}$$

Generally θ_s is estimated from *BD*, assuming a soil particle density of 2.65 g cm^{-3} [34,57], while K_s and the soil sorptivity, *S*, are estimated by fitting the explicit three-dimensional infiltration model proposed by Haverkamp et al. [55] to the transient cumulative infiltration data.

Structure-dependent scale parameters were estimated by a falling head infiltration experiment of Beerkan type. In detail, the 3D cumulative infiltration, *I* (L), and the infiltration rate, *i* (L T^{-1}), can be approached by the following explicit transient (Equations (6a) and (6b)) and steady-state (Equations (6c) and (6d)) expansions [32,55]:

$$I(t) = S\sqrt{t} + \left(AS^2 + BK_s\right)t \tag{3a}$$

$$i(t)\frac{S}{2\sqrt{t}} + \left(AS^2 + BK_s\right) \tag{3b}$$

$$I_{+\infty}(t) = \left(AS^2 + K_s\right)t + C\frac{S^2}{K_s} \tag{3c}$$

$$i_s = AS^2 + K_s \tag{3d}$$

where i_s (L T^{-1}) is the steady-state infiltration rate, *A* (L^{-1}), *B* and *C* are constants defined taking into account initial conditions as [55]:

$$A = \frac{\gamma}{r(\theta_s - \theta_i)} \tag{4a}$$

$$B = \frac{2-\beta}{3}\left[1 - \left(\frac{\theta_i}{\theta_s}\right)^{\eta}\right] + \left(\frac{\theta_i}{\theta_s}\right)^{\eta} \tag{4b}$$

$$C = \frac{1}{2\left[1 - \left(\frac{\theta_i}{\theta_s}\right)^{\eta}\right](1-\beta)}\ln\left(\frac{1}{\beta}\right) \tag{4c}$$

where *r* (L) is the radius of the disk source, γ and β are the infiltration constants [55].

Finally, the retention curve scale parameter, h_g, was estimated by the relationship:

$$h_g = -\frac{S^2}{c_p(\theta_s - \theta_i)\left[1 - (\theta_i/\theta_s)^{\eta}\right]K_s} \tag{5}$$

where θ_i [L^3 L^{-3}] is the soil water content at the time of sampling, and c_p is a coefficient dependent on *n*, *m*, and *h* according to Lassabatère et al. [32] (Equation (6b)).

Three alternative algorithms, i.e., BEST-slope [32], BEST-intercept [58], and BEST-steady [59] were applied in this investigation, differing only in the way the infiltration model is fitted to the experimental data [59]. However, in accordance with the reasoning given in the data analysis section, only one of the applied algorithms was selected for comparison purposes (CT vs. NT). Finally, the infiltration constants, β and γ, were fixed at the reference values of literature ($\beta = 0.6$ and $\gamma = 0.75$) [32]. An updated version of the workbook by Di Prima et al. [60] that considers all the available algorithms was used to analyse the experimental cumulative infiltrations, *I(t)*, by the three BEST-algorithms.

Two soil water dynamic based indicators were considered in this investigation to compare alternative soil management strategies (CT and NT), namely the flow-weighted mean pore size λ_m [L] and the number of hydraulically active pores per unit area $C_{\lambda m}$ [L^{-2}] [36,42,43].

The flow-weighted mean pore size λ_m was calculated by the following relationship according to Mubarak et al. [42]:

$$\lambda_m = \frac{\sigma_w}{\rho_w g \alpha h} \tag{6}$$

where σ_w [M T^{-2}] is the surface tension of water, ρ_w [M L^{-3}] is the density of water, *g* [L T^{-2}] is the acceleration due to gravity, and αh [L] is the capillary length, calculated from the scale parameter h_g as $\alpha h = - h_g$ [43,54].

The value of $C_{\lambda m}$ was calculated according to Watson and Luxmoore [61]:

$$C_{\lambda m} = \frac{8\mu K_s}{\rho_w g \pi \lambda_m{}^4} \tag{7}$$

where μ [M L^{-1}T^{-}1] stands for the dynamic viscosity of water.

With regard to the aforementioned indicators, it may be useful to underline that λ_m represents the pore size that contributes to water infiltration under ponding condition, thus providing an estimation of the relative importance of gravity and capillary forces on the total flow; consequently, $C_{\lambda m}$ provides an estimation of the number of soil pores, for unit of area, having a mean size equal to λ_m [46].

2.4. Capacitive-Based Indicators

For each soil water retention curve determined in the lab, four capacity-based indicators were estimated, and the impact of CT and NT on soil water conservation was assessed with reference to the optimal values of literature [50,62]. In detail, estimated soil water retention values corresponding to soil pressure head, $h = 0$, 10, 100, and 15,300 cm were used to calculate the macroporosity (P_{mac}), air capacity (AC), relative field capacity (RFC), and plant available water capacity ($PAWC$), as summarized in Table 1. Therefore, the impact of soil management strategy was established as the difference between mean measured and optimal values (Table 1).

Table 1. Selected capacity-based indicators and corresponding optimal ranges or critical limits according to Reynolds et al. [62] and Castellini et al. [50].

Soil Physical Indicator	Reference Value	Mean Optimal Value
Macroporosity, P_{mac} (cm^3 cm^{-3}) $P_{mac} = \theta_s - \theta_m$	$0.04 \leq P_{mac} \leq 0.10$ optimal $P_{mac} < 0.04$ aeration limited soil $P_{mac} > 0.10$ water limited soil	$P_{mac} = 0.07$
Air capacity, AC (cm^3 cm^{-3}) $AC = \theta_s - \theta_{FC}$	$0.10 \leq AC \leq 0.26$ optimal $AC < 0.10$ aeration limited soil $AC > 0.26$ water limited soil	$AC = 0.18$
Relative field capacity, RFC (dimensionless) $RFC = \frac{\theta_{FC}}{\theta_s} = \left[1 - \left(\frac{AC}{\theta_s}\right)\right] = \left(\frac{PAWC + \theta_{PWP}}{\theta_s}\right)$	$0.6 \leq RFC \leq 0.7$ optimal $RFC < 0.6$ water limited soil $RFC > 0.7$ aeration limited soil	$RFC = 0.65$
Plant available water capacity, $PAWC$ (cm^3 cm^{-3}) $PAWC = \theta_{FC} - \theta_{PWP}$	$PAWC \geq 0.20$ ideal $0.15 \leq PAWC < 0.20$ good $0.10 \leq PAWC < 0.15$ limited $PAWC < 0.10$ poor	$PAWC = 0.20$

θ_s = saturated soil water content; θ_m = water content of the soil matrix ($h = -10$ cm); θ_{FC} = soil water content at the field capacity ($h = -100$ cm); θ_{PWP} = soil water content at the permanent wilting point ($h = -15{,}300$ cm).

2.5. Data Analysis

For each variable considered in this investigation (cl, si, sa, θ_i, θ_s, BD, TOC, h_g, S, K_s, λ_m, $C_{\lambda m}$, P_{mac}, AC, RFC, $PAWC$), a given dataset was summarized by calculating the mean and the associated coefficient of variation (CV). Arithmetic means were calculated for cl, si, sa, θ_i, θ_s, BD, TOC, λ_m, $C_{\lambda m}$, P_{mac}, AC, RFC, $PAWC$, since they were assumed to be normally distributed, as commonly suggested in the literature [36,62], while h_g and soil variables directly linked to the infiltration experiment, S and K_s, were assumed to be logarithmically distributed [63,64], and geometric means and associated CVs were calculated using the appropriate lognormal equations [65,66].

The impact of alternative soil management strategies on soil properties, i.e. CT and NT, was assessed in terms of cumulative infiltration and infiltration rate ($I(t)$ and $q(t)$), because it is expected that such different soil management can have a significant influence on the hydrodynamic properties of the soil. For the aforementioned soil systems, BD and TOC were considered to give account of possible effects on soil compaction or soil structure, respectively. Parameters directly obtained by

BEST (i.e., scale parameter, soil sorptivity, and saturated hydraulic conductivity, h_g, S, K_s) or those estimated from BEST (dimension and number of hydraulically active pores, λ_m and $C_{\lambda m}$) were also considered because the former represent fundamental hydraulic properties, and the latter are useful indicators for comparison purposes [41,67]. The hydraulic functions, that is, the hydraulic conductivity function, $K(\theta)$, and the water retention curve, $\theta(h)$, were respectively estimated or measured to show any differences in the saturated/unsaturated soil domain. Finally, capacity-based indicators (P_{mac}, AC, RFC, $PAWC$) were selected to assess the optimal balance between air and water into the soil [50].

Regarding the BEST application, we would like to point out that -intercept or -steady provide the same h_g estimation [34] and, consequently, a unique λ_m value for these algorithms can be obtained, whilst different estimations of S and K_s values can be obtained, according to the applied BEST-algorithm. Therefore, the impact of BEST-algorithms on S and K_s estimations was evaluated, but only one was selected to compare soil management strategies. For this purpose, as BEST-algorithms mainly consider the transient or the steady-state phase of the infiltration process [59], an analysis of the cumulative infiltration curves was carried out to establish the possible non-attainment of steady flow. Then, the equilibration time, t_s, namely, the time needed to reach steady-state conditions, the infiltrated depth at the equilibration time, $I(t_s)$, and the total duration, for the infiltration runs were determined and used to select the most appropriate algorithm for comparison purposes (CT vs. NT). Specifically, the t_s value was determined according to Angulo-Jaramillo et al. [68] as the first value for which:

$$\frac{|I(t) - I_{reg}(t)|}{I(t)} \cdot 100 \leq E \tag{8}$$

where $I(t)$ is the cumulative infiltration during time t, $I_{reg}(t)$ is the cumulative infiltration estimated from the regression analysis of the $I(t)$ vs. t plot, while E is the criterion for establishing the onset of linearity. Equation (8) is applied starting from $t = 0$ and progressively excluding the first data points until a value of $E \leq 2\%$ was reached [68].

Finally, the statistical significance between CT and NT or between soil properties was performed according to a two-tailed t-test ($P = 0.05$).

3. Results

3.1. Basic Soil Properties

According to the USDA classification, the texture of the upper layers of the soil (0–10 cm) was always clay in Candela (i.e., both for CT and NT) while it was both clay-loam or clay in Gravina site, depending on the soil management (CT or NT, respectively) (Table 2). Therefore, if compared to CT, the long-term NT showed an increase in the clay and sand fractions of the upper layer of Gravina soil. *TOC*, *BD*, θ_s, and θ_i, values of investigated soils were reported in Table 3. In detail, similar *TOC* values were generally detected, except for Candela where CT showed significantly lower values. Significantly higher values of *BD* were observed only under NT of Gravina as they were 1.15 times higher than CT, while no significant differences were detected in Candela. Since in BEST the saturated soil water content, θ_s, is estimated from the value of *BD*, significantly lower values of θ_s were also detected in the former case on the no-tillage plot (Table 3). Finally, relatively comparable θ_i values were detected in Candela, while higher differences between soil management strategies were detected in Gravina (Table 3). The impact of θ_i values on soil hydraulic characterization is reported in the following section.

Table 2. Clay (*cl*), silt (*si*), and sand (*sa*) contents (0–0.1-m depth) according to the USDA classification for sites (Gravina and Candela) and soil managements (conventional tillage, CT and no-tillage, NT) considered.

Site (Soil Management)	cl (%)	si (%)	sa (%)	USDA
Gravina (CT)	34.9	34.7	30.4	clay-loam
Gravina (NT)	42.9	21.5	35.5	clay
Candela (CT)	50.3	21.8	27.8	clay
Candela (NT)	55.0	24.5	20.5	clay

Table 3. Mean and associated coefficient of variation (CV, in parenthesis) of initial and saturated volumetric soil water content, θ_i and θ_s (cm^3/cm^3), bulk density, *BD* (g/cm^3), and total organic carbon, *TOC* (%).

	TOC	*BD*	θ_s	θ_i
Gravina CT	1.518 a	1.1396 a	0.5699 a	0.1781 a *
	(7.6)	(9.7)	(9.7)	(9.3)
Gravina NT	1.614 a	1.3007 b	0.5092 b	0.2156 a
	(13.5)	(7.4)	(7.4)	(13.8)
Candela CT	1.248 a	1.2876 a	0.5141 a	0.2489 a
	(4.7)	(16.2)	(16.2)	(23.3)
Candela NT	1.514 b	1.3299 a	0.4982 a	0.2807 a
	(14.1)	(15.0)	(15.0)	(26.1)

* For a given experimental site (i.e., Gravina or Candela), mean values followed by the same letter are not significantly different according to a two-tailed *t*-test ($P = 0.05$).

3.2. Comparison and Choice among BEST-Algorithms

The volumetric soil water content at the time of experiments, θ_i, has exceeded the suggested condition by Lassabatere et al. [32] according to which BEST experiments should be carried out in relatively dry soil conditions (i.e., $\theta_i \leq 0.25\,\theta_s$); this soil condition should ensure a greater success rate to obtain, using all the available BEST-algorithms (slope, intercept and steady), physically plausible estimates of soil hydraulic properties (i.e., positive values of K_s and S) [59]. Relatively high θ_i values, in fact, strongly affected Candela sites results, as relatively low percentages of success were obtained using BEST-slope under CT (1/12; 8% of success) and under NT (27%); relatively higher success rates however were obtained with the same algorithm at Gravina where BEST-slope provided about 67% and 69% of successes for CT and NT, respectively. This success deficit of BEST-slope was not completely surprising as several references of literature suggest such a response and Castellini et al. [36], for example, recently discussed this topic for most of the available investigations. On the contrary, since the other two BEST-algorithms, i.e. intercept and steady, always provided analyzable results in 100% of considered cases, they were considered to establish a comparison between conventional and alternative soil management strategies (CT vs. NT).

Table 4 shows the comparison results of S and K_s carried out by the alternative BEST-algorithms. Application of -intercept or -steady showed differences for each considered variable since discrepancies by a factor of 1.1–1.3 for S and 1.2–1.7 for K_s, were detected. According to a paired two-tailed *t*-test ($p = 0.05$), the differences between BEST-algorithms were always significant (Table 4). Results of this investigation are in line with the main references of literature. Bagarello et al. [59], for example, using a large sample size (N = 401), detected lower values of S and K_s, using BEST-steady as compared to BEST-intercept; observed K_s discrepancies between algorithms were at most equal to a factor of 1.5 (1.1 as a mean), and can be considered practically negligible for many agronomic applications [67,69]. Consequently, an analysis of the cumulative infiltration curves was made to verify if the steady flow conditions were always reached during field experiments, and it was used as a criterion for choosing the most suitable calculation algorithm, and to compare selected soil treatments.

Table 4. Geometric mean (GM) and associated coefficient of variation (CV%) of soil sorptivity (*S*) and saturated hydraulic conductivity (K_s) obtained for each BEST-algorithm (Intercept and Steady), experimental site (Gravina and Candela) and soil management (conventional tillage, CT and no-tillage, NT).

Variable	S (mm s^{-1})				K_s (mm h^{-1})			
Algorithm	Intercept		Steady		Intercept		Steady	
Statistic	GM	CV	GM	CV	GM	CV	GM	CV
Gravina CT	1.429 aA	38.3	1.277 bA	35.7	96.932 aA	53.8	77.496 bA	53.7
Gravina NT	1.589 aA	53.7	1.460 bA	53.5	136.112 aA	113.0	114.945 bA	115.4
Candela CT	2.759 aA	41.9	2.130 bA	44.7	230.167 aA	104.5	137.105 bA	121.3
Candela NT	2.771 aA	31.2	2.299 bA	32.4	297.889 aA	85.6	205.223 bA	100.4

* For a given variable (*S*, K_s) and combination (i.e., experimental site–soil management), mean values of BEST-algorithm followed by the same lowercase letter are not significantly different according to a paired two-tailed *t*-test (*P* = 0.05). For a given variable and BEST-algorithm, mean values of the different combinations (i.e., Gravina, CT vs. NT) followed by the same capital letter are not significantly different according to a *two tailed* t-test (*P* = 0.05).

Cumulative infiltrations showed the expected shapes with a concave part (concavity downwards) corresponding to the transient state and a linear part at the end of the curves related to the steady-state (Figure 2). Relatively flat cumulative infiltrations, i.e., relatively lower infiltration rates, were generally observed in Gravina more than in Candela (Figure 2a,b), showing for the former site relatively higher total time of experiments (i.e., infiltration experiments that lasted up to 2.5 h). However, a visual inspection of *I*(*t*) relationships suggests no substantial difference between soil management strategies. This finding was also showed in terms of infiltration rates (Figure 2c,d); for a given infiltration time, differences in mean values of infiltration rate between sites did not exceed an order of magnitude.

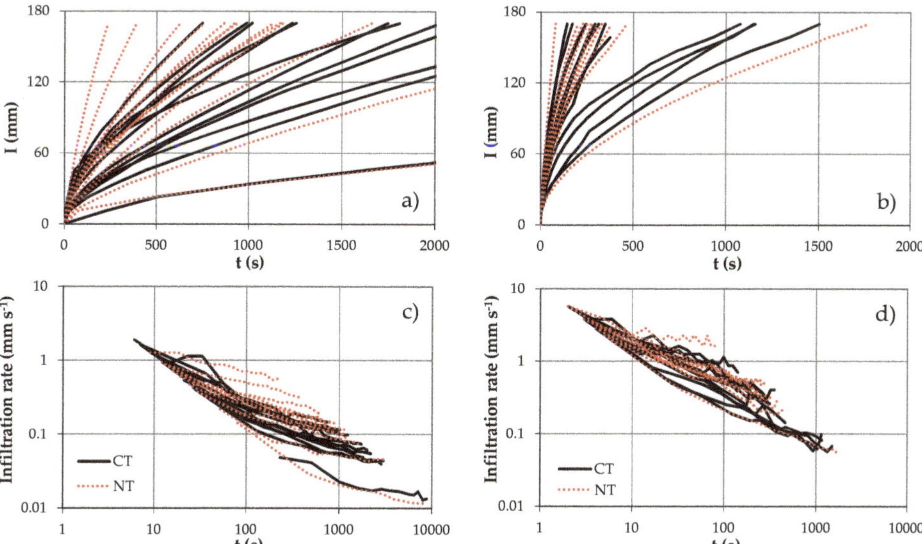

Figure 2. Cumulative infiltrations (**a,b**) and infiltration rates (**c,d**) at the study sites, Gravina (on the left) and Candela (on the right), under conventional tillage (CT) and no-tillage (NT). Note that *I*(*t*) relationship of Gravina was truncated at t = 2000 s; lost information can be assumed from Table 4.

Table 5 summarizes the results of steady-state flow analysis. Equilibration time, t_s, i.e. the time needed to reach steady-state conditions or, similarly, the duration of the transient phase, was different among soils ranging by two orders of magnitude (from a minimum of 65 s to a maximum of 6415 s). In particular, for a given site, higher mean t_s values were obtained under CT rather than NT, with differences that were equal to a factor of 1.6 or 1.3, respectively, for Gravina or Candela. The infiltrated depth at the equilibration time was 70% of the total infiltrated depth (I_{tot} = 170 mm) for Gravina and 80% of I_{tot} for Candela (Table 5). This suggests that clear steady-state conditions were always reached before the end of infiltration experiments, and both BEST-intercept and BEST-steady could be used considering the last data points, according to the specific infiltration curve. Consequently, as steady-state conditions were clearly highlighted, the comparison between soil management in terms of S, K_s, and $C_{\lambda m}$, and in terms of h_g and λ_m, will be referred only to the BEST-steady algorithm.

Table 5. Minimum (Min), maximum (Max), mean, and coefficient of variation (CV%), of the equilibration time, t_s (s), infiltrated depth at the equilibration time, $I(t_s)$ (mm), and total duration, t_{end} (s) for the infiltration runs.

Variable	Site	Statistics			
		Min	Max	Mean	CV%
t_s	Gravina CT	458	6415	1417	113.9
	Gravina NT	153	3962	868	111.0
	Candela CT	87	811	345	76.9
	Candela NT	65	1182	259	119.3
$I(t_s)$	Gravina CT	102	136	119	7.6
	Gravina NT	68	136	118	17.8
	Candela CT	102	147	133	9.7
	Candela NT	102	147	132	10.4
t_{end}	Gravina CT	750	8795	2311	93.8
	Gravina NT	234	8750	1683	133.6
	Candela CT	141	1501	590	81.8
	Candela NT	79	1775	414	111.6

3.3. Comparison between CT and NT

Comparison between CT and NT showed relatively low differences between the soil management strategies, since a general equivalence in terms of hydrodynamic soil properties was detected. In fact, according to the results discussed before and regardless of the site considered, no significant difference between CT and NT was detected in terms of S (differences not higher than a factor of 1.1) and K_s (factor of 1.5) (Table 4). This finding suggests that, despite the substantial differences between farms about the time elapsed in the conversion from CT to NT (six to twenty-four years), hydrodynamic soil properties under CT and NT were comparable both in Candela and Gravina. The negligible differences between the soil management strategies are clearly shown in Figure 3 since, neglecting some points of the mean cumulative water infiltration curve (that is, the first or last two-three data points) for which sorptivity or some discrepancies in the mean flows may have occurred, a ratio of Δt-CT/Δt-NT practically equal to one was observed for more than half (up to about the ninth applied water volume) of the infiltration process. Therefore, detected differences between Δt-CT and Δt-NT were also expressive of a substantial equivalence of the compared soil management strategies over infiltration time.

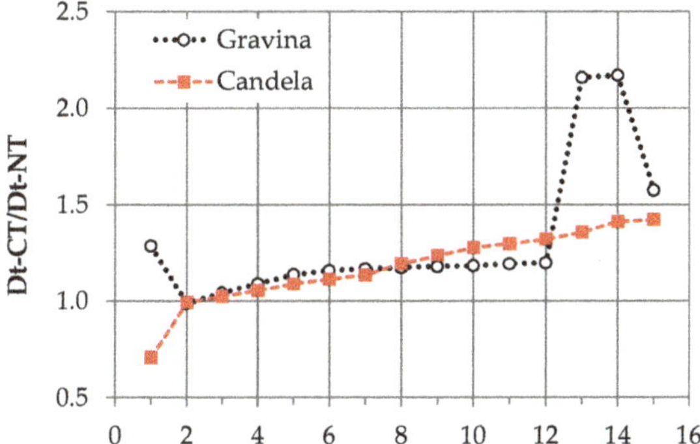

Figure 3. Ratio between the mean infiltration time for water infiltration in CT and NT during the Beerkan runs against the number of the applied volumes of water.

Figure 4 shows the interquartile range of the water retention curve scale parameter, h_g, of the flow-weighted mean pore size, λ_m and of the number of hydraulically active pores, $C_{\lambda m}$. Lower mean values of h_g (higher in absolute value) were generally detected at Candela than Gravina (173–155 mm and 79–98 mm, respectively). For a given soil management strategy, Candela showed a higher variation range by a factor of 1.5–1.8 as compared with Gravina. These findings support bulk density values, on average higher in Candela than in Gravina. However, for a given site, no significant differences were detected between CT and NT (Figure 4). Accordingly, the flow-weighted mean pores size, λ_m, was higher in Gravina (0.097–0.077 mm) than in Candela (0.044–0.052 mm), respectively under CT and NT, but no statistical significance between treatments was detected (Figure 4). Differences significantly higher were instead detected in terms of $C_{\lambda m}$ in Gravina, since NT showed a higher number of hydraulically active pores (by a factor of 3.1) as compared to CT (Figure 4). This result is quite surprising as, although the expected inverse relationship between λ_m and $C_{\lambda m}$ was detected, i.e., few pores larger or many smaller ones, a long-term undisturbed soil would have let suppose the first condition, as reported for marginal soils with natural vegetation [46]. However, $C_{\lambda m}$ depends on both λ_m and K_s and, although discrepancies in soil texture classes were detected (i.e., higher percentages of clay and sand fractions were detected under NT), results of this investigation suggests that a long-term no-tilled soil with relatively small pores, but more numerous and probably well-connected, may be more conductive as compared to CT; this can mitigate the increase of bulk density in long-term no-tilled soils. On the other hand, Candela showed only negligible differences on number of conductive pores, and NT findings do not seem to be all consistent with each other (i.e., larger pore size corresponded to a larger number of conductive pores, or higher BD values did not determine higher values of h_g; the latter was also extremely variable under NT). However, although no statistical significance was found for the mentioned soil properties, observed inconsistencies probably can be considered as a signal that soil conservation practices, repeated for a relatively short time of six years, do not allow to detect clear relationships between soil properties. Moreover, even when soil data of two farms were poled together to give account only for the effect of soil management (Gravina + Candela), the comparison in terms of both λ_m and $C_{\lambda m}$ provided consistent findings, since CT and NT were not different according to a two-tailed *t*-test (Figure 5), and the expected decreasing relationship between these two variables was verified. As a consequence, we are inclined to suppose that a soil compaction under Candela-CT may have occurred, as evidenced by the relatively low variability (and concomitant extreme outlier) of λ_m (Figure 4). This reasoning finds support in a similar soil behavior showed by Souza et al. [43], since an increasing relationship between λ_m and $C_{\lambda m}$ was detected due to the formation of superficial crust that locally may have reduced the size of conductive pores, as compared to non-crusted soils.

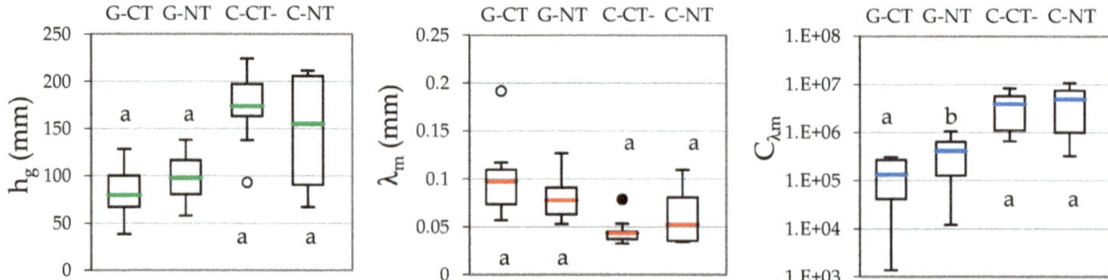

Figure 4. Mean values and interquartile range (min, max, 1st and 3rd quartile) of retention curve scale parameter, h_g (in absolute value), flow-weighted mean pore size, λ_m, and number of hydraulically active pores per unit area ($C_{\lambda m}$) obtained with BEST-steady. Open circles represent outliers (closed circles extreme outliers). For a given site, Gravina and Candela (G and C), soil management strategies (CT or NT) followed by the same letter are not significantly different according to a two-tailed *t*-test ($p = 0.05$).

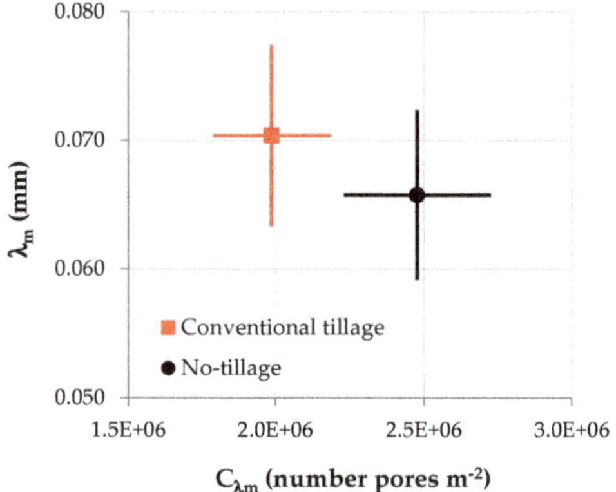

Figure 5. Comparison between of CT and NT in terms of mean values of flow-weighted mean pore size, λ_m, and number of hydraulically active pores per unit area, $C_{\lambda m}$ (i.e., Gravina + Candela dataset). Bars represent $\pm$ 10% deviation on the mean values.

The comparison in terms of hydraulic functions is reported in Figure 6. For two selected θ values (i.e., 0.4 and 0.7), differences in the $K(\theta)$ values between the soil management strategies were within an order of magnitude. Accordingly, similar results for shape parameters of BEST were detected ($m = 0.05 - 0.06$, $n = 2.11 - 2.13$, and $\eta = 21 - 19$, respectively for Candela and Gravina). In particular, higher K differences were generally detected for Gravina (i.e., differences by a factor of 2.9–10.3 at $\theta = 0.4$ and 0.7, respectively) than Candela (3.2–4.3). However, slightly higher differences were detected in the former case close to water saturation as compared to unsaturated ones, suggesting a more efficient conductive system under relatively higher moisture conditions (Figure 6).

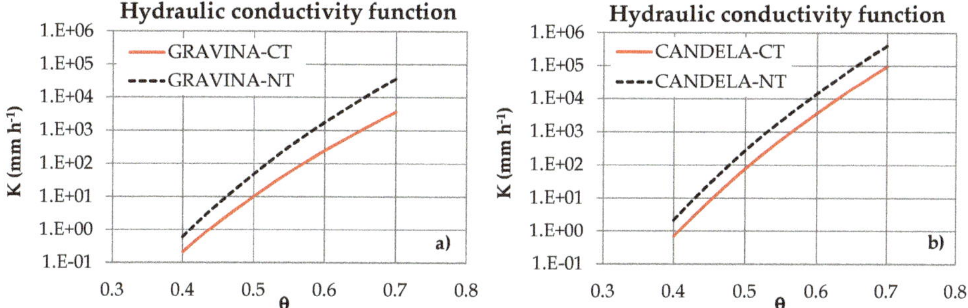

Figure 6. Hydraulic conductivity functions obtained with BEST-steady at Gravina (**a**) and Candela (**b**) sites.

The comparison between CT and NT based on measured soil water retention is depicted in Figure 7. In general, measured soil water retention data showed a relatively low variability as coefficients of variation (CVs) of θ values ranged from about 15–20% to 2–3%, decreasing as the potential decreases (i.e., -5 to $-15{,}300$ cm). As shown, noticeable differences in soil water retention were detected only in the Candela site, as relatively low differences were detected between CT and NT in Gravina only close to water saturation (e.g., 0.06 cm^3 cm^{-3} at $h = -5$ cm) (Figure 7). In other words, the water retention curve of Candela-CT highlighted the typical features of a tilled soil, namely: (i) relatively lower bulk density value, and (ii) relatively low value of the pressure head at the inflection point (i.e., 10 cm against 25–26 cm for the remaining ones). This behavior is adequately represented in Figure 8, where the comparison among pore size distributions clearly shows a higher modal diameter for Candela-CT (298 µm) as compared to other sites (about 115 µm); for Candela-CT, this resulted in lower degrees of saturation at the inflection point of the water retention curve [62], by 4% as compared to Candela-NT, or by 15%–29% as compared to Gravina-CT or Gravina-NT. Therefore, as compared to hydrodynamic soil properties, a more evident impact of soil management strategy (CT or NT) can be quantified in terms of capacity-based indicators.

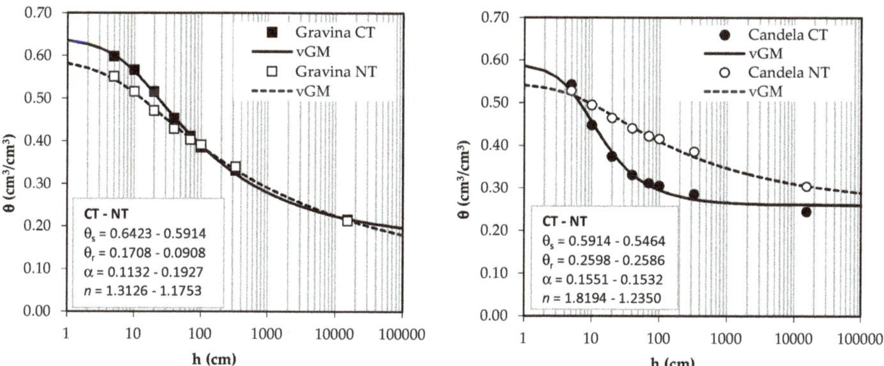

Figure 7. Measured (points) and modelled (lines) soil water retention data (van Genuchten model, vGM) for conventional tillage (CT) and no-tillage (NT) of Gravina (on the left) and Candela (on the right) sites. Parameters of vGM (θ_s, θ_r, α, and n) are also reported ($\alpha = 1/$cm).

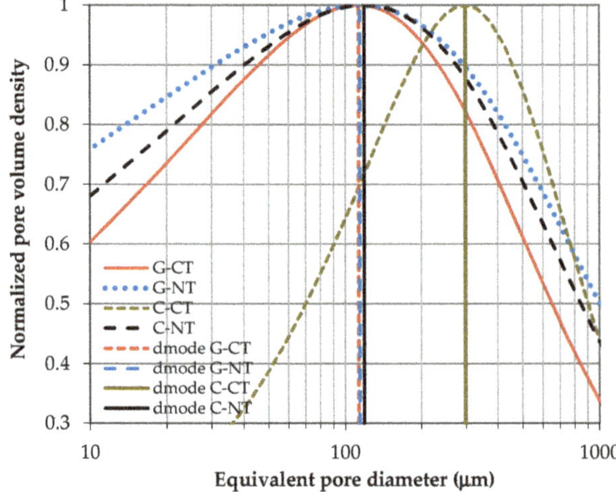

Figure 8. Normalized pore volume distributions and the corresponding modal diameters (d_{mod}) of the four site–soil management combinations (G, Gravina; C, Candela; CT, conventional tillage; NT, no-tillage), calculated according to Reynolds et al. [62].

In accordance with the literature assessing optimal physical conditions of agricultural soils (Table 1), a better soil physical quality was observed for Gravina rather than for Candela, since optimal values of macroporosity, air capacity, and relative field capacity, as well as good levels of plant available water, were always detected both under CT and NT (Table 6). Conversely, the Candela site showed some discrepancies between soil management as nonoptimal soil conditions were always detected under CT, while optimal values of macroporosity or of air capacity were detected under NT. As summarized by *RFC*, opposite conditions generally were identified between CT and NT, as too porous (*RFC* < 0.6) or too compact (*RFC* > 0.7) soil conditions were detected in the former and in the latter treatment of Candela site, respectively (Table 6). For a given soil management strategy, CT or NT, the impact of the time that has elapsed since the conversion to NT on capacity-based indicators (24 and 6 years, for respectively for Gravina and Candela) was shown in Figure 9. For a given soil indicator (i.e., P_{mac}, *AC*, *RFC*, *PAWC*), lower discrepancies between CT and NT than optimal mean values were always detected under the long-term conversion of Gravina (Figure 9). For this site, moreover, no-tillage showed always results closer to the optimum (in terms of AC and RFC), or virtually equivalent (P_{mac} and *PAWC*), as compared to conventional tillage, reinforcing the hypothesis that sufficiently long periods of time positively affects the soil physical properties' stabilization.

Table 6. Mean values of capacity-based indicators measured at Gravina (G) and Candela (C) sites under conventional tillage (CT) and no-tillage (NT).

Site-Soil Management	P_{mac}	AC	RFC	PAWC
G-CT	0.080	0.253	0.606	0.173
G-NT	0.079	0.204	0.655	0.173
	=	=	=	=
C-CT	**0.136**	**0.297**	**0.498**	**0.035**
C-NT	0.049	0.137	**0.749**	0.104
	≠	≠	=	=

P_{mac} = macroporosity (cm^3 cm^{-3}); *AC* = air capacity (cm^3 cm^{-3}); *RFC* = relative field capacity (-); plant available water capacity = *PAWC* (cm^3 cm^{-3}). The symbols indicate that, for a given comparison, soil indicators provided an equal (=) or a different (≠) evaluation if compared to the reference values of Table 1; non-optimal values were highlighted in bold.

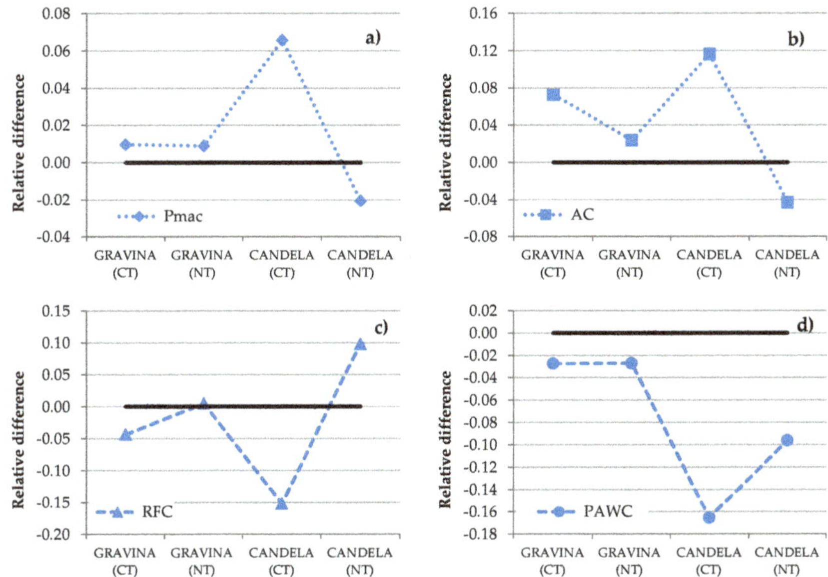

Figure 9. Relative differences between measured mean values and optimal mean values for macroporosity (**a**), air capacity (**b**), relative field capacity (**c**) and plant available water capacity (**d**), according to the Table 1.

LAI and biomass were higher under CT than NT at Gravina by a factor of 1.2 (Figure 10), but observed differences were not statistically significant. Therefore, lower grain yields by 10% were obtained under NT (3.0–2.7 t ha^{-1}, respectively for CT and NT). No information was reported for Candela as some crop damages were observed after germination due to extreme weather conditions; this resulted in noncomparable plant densities between the alternative soil management strategies.

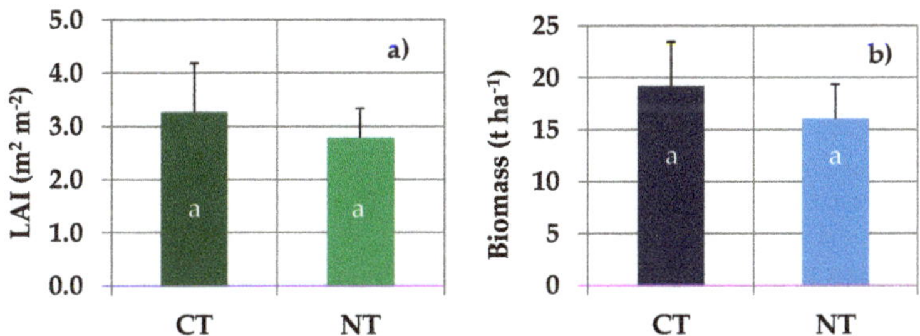

Figure 10. Leaf area index (LAI) (**a**) and fresh biomass (**b**) measured at Gravina site. Bars represent standard deviations. For each diagram, values followed by the same letter are not significantly different according to a two-tailed *t*-test (*p* = 0.05).

4. Discussion

Although the literature suggests that seasonal changes in soil physical properties may be caused by several physical factors, for example, the arrangement of soil particles, pore system configuration and compaction due to raindrop impact [40,70,71], and contrasting results may be summed according to selected investigation (among others, [70–73]), findings basically agree that soil tillage represents the main factor that can sharply reduce soil density and increase hydrodynamic properties [70]. The main effects of CT on soil physical properties run out during an annual or two-year crop cycle [70–73], as soil is gradually compacted due to natural reconsolidation by gravity and raindrop impact [70,74,75]. However, cases where the above-mentioned natural stabilization of soil properties are not uncommon as a natural soil compaction was reported even at the end of an annual growth cycle [67,73]. Past this time, if a soil is no longer tilled, the literature agrees that the above-mentioned natural stabilization of soil properties may occur after about five to six years [17,76].

In this investigation, the relative short/long term impact (i.e., six or twenty-four years) of selected soil management, CT or NT, was evaluated mainly in terms of soil bulk density, hydrodynamic soil properties, soil hydraulic conductivity functions, and capacity-based indicators, and discussed with reference to the literature. Contrasting results of NT effects on BD (i.e., NT≈CT or NT»CT) were generally reported in the literature [21,27] as several main factors (soil texture, organic matter, climatic environment, and agronomy) are involved in the soil reconsolidation processes. Similar conclusions were drawn by [77] for fine-textured soils (i.e., clay loam or silty clay loam) in a Mediterranean environment very similar to those considered in this investigation (i.e. northern Apulia), because both significant (by a factor of 1.22) and negligible (1.03) differences between NT and CT were observed. Our results showed that only long-term NT resulted in a significant increase in BD, but, according to the literature guidelines to establish acceptable levels of soil compaction (i.e., $BD \leq 1.30$ g cm^{-3}) above which yield loss could occur due to inadequate soil aeration [62], our results also demonstrate that 24 years of continuous no-tillage has not degraded the soil and acceptable levels of soil density remain. Therefore, according to Soane et al. [1], soil bulk density may be recommended as the main, and easily determinable, soil parameter to detect soil physical deterioration due to a continuous no-tillage.

Infiltration rate or hydraulic conductivity of NT soils is sometimes, but not always, found to be appreciably higher than in ploughed soils [1]. Vogeler et al. [21] detected higher K_s values under NT rather than CT in a German soil, while no differences were reported under unsaturated conditions by Moret and Arrúe [71]. Villarreal et al. [44], comparing CT and NT for a crop rotation including maize and soybean for the last 15 years, detected higher values of S under CT (a factor within the range 1.1–2.4) or variable differences on K_s depending on the sampling time, namely higher under CT before seeding or six leaf stage (a factor of 2.4–1.4, respectively), or higher under NT between the six-leaf stage and physiological maturity (a factor of 1.8–1.9). Also, even minor mean differences were observed in terms of soil porosity (total, macro, meso, or microporosity) [44]. Although this investigation represents a knowledge contribution for a Mediterranean agro-environment for which a lack of information exists, it provides specific information of short- and long-term effects of NT practice on hydrodynamic and hydrostatic soil properties. Although a seasonal variability in soil properties cannot be excluded a priori, since literature suggests prudence on this, the general result obtained in this study was a null effect of long-term NT on hydrodynamic soil properties. At least four main results can be summarized for the investigated fine textured-soils: (i) long-term no-tillage significantly increased BD but this does not result in a reduction in soil permeability; (ii) short-term (six years) no-tillage did not result in significant changes in selected soil properties, possibly because this was a transition phase; (iii) although short-term and long-term NT systems showed similar hydrodynamic soil properties, they highlighted different characteristics of the conductive pore system; (iv) a more conductive pores network, i.e., consisting of relatively smaller pores that are more numerous and probably better interconnected, has been identified under long-term NT compared to CT. Conversely, application of capacity-based indicators revealed differences between treatments suggesting that, when well-managed, NT systems can give rise to an optimal balance between the liquid and gaseous phases of the soil.

Overall, the literature suggests that a different NT response on grain yield reduction is expected, depending on the considered agro-environment climate, i.e. humid or dry [11]; comparable yields between CT and NT were highlighted only under dry climates, as higher grain yield reductions are generally expected under humid ones (i.e., not higher than 10%). However, findings by Van de Putte et al. [8] based on 47 European studies have shown that adoption of conservation agriculture may decrease the yield from 4.5 to 8.5%, also in drier climatic conditions. Results of our investigation, carried out in a relatively dry climatic environment, showed differences in grain yields not higher than 10%. Factors that plausibly have influenced this result were mainly meteorological, namely an adequate water supply during pre-germination and tilling stages, a higher water retention of tilled soils, and a corresponding greater plants density. Therefore, although our results seem to be more in agreement with the thesis by Pittelkow et al. [11], the findings of this investigation should be taken with relative caution since they account for only one year of investigation. In order to mitigate both the yield losses of no-tillage and an excessive, potentially harmful, increase in soil compaction, suggested actions can contemplate: (i) a periodic minimum tillage of the soil or (ii) regularly alternating CT and NT within a rotation of different crops [8,78].

Finally, as compared to the methods commonly applied for similar investigations, mainly point-measurements, BEST represents a parametric procedure which allows the soil hydraulic functions to be obtained. In particular, as discussed above, soil water availability determination represents key information for the economic sustainability assessment of NT compared to CT under rainfed environments and differences in soil water retention were shown both for long-term and short-term conversion. Therefore, application of physical-based models that make use of the Richards equation [79,80] enables an accurate assessment of the long-term no-tillage sustainability both under present [81] or under future climate change scenarios [82] in Mediterranean regions.

5. Conclusions

In this study, the comparison between alternative soil management strategies for durum wheat cultivation, CT and NT, has shown a substantial equivalence in terms of soil hydrodynamic properties in the two farms investigated. Although the conversion time from CT to NT was very dissimilar between farms, i.e. 24/6 years in Gravina/Candela farms, negligible discrepancies in hydrodynamic soil properties were detected. In fact, differences between CT and NT were equal with factors of 1.3 and 1.4 (cumulative infiltration or infiltration rate), 2.1 and 1.7 (soil sorptivity), 0.6 and 0.6 (saturated hydraulic conductivity), and 1.3 and 0.8 (conductive pore size and soil hydraulic functions) for the Gravina and Candela sites, respectively. Although sporadic significant differences between soil management strategies were detected (NT soil was either more compact or more organic), NT soil management does not seem to have worsened the main physical properties, i.e., bulk density or structure, of investigated clay soils. However, significant differences in the number of hydraulically active pores were detected for Gravina-NT as compared to Gravina-CT so that, for this farm, long-term no-tilled soil can be presented as a functionally similar hydrodynamic system to CT but characterized by a prevalence of a relatively higher number of small pores. However, as this result was not obtained in Candela and some inconsistencies in soil relationships for this farm were highlighted, we cannot rule out that the short time from the conversion, i.e., only six years, may have made more some results uncertain. Therefore, for considered agro-environment, a logical implication of this investigation is also that until six years of no-tillage, it is not possible to state that steady soil properties were reached and that the investigated soil of Candela was in the final period of transition. BEST-procedure alone has made it possible to highlight suspiciously low and non-variable λ_m values due to soil compaction under CT at Candela, while capacity-based indicators application, obtained from the measured retention curve, allowed verification of the agronomic and environmental sustainability of no-tillage treatment for durum wheat cultivation. Therefore, the integrated approach adopted seems to be applicable to assess the impact of soil management strategies on the physical and hydraulic properties of the soil.

Further research on this topic should conduct new long-term field experiments in Mediterranean environments to accurately monitor the main variables of the soil–plant–atmosphere system (climate, soil, and agronomic management), to establish the long-term agronomic and environmental sustainability of the no-tillage strategy under wheat. As wheat straw cover may protect the no-tilled soil surface from rainfall impact, research aimed at quantifying the soil sealing reduction will be carried out in the future, as it represents a main topic both for hydrology and management of the soil.

Supplementary Materials: The following are available online at http://www.mdpi.com/2073-4441/11/3/484/s1, Table S1: Summary of the main cultivation practices carried out at the investigated sites during the growth season 2016–2017.

Author Contributions: M.C. outlined the investigation, has carried out data analysis and wrote the manuscript. F.F, P.G, L.G, C.V. and A.V.V. have carried out the experimental work. All authors contributed to critically discuss the results and review the manuscript.

Acknowledgments: The work was supported and funded by the project "STRATEGA, Sperimentazione e TRAsferimento di TEcniche innovative di aGricoltura conservativA", funded by Regione Puglia–Dipartimento Agricoltura, Sviluppo Rurale ed Ambientale, CUP: B36J14001230007. Authors would like to thank the owners of investigated farms, Mr. Piero Giglio and Mr. Giovanni Zingariello, for availability and support provided in the field, respectively at Gravina and Candela.

Conflicts of Interest: The authors declare no conflict of interest.

References

1. Soane, B.D.; Ball, B.C.; Arvidsson, J.; Basch, G.; Moreno, F.; Roger-Estrade, J. No-till in northern, western and south-western Europe: A review of problems and opportunities for crop production and the environment. *Soil Tillage Res.* **2012**, *118*, 66–87. [CrossRef]

2. Blanco-Canqui, H.; Ruis, S.J. No-tillage and soil physical environment. *Geoderma* **2018**, *326*, 164–200. [CrossRef]

3. Prosdocimi, M.; Jordán, A.; Tarolli, P.; Keesstra, S.; Novara, A.; Cerdà, A. The immediate effectiveness of barley straw mulch in reducing soil erodibility and surface runoff generation in Mediterranean vineyards. *Sci. Total Environ.* **2016**, *547*, 323–330. [CrossRef] [PubMed]

4. Bogunovic, I.; Pereira, P.; Kisic, I.; Sajko, K.; Sraka, M. Tillage management impacts on soil compaction, erosion and crop yield in Stagnosols (Croatia). *Catena.* **2018**, *160*, 376–384. [CrossRef]

5. Bottinelli, N.; Jouquet, P.; Capowiez, Y.; Podwojewski, P.; Grimaldi, M.; Peng, X. Why is the influence of soil macrofauna on soil structure only considered by soil ecologists? *Soil Tillage Res.* **2015**, *146*, 118–124. [CrossRef]

6. Tebrügge, F.; Düring, R. Reducing tillage intensity–a review of results from a long-term study in Germany. *Soil Tillage Res.* **1999**, *53*, 15–28. [CrossRef]

7. Colecchia, S.A.; Rinaldi, M.; De Vita, P. Effects of tillage systems in durum wheat under rainfed Mediterranean conditions. *Cereal Res. Commun.* **2015**, *43*, 704–716. [CrossRef]

8. Van de Putte, A.; Govers, G.; Diels, J.; Gillijns, K.; Demuzere, M. Assessing the effect of soil tillage on crop growth: A meta-regression analysis on European crop yields under conservation agriculture. *Eur. J. Agron.* **2010**, *33*, 231–241. [CrossRef]

9. Marandola, D.; De Maria, M. La semina su sodo: Numeri e situazione in Italia. *L'Inf. Agrar.* **2013**, *27*, 42–46. (In Italian)

10. Stubbs, T.L.; Kennedy, A.C.; Schillinger, W.F. Soil ecosystem change during the transition to no-till cropping. *J. Crop Improv.* **2004**, *11*, 105–135. [CrossRef]

11. Pittelkow, C.M.; Linquist, B.A.; Lundy, M.E.; Liang, X.; Van Groenigen, K.J.; Lee, J.; Van Gestel, N.; Six, J.; Venterea, R.T.; Van Kessel, C. When does no-till yield more? A global meta-analysis. *Field Crops Res.* **2015**, *183*, 156–168. [CrossRef]

12. Kertész, A.; Madarász, B. Conservation Agriculture in Europe. *Int. Soil Water Conserv. Res.* **2014**, *2*, 91–96. [CrossRef]

13. Somasundaram, J.; Chaudhary, R.S.; Awanish Kumar, D.; Biswas, A.K.; Sinha, N.K.; Mohanty, M.; Hati, K.M.; Jha, P.; Sankar, M.; Patra, A.K.; et al. Effect of contrasting tillage and cropping systems on soil aggregation, carbon pools and aggregate-associated carbon in rainfed Vertisols. *Eur J Soil Sci.* **2018**. [CrossRef]

14. Giambalvo, D.; Amato, G.; Badagliacca, G.; Ingraffia, R.; Di Miceli, G.; Frenda, A.S.; Plaia, A.; Venezia, G.; Ruisi, P. Switching from conventional tillage to no-tillage: Soil N availability, Nuptake,15N fertilizer recovery, and grain yield of durum wheat. *Field Crops Res.* **2018**, *218*, 171–181. [CrossRef]

15. Troccoli, A.; Maddaluno, C.; Mucci, M.; Russo, M.; Rinaldi, M. Is it appropriate to support the farmers for adopting Conservation Agriculture? Economic and environmental impact assessment. *Ital. J. Agron.* **2015**, *10*, 169–177. [CrossRef]

16. Marandola, D.; Monteleone, A. I PSR 2014-2020 puntano sulla semina su sodo. *L'Inf. Agrar.* **2016**, *2*, 59–66. (In Italian)

17. Lozano, L.A.; Soracco, C.G.; Buda, V.S.; Sarli, G.O.; Filgueira, R.R. Stabilization of soil hydraulic properties under a long term no-till system. *Rev. Bras. Ciênc. Solo* **2014**, *38*, 1281–1292. [CrossRef]

18. Ferrara, R.M.; Mazza, G.; Muschitiello, C.; Castellini, M.; Stellacci, A.M.; Navarro, A.; Lagomarsino, A.; Vitti, C.; Rossi, R.; Rana, G. Short-term effects of conversion to no-tillage on respiration and chemical-physical properties of the soil: A case study in a wheat cropping system in semi-dry environment. *Ital. J. Agrometeorol.* **2017**, *1*, 47–58.

19. Strudley, M.W.; Green, T.R.; Ascough, J.C., II. Tillage effects on soil hydraulic properties in space and time: State of the science. *Soil Tillage Res.* **2008**, *99*, 4–48. [CrossRef]

20. Chandrasekhar, P.; Kreiselmeier, J.; Schwen, A.; Weninger, T.; Julich, S.; Feger, K.-H.; Schwärzel, K. Why We Should Include Soil Structural Dynamics of Agricultural Soils in Hydrological Models. *Water* **2018**, *10*, 1862. [CrossRef]

21. Vogeler, I.; Rogasik, J.; Funder, U.; Panten, K.; Schnug, E. Effect of tillage systems and P-fertilization on soil physical and chemical properties, crop yield and nutrient uptake. *Soil Tillage Res.* **2009**, *103*, 137–143. [CrossRef]

22. Reichert, J.M.; Rosa, V.T.; Vogelmann, E.S.; Rosa, D.P.; Horn, R.; Reinert, D.J.; Sattler, A.; Denardin, J.E. Conceptual framework for capacity and intensity physical soil properties affected by short and long-term (14 years) continuous no-tillage and controlled traffic. *Soil Tillage Res.* **2016**, *158*, 123–136. [CrossRef]

23. Johnston, A.E.; Poulton, P.R. The importance of long-term experiments in agriculture: Their management to ensure continued crop production and soil fertility; the Rothamsted experiment. *Eur. J. Soil Sci.* **2018**, *69*, 113–125. [CrossRef] [PubMed]

24. Kumar, S.; Kadono, A.; Lal, R.; Dick, W. Long-term tillage and crop rotations for 47–49 years influences hydrological properties of two soils in Ohio. *Soil Sci. Soc. Am. J.* **2012**, *76*, 2195–2207. [CrossRef]

25. Ventrella, D.; Stellacci, A.M.; Castrignanò, A.; Charfeddine, M.; Castellini, M. Effects of crop residue management on winter durum wheat productivity in a long term experiment in Southern Italy. *Eur. J. Agron.* **2016**, *77*, 188–198. [CrossRef]

26. Blanco-Canqui, H.; Wienhold, B.J.; Jin, V.L.; Schmer, M.R.; Kibet, L.C. Long-term tillage impact on soil hydraulic properties. *Soil Tillage Res.* **2017**, *170*, 38–42. [CrossRef]

27. Chang, C.; Lindwall, C.W. Effects of tillage and crop rotation on physical properties of a loam soil. *Soil Tillage Res.* **1992**, *22*, 383–389. [CrossRef]

28. Lipiec, J.; Kuś, J.; Słowińska-Jurkirwicz, A.; Nosalewicz, A. Soil porosity and water infiltration as influenced by tillage methods. *Soil Till. Res.* **2006**, *89*, 210–220. [CrossRef]

29. Azooz, R.H.; Arshad, M.A. Soil water drying and recharge rates as affected by tillage under continuous barley and barley–canola cropping systems in northwestern Canada. *Can. J. Soil Sci.* **2001**, *81*, 45–52. [CrossRef]

30. Castellini, M.; Pirastru, M.; Niedda, M.; Ventrella, D. Comparing physical quality of tilled and no-tilled soils in an almond orchard in southern Italy. *Ital. J. Agron.* **2013**, *8*, 149–157. [CrossRef]

31. Ciollaro, G.; Lamaddalena, N. Effect of tillage on the hydraulic properties of a vertic soil. *J. Agric. Eng. Res.* **1998**, *71*, 147–155. [CrossRef]

32. Lassabatère, L.; Angulo-Jaramillo, R.; Ugalde, J.M.S.; Cuenca, R.; Braud, I.; Haverkamp, R. Beerkan estimation of soil transfer parameters through infiltration experiments: BEST. *Soil Sci. Soc. Am. J.* **2006**, *70*, 521–532. [CrossRef]

33. Bagarello, V.; Di Prima, S.; Iovino, M.; Provenzano, G.; Sgroi, A. Testing different approaches to characterize Burundian soils by the BEST procedure. *Geoderma* **2011**, *162*, 141–150. [CrossRef]

34. Castellini, M.; Di Prima, S.; Iovino, M. An assessment of the BEST procedure to estimate the soil water retention curve: A comparison with the evaporation method. *Geoderma* **2018**, *320*, 82–94. [CrossRef]

35. Siltecho, S.; Hammecker, C.; Sriboonlue, V.; Clermont-Dauphin, C.; Trelo-ges, V.; Antonino, A.C.D.; Angulo-Jaramillo, R. Use of field and laboratory methods for estimating unsaturated hydraulic properties under different land uses. *Hydrol. Earth Syst. Sci.* **2015**, *19*, 1193–1207. [CrossRef]

36. Castellini, M.; Iovino, M.; Pirastru, M.; Niedda, M.; Bagarello, V. Use of BEST procedure to assess soil physical quality in the Baratz Lake catchment (Sardinia, Italy). *Soil Sci. Soc. Am. J.* **2016**, *80*, 742–755. [CrossRef]

37. Souza, R.; Souza, E.; Netto, A.M.; de Almeida, A.Q.; Júnior, G.B.; Silva, J.R.I.; de Sousa Lima, J.R.; Antonino, A.C.D. Assessment of the physical quality of a Fluvisol in the Brazilian semiarid region. *Geoderma Reg.* **2017**, *10*, 175–182. [CrossRef]

38. Cullotta, S.; Bagarello, V.; Baiamonte, G.; Gugliuzza, G.; Iovino, M.; La Mela Veca, D.S.; Maetzke, F.; Palmeri, V.; Sferlazza, S. Comparing different methods to determine soil physical quality in a mediterranean forest and pasture land. *Soil Sci. Soc. Am. J.* **2016**, *80*, 1038–1056. [CrossRef]

39. Lozano-Baez, S.E.; Cooper, M.; Ferraz, S.F.B.; Ribeiro Rodrigues, R.; Pirastru, M.; Di Prima, S. Previous land use affects the recovery of soil hydraulic properties after forest restoration. *Water* **2018**, *10*, 453. [CrossRef]

40. Bagarello, V.; Castellini, M.; Di Prima, S.; Iovino, M. Soil hydraulic properties determined by infiltration experiments and different heights of water pouring. *Geoderma* **2014**, *213*, 492–501. [CrossRef]

41. Di Prima, S.; Concialdi, P.; Lassabatere, L.; Angulo-Jaramillo, R.; Pirastru, M.; Cerdà, A.; Keesstra, S. Laboratory testing of Beerkan infiltration experiments for assessing the role of soil sealing on water infiltration. *Catena* **2018**, *167*, 373–384. [CrossRef]

42. Mubarak, I.; Mailhol, J.C.; Angulo-Jaramillo, R.; Ruelle, P.; Boivin, P.; Khaledian, M. Temporal variability in soil hydraulic properties under drip irrigation. *Geoderma* **2009**, *150*, 158–165. [CrossRef]

43. Souza, E.S.; Antonino, A.C.D.; Heck, R.J.; Montenegro, S.M.G.L.; Lima, J.R.S.; Sampaio, E.V.S.B.; Jaramillo, R.A.; Vauclin, M. Effect of crusting on the physical and hydraulic properties of a soil cropped with castor beans (*Ricinus communis* L.) in the north eastern region of Brazil. *Soil Tillage Res.* **2014**, *141*, 55–61. [CrossRef]

44. Villarreal, R.; Soracco, C.G.; Lozano, L.A.; Melani, E.M.; Sarli, G.O. Temporal variation of soil sorptivity under conventional and no-till systems determined by a simple laboratory method. *Soil Tillage Res.* **2017**, *168*, 92–98. [CrossRef]

45. Somasundaram, J.; Reeves, S.; Wang, W.; Heenan, M.; Dalal, R. Impact of 47 years of no tillage and stubble retention on soil aggregation and carbon distribution in a Vertisol. *Land Degrad. Dev.* **2017**, *28*, 1589–1602. [CrossRef]

46. Iovino, M.; Castellini, M.; Bagarello, V.; Giordano, G. Using static and dynamic indicators to evaluate soil physical quality in a Sicilian area. *Land Degrad. Dev.* **2016**, *27*, 200–210. [CrossRef]

47. Di Prima, S.; Rodrigo-Comino, J.; Novara, A.; Iovino, M.; Pirastru, M.; Keesstra, S.; Cerda, A. Assessing soil physical quality of citrus orchards under tillage, herbicide and organic managements. *Pedosphere* **2018**, *28*, 463–477. [CrossRef]

48. Gee, G.W.; Or, D. Particle-size analysis. Methods of Soil Analysis, Physical Methods. *Soil Sci. Soc. Am.* **2002**, 255–293.

49. Vitti, C.; Stellacci, A.M.; Leogrande, R.; Mastrangelo, M.; Cazzato, E.; Ventrella, D. Assessment of organic carbon in soils: A comparison between the Springer–Klee wet digestion and the dry combustion methods in Mediterranean soils (Southern Italy). *Catena* **2016**, *137*, 113–119. [CrossRef]

50. Castellini, M.; Stellacci, A.M.; Barca, E.; Iovino, M. Application of multivariate analysis techniques for selecting soil physical quality indicators: A case study in long-term field experiments in Apulia (southern Italy). *Soil Sci. Soc. Am. J.* **2019**. [CrossRef]

51. Van Genuchten, M.T. A closed-form equation for predicting the hydraulic conductivity of unsaturated soils. *Soil Sci. Soc. Am. J.* **1980**, *44*, 892–898. [CrossRef]

52. Brooks, R.H.; Corey, T. Hydraulic properties of porous media. In *Hydrology Papers 3*; Colorado State University: Fort Collins, Colorado, 1964; 27p.

53. Burdine, N.T. Relative permeability calculation from pore size distribution data. *Petr. Trans. Am. Inst. Min. Metall. Eng.* **1953**, *198*, 71–77. [CrossRef]

54. Haverkamp, R.; Debionne, S.; Viallet, P.; Angulo-Jaramillo, R.; de Condappa, D. Soil properties and moisture movement in the unsaturated zone. In *The Handbook of Groundwater Engineering*; Delleur, J.W., Ed.; CRC Press: Boca Raton, FL, USA, 2006; pp. 1–59.

55. Haverkamp, R.; Ross, P.J.; Smettem, K.R.J.; Parlange, J.Y. Threedimensional analysis of infiltration from the disc infiltrometer: 2. Physically based infiltration equation. *Water Resour. Res.* **1994**, *30*, 2931–2935. [CrossRef]

56. Minasny, B.; McBratney, A.B. Estimating the water retention shape parameter from sand and clay content. *Soil Sci. Soc. Am. J.* **2007**, *71*, 1105–1110. [CrossRef]

57. Xu, X.; Kiely, G.; Lewis, C. Estimation and analysis of soil hydraulic properties through infiltration experiments: Comparison of BEST and DL fitting methods. *Soil Use Manag.* **2009**, *25*, 354–361. [CrossRef]

58. Yilmaz, D.; Lassabatère, L.; Angulo-Jaramillo, R.; Deneele, D.; Legret, M. Hydrodynamic characterization of basic oxygen furnace slag through an adapted BEST method. *Vadose Zone J.* **2010**, *9*, 107–116. [CrossRef]

59. Bagarello, V.; Di Prima, S.; Iovino, M. Comparing alternative algorithms to analyze the Beerkan infiltration experiment. *Soil Sci. Soc. Am. J.* **2014**, *78*, 724–736. [CrossRef]

60. Di Prima, S. Automatic analysis of multiple Beerkan infiltration experiments for soil Hydraulic Characterization. In Proceedings of the 1st CIGR Inter-Regional Conference on Land and Water Challenges, Bari, Italy, 10–14 September 2013; p. 127. [CrossRef]

61. Watson, K.; Luxmoore, R. Estimating macroporosity in a forest watershed by use of a tension infiltrometer. *Soil Sci. Soc. Am. J.* **1986**, *50*, 578–582. [CrossRef]

62. Reynolds, W.D.; Drury, C.F.; Tan, C.S.; Fox, C.A.; Yang, X.M. Use of indicators and pore volume-function characteristics to quantify soil physical quality. *Geoderma* **2009**, *152*, 252–263. [CrossRef]

63. Mohanty, B.P.; Ankeny, R.; Horton, M.D.; Kanwar, R.S. Spatial analysis of hydraulic conductivity measured using disc infiltrometers. *Water Resour. Res.* **1994**, *30*, 2489–2498. [CrossRef]

64. Warrick, A.W. Appendix 1: Spatial variability. In *Environmental Soil Physics*; Hillel, D., Ed.; Academic Press: San Diego, CA, USA, 1998; pp. 655–675.

65. Lee, D.M.; Reynolds, W.D.; Elrick, D.E.; Clothier, B.E. A comparison of three field methods for measuring saturated hydraulic conductivity. *Can. J. Soil Sci.* **1985**, *65*, 563–573. [CrossRef]

66. Bagarello, V.; Castellini, M.; Iovino, M.; Sgroi, A. Testing the concentric-disk tension infiltrometer for field measurements of soil hydraulic conductivity. *Geoderma* **2010**, *158*, 427–435. [CrossRef]

67. Castellini, M.; Ventrella, D. Impact of conventional and minimum tillage on soil hydraulic conductivity in typical cropping system in southern Italy. *Soil Tillage Res.* **2012**, *124*, 47–56. [CrossRef]

68. Angulo-Jaramillo, R.; Bagarello, V.; Iovino, M.; Lassabatere, L. (Eds.) Saturated soil hydraulic conductivity. In *Infiltration Measurements for Soil Hydraulic Characterization*; Springer: Cham, Switzerland, 2016; pp. 43–180.

69. Elrick, D.E.; Reynolds, W.D. Methods for analyzing constant-head well permeameter data. *Soil Sci. Soc. Am. J.* **1992**, *56*, 320–323. [CrossRef]

70. Hu, W.; Shao, M.G.; Wang, Q.J.; Fan, J.; Horton, R. Temporal changes of soil hydraulic properties under different land uses. *Geoderma* **2009**, *149*, 355–366. [CrossRef]

71. Moret, D.; Arrúe, J.L. Dynamics of soil hydraulic properties during fallow as affected by tillage. *Soil Tillage Res.* **2007**, *96*, 103–113. [CrossRef]

72. Alletto, L.; Coquet, Y. Temporal and spatial variability of soil bulk density and near-saturated hydraulic conductivity under two contrasted tillage management systems. *Geoderma* **2009**, *152*, 85–94. [CrossRef]

73. Schwen, A.; Bodner, G.; Scholl, P.; Buchan, G.D.; Loiskandl, W. Temporal dynamics of soil hydraulic properties and the water-conducting porosity under different tillage. *Soil Tillage Res.* **2011**, *113*, 89–98. [CrossRef]

74. Castellini, M.; Niedda, M.; Pirastru, M.; Ventrella, D. Temporal changes of soil physical quality under two residue management systems. *Soil Use Manag.* **2014**, *30*, 423–434. [CrossRef]

75. Castellini, M.; Giglio, L.; Niedda, M.; Palumbo, A.D.; Ventrella, D. Impact of biochar addition on the physical and hydraulic properties of a clay soil. *Soil Tillage Res.* **2015**, *154*, 1–13. [CrossRef]

76. Rhoton, F.E. Influence of time on soil response to no-till practices. *Soil Sci. Soc. Am. J.* **2000**, *64*, 700–709. [CrossRef]

77. De Vita, P.; Di Paolo, E.; Fecondo, G.; Di Fonzo, N.; Pisante, M. No-tillage and conventional tillage effects on durum wheat yield, grain quality and soil moisture content in southern Italy. *Soil Till. Res.* **2007**, *92*, 69–78. [CrossRef]

78. Dang, Y.P.; Moody, P.W.; Bell, M.J.; Seymour, N.P.; Dalal, R.C.; Freebairn, D.M. Strategic tillage in no till farming systems in Australia's northern grains-growing regions: II. Implications for agronomy, soil and environment. *Soil Tillage Res.* **2015**, *152*, 115–123. [CrossRef]

79. Ma, Y.; Feng, S.; Huo, Z.; Song, X. Application of the SWAP model to simulate the field water cycle under deficit irrigation in Beijing, China. *Math. Comp. Model.* **2011**, *54*, 1044–1052. [CrossRef]

80. Shelia, V.; Šimůnek, J.; Boote, K.; Hoogenbooom, G. Coupling DSSAT and HYDRUS-1D for simulations of soil water dynamics in the soil-plant-atmosphere system. *J. Hydrol. Hydromech.* **2018**, *66*, 232–245. [CrossRef]

81. Ventrella, D.; Charfeddine, M.; Giglio, L.; Castellini, M. Application of DSSAT models for an agronomic adaptation strategy under climate change in Southern Italy: Optimum sowing and transplanting time for winter durum wheat and tomato. *Ital. J. Agron.* **2012**, *7*, 109–115. [CrossRef]
82. Ventrella, D.; Giglio, L.; Charfeddine, M.; Lopez, R.; Castellini, M.; Sollitto, D.; Castrignanò, A.; Fornaro, F. Climate change impact on crop rotations of winter durum wheat and tomato in southern Italy: Yield analysis and soil fertility. *Ital. J. Agron.* **2012**, *7*, 100–108. [CrossRef]

Article

Recovery of Soil Hydraulic Properties for Assisted Passive and Active Restoration: Assessing Historical Land Use and Forest Structure

Sergio Esteban Lozano-Baez [1,*], Miguel Cooper [2], Silvio Frosini de Barros Ferraz [3], Ricardo Ribeiro Rodrigues [1], Mirko Castellini [4] and Simone Di Prima [5]

[1] Laboratory of Ecology and Forest Restoration (LERF), Department of Biological Sciences, "Luiz de Queiroz" College of Agriculture, University of São Paulo, Av. Pádua Dias 11, Piracicaba SP 13418-900, Brazil; rrresalq@usp.br

[2] Department of Soil Science, "Luiz de Queiroz" College of Agriculture, University of São Paulo, Av. Pádua Dias 11, Piracicaba SP 13418-900, Brazil; mcooper@usp.br

[3] Forest Hydrology Laboratory, "Luiz de Queiroz" College of Agriculture, University of São Paulo, Av. Pádua Dias 11, Piracicaba SP 13418-900, Brazil; silvio.ferraz@usp.br

[4] Council for Agricultural Research and Economics, Research Centre for Agriculture and Environment (CREA-AA), Via C. Ulpiani 5, 70125 Bari, Italy; mirko.castellini@crea.gov.it

[5] Université de Lyon, UMR5023 Ecologie des Hydrosystèmes Naturels et Anthropisés, CNRS, ENTPE, Université Lyon 1, 3 rue Maurice Audin, 69518 Vaulx-en-Velin, France; simone.diprima@entpe.fr

* Correspondence: sergio.lozano@usp.br; Tel.: +55-19-3429-4100

Received: 22 November 2018; Accepted: 28 December 2018; Published: 7 January 2019

Abstract: Tree planting and natural regeneration are the main approaches to achieve global forest restoration targets, affecting multiple hydrological processes, such as infiltration of rainfall. Our understanding of the effect of land use history and vegetation on the recovery of water infiltration and soil attributes in both restoration strategies is limited. Therefore, we investigated the recovery of top-soil saturated soil hydraulic conductivity (K_s), soil physical and hydraulic properties in five land use types: (i) a secondary old-growth forest; (ii) a forest established through assisted passive restoration 11 years ago; (iii) an actively restored forest, with a more intensive land use history and 11 years of age; (iv) a pasture with low-intensity use; and (v) a pasture with high-intensity use, in the Brazilian Atlantic Forest. For these land use types, we determined the historical land use patterns and conducted soil sampling, using the Beerkan method to determine K_s values in the field. We also measured tree basal area, canopy cover, vegetation height, tree density and species richness in forest covers. The K_s decreased when land use was more intense prior to forest restoration actions. Our results indicate that land use legacy is a crucial factor to explain the current difference in soil and vegetation attributes among study sites.

Keywords: Beerkan method; forest restoration; infiltration; natural regeneration; pasture

---

1. Introduction

Forest restoration strategies are being implemented around the world through ambitious international (e.g., Bonne Challenge and New York Declaration on Forests), regional (e.g., Initiative 20 × 20 and AFR100) and national initiatives such as restoration plans in many countries [1]. Consequently, secondary forests have expanded in tropical regions [1,2]. In Brazil, the location of our study area, the "Atlantic Forest Restoration Pact" aims by 2050 to increase the current forest cover from 17% to at least 30%, with a restoration target of 15 million hectares [3]. These initiatives include both passive and active restoration strategies. Passive ecological restoration refers to spontaneous recovery of tree species in an ecosystem that has been damaged, while assisted passive restoration involves human interventions to assist natural regeneration [4,5]. This can include introduction of propagules and removal of invasive species and persistent disturbances, for example, fire or livestock grazing [4]. On the other hand, active restoration requires a higher human intervention through planting of tree seedlings to accelerate the recovery process [6,7].

Both restoration approaches have been shown to impact positively the provision of ecosystem services, as well recovering biodiversity and ecosystem functions [8]. However, most restoration research around the world has focused on aboveground plant communities, whilst the belowground environment (e.g., soil physical and hydraulic properties) has been poorly studied [9,10]. For example, the response of the infiltration process, and soil physical and hydraulic properties after forest restoration is virtually unknown [11]. A crucial parameter in the infiltration process is the soil saturated hydraulic conductivity (K_s), which influences water percolation through the soil matrix [12,13]. It is well known from previous studies that K_s is highly variable compared to other soil physical properties [14,15]. In fact, the K_s depends strongly on the highly variable soil structure, and it is known to vary several orders of magnitude [16,17], especially on forested soils [18,19]. In general, K_s recovery and soil hydraulic properties have been reported separately in passive [20–24] and active [25–28] restoration, but few comparisons between both restoration strategies have been conducted. Lozano-Baez et al. [28] investigated the surface K_s recovery under a nine-year-old actively restored forest in the Atlantic Forest of Brazil and observed that the land use prior to forest restoration influences the K_s recovery. Moreover, the few recent comparisons between active and passive restoration show contradictory results. For instance, K_s at 12.5 cm depth in Brazilian Amazônia was higher under a 15-year-old passively restored forest than a 10-year-old tree plantation [11]. In contrast, other authors in Madagascar found much lower surface K_s in 2–10-year-old naturally regenerating fallow than actively restored forest of 6–9 years of age [29].

Most previous studies have assessed the recovery of soil physical and hydraulic properties without addressing the relationships among soil, vegetation and land use history. These relationships are fundamental to better understand the recovery process (e.g., resilience of the ecosystem) and successional trajectories after forest restoration [30,31]. Foster et al. [32] argued that the imprints of past land use on ecosystems may persist for decades to centuries. In particular, after forest restoration, such imprints of past land use on soil (e.g., K_s, soil physical and hydraulic properties) may persist for a time frame of more than a decade, as suggested by several studies [12,26,33]. However, the above-mentioned mechanisms and relationships that affect the recovery process are poorly understood.

As part of a larger research effort investing the effects of forest restoration on K_s, this study aimed to extend the work of Lozano-Baez et al. [28] at a new location. Apart from presenting new K_s data for pastures with different land use intensities and a secondary old-growth forest, this paper includes the first measurements of K_s for a forest established through assisted passive restoration in the Brazilian Atlantic Forest. We further quantified and compared the K_s, soil physical and hydraulic properties recovery of active vs. assisted passive restoration strategies from the same restoration program described by Lozano-Baez et al. [28]. We examined whether differences in land use history led to differences in these soil attributes (e.g., K_s, bulk density, soil organic carbon content, soil porosity, initial and saturated soil water content) and vegetation attributes (e.g., basal area, canopy cover, vegetation height, tree density and species richness). We studied five land-cover types: (i) a secondary old-growth forest, used as a reference forest (hereafter, RF); (ii) a forest established through assisted passive restoration (hereafter, APR); (iii) an actively restored forest (hereafter, AR); (iv) a pasture with low-intensity use (hereafter, LiP); and (v) a pasture with high-intensity use (hereafter, HiP). In forest stands, we associated the recovery of K_s, soil physical and hydraulic properties with the vegetation attributes. We hypothesized that K_s would vary with intensity of land use in the past among land-cover types as follows: RF > APR > AR > LiP > HiP. As the AR site had a more intensive land use history, we expected that K_s recovery and vegetation attributes would be higher in the APR.

2. Materials and Methods

2.1. Study Area

The study area is located in the county of Campinas (22°53' S, 46°54' W), São Paulo State, Southeast Brazil (Figure 1). The climate in this region is classified as Cwa according to the Köppen classification mean annual precipitation is 1700 mm and mean annual temperature is 20 °C, with dry winters and wet summers [34]. Our study sites are located at the transition between the Atlantic Plateau and the Peripheral Depression geomorphological provinces [35]. The soils are classified as Ultisols [36] and the original vegetation in this area is a seasonal semi-deciduous forest, belonging to the Atlantic Forest biome. This region is highly fragmented, because of 200 years of historical landscape changes [37]. In particular, our study area is located inside the sub-basin of the Atibaia River where the main land uses are native vegetation and pastureland, occupying 33% and 30% of the sub-basin, respectively. The native vegetation includes Atlantic Forest remnants with different sizes and ages [38].

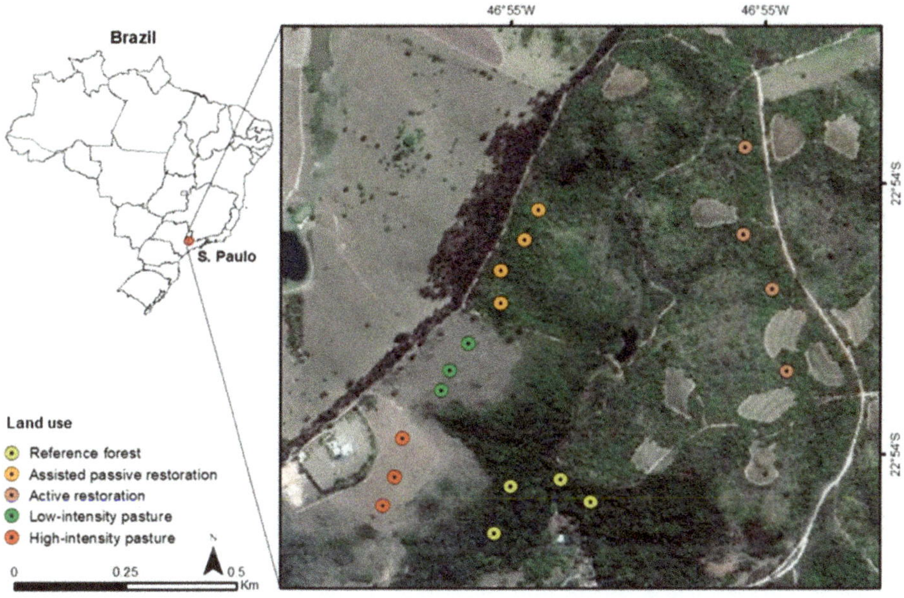

Figure 1. Location of the 18 study plots in the state of São Paulo, Southeast Brazil.

Within this area, we selected five land uses to measure soil physical and hydraulic properties, vegetation structure and diversity (Figure 2). In general, the deforestation of our study area already existed at the beginning of the 19th century, with the objective of introducing coffee (*Coffea arabica*) plantations. However, after the crisis in coffee cultivation during the early 20th century, the plantations were gradually replaced by pastures.

Land use history for the study sites was reconstructed based on interviews with the local population and aerial photographs taken in 1968, 1978, 1994, 2005 and 2017. The site RF is a secondary old-growth forest characterized by having the highest slope between the study sites, which was 28.8 ± 4.9% (SD). The slope was measured in the study plots with laser distance meter. Site RF was used as a control area to assess reference values for soil physical and hydraulic properties. According to interviews, in the early 20th century, this site was affected by natural fire disturbances and it was partially cleared at least once in the past for agricultural purposes. Moreover, aerial photographs showed that forest cover in most of the area was established and has increased since 1968. In this context, all RF plots were located in sites with forest cover in the last 40 years.

Figure 2. Pictures showing the vegetation cover and the Ultisol top-soil profile for each study site. The black lines in each top-soil profile represent 0.2 m scale. RF, Reference Forest; APR, Assisted Passive Restoration; AR, Active Restoration; LiP, Low-intensity Pasture; HiP, High-intensity Pasture.

Site APR is located adjacent to the LiP. The slope (28.1 ± 2.8%) was similar to the RF. From 1968 to 1994, it was used for milk cattle grazing. Then, the area was abandoned and remained without a specific land use until 2007, leaving the forest to naturally regrow over 12 years. In 2007, to decide the best restoration strategy for the area, the "Diagnostic" protocol proposed by Rodrigues et al. [37] was implemented. This protocol allowed identifying the initial environmental situation and evaluating the potential of autogenic restoration of the area. Considering that this site evidenced favorable abiotic and biotic conditions (e.g., naturally regenerating native plants) for native plant establishment, the restoration diagnosis was of fair potential for autogenic restoration. Thus, forest restoration techniques included the encouragement of regenerating individual native trees and shrubs by manual and chemical control of invasive grasses. Moreover, enrichment plantings with native tree species were also implemented in patches without natural regeneration. In this regard, our measurements reflect the effect of 11 years of APR on a soil with a previous second-growth forest.

At AR site, the slope (22.8 ± 1.7%) was lower than the RF and APR. Initially, this site was used for dairy cattle grazing from 1968 to 1986. Later, it was replaced by coffee (*C. arabica*) plantations until 1994. It is important to emphasize that, at the beginning of the coffee plantation phase, widespread terracing was implemented. Then, in 1994, the coffee was replaced by pastures with *Urochloa brizantha* for beef cattle grazing. In 2007, the "Diagnostic" protocol mentioned previously was implemented. Given that AR site evidenced very few spontaneously regenerating seedlings and degraded environmental conditions that limited the passive restoration strategy, the restoration diagnosis in this area was of very low potential for autogenic restoration. Thus, AR was implemented through a restoration model that aimed to provide economical insurance and ensure successional processes to landowners [37]. Restoration plantings were implemented as mixed plantation with high-diversity-mix of seedlings (>50 native trees species). During the planting, these species were organized in fourth groups (e.g., initial, filling, middle and final species) according to the rate of growth and commercial value. Initial species (e.g., *Acacia polyphilla*, *Croton floribundus* and *Schinus terebinthifolius*) can be harvested for fuel production in 10–15 years, and are characteristically fast-growing, providing fast soil coverage and beneficial initial conditions for other species growth. Filling species (e.g., *Croton urucurana*, *Gochnatia polymorpha* and *Trema micrantha*) are also fast-growing species planted in the same line as the species. Middle species (e.g., *Astronium graveolens*, *Gallesia integrifolia* and *Machaerium stipitatum*) can be harvested during Years 20–25, and are more valuable wood species that will replace the initial and filling species. Final species (e.g., *Aspidosperma polyneuron*, *Cariniana estrellensis* and *Cariniana legalis*) are narrow canopy and slow-growing species that can be harvested during Years 40–45 for luxury and finished carpentry. The species planted are listed in Table S1. The total density of seedling was

1660 ind·ha^{-1}, in a 3 m × 2 m spacing, using mechanized soil preparation. Before planting, invasive grasses were controlled through herbicide application. Fertilizers and irrigation were applied at the time of planting and during the first year [37,39]. As a result, our measurement in this restoration site represent the effect of 11 years of active restoration on highly degraded soil, with an intense land use history.

Site LiP with a slope of 22.7 ± 2.1% is located adjacent to the HiP, and both sites share a similar land use history until 2008. Since this year, in the LiP, grazing has been intermittent and with low productivity (e.g., stocking rate lower than two livestock units per hectare). During our field campaign, the vegetative cover in the LiP was dominated by the same grass specie (*U. brizantha*), with a mean height about 50 cm and isolated native trees, shrub species and nonnative grasses scattered in the area were also evident (Figure 2). Consequently, our results reflect the influence of 40 years of grazing, with a lower land use intensity in the last decade.

At site HiP, the slope was 23.3 ± 3.2%. This site was covered by a coffee (*C. arabica*) plantation until 1968. Afterwards, the coffee was replaced by pasture, planting *U. brizantha* as grass species. Since 1978, this area has been heavily grazed with dairy cattle, supporting a stocking rate greater than two livestock units per hectare, with regular application of fertilizers and other inputs. As a result, our measurements at this site represent the effect of 40 years of continuous grazing.

A graphical summary of the land use history for the five land uses described previously is provided in Figure 3.

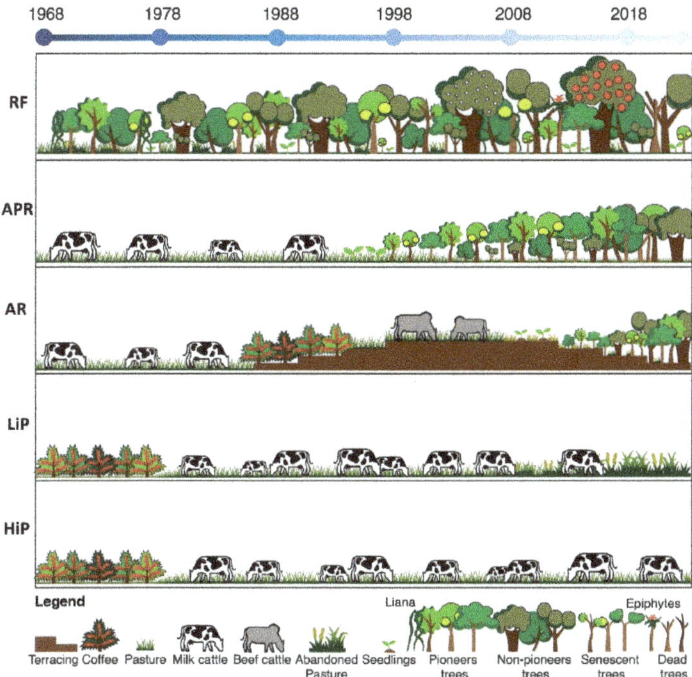

Figure 3. Land use history for each land use type. RF, Reference Forest; APR, Assisted Passive Restoration; AR, Active Restoration; LiP, Low-intensity Pasture; HiP, High-intensity Pasture.

2.2. Experimental Design

The study sites were located in a similar landscape position along the hillslope gradient and were selected to have the same soil type following Zwartendijk et al. [23]. In forest stands, we established four plots, and for the pasture sites three plots were established. Sampling the same number of plots per land uses was impossible due the restricted accessibility in pasture sites, resulting in 18 plots altogether (Figure 1). For sampling vegetation and soil attributes, the size of each plot was 500 m^2 (50 m long and 10 m wide), a total area of 2000 m^2 for each site. The size and number of plots were chosen according to similar investigations aimed at evaluating vegetation structure and composition in tropical forest restoration projects [40–42].

2.3. Vegetation Sampling

Vegetation sampling was conducted from September to November 2017 in the RF, APR and AR plots. In each 500 m^2 plot, we identified and sampled all living trees and shrubs with height $\geq$ 50 cm and diameter at breast height (DBH) > 5 cm. Additionally, we installed a 200 m^2 (50 m long and 4 m wide) subplot at the center of each plot, to identify and measure all trees and shrubs with DBH < 5 cm and height $\geq$ 50 cm. For all sites considered in this investigation, we measured the following vegetation attributes: (1) tree basal area; (2) canopy cover; (3) vegetation height; (4) tree density; and (5) species richness. These are key ecological indicators, useful to evaluate vegetation structure and composition in tropical forest restoration projects, also, they are being recommend in Atlantic Forest monitoring protocols [42,43]. As suggested by Viani et al. [42], the percentage of canopy cover in each plot was measured by an adaptation of the line interception method [44], installing a 50 m line transect in the middle of each study plot. Vegetation height was measured with a 5 m measuring stick, and the remaining height of trees taller than this was estimated visually. As suggested by Suganuma and Durigan [40], for tree density, we analyzed: (1) density of trees (DBH > 5 cm); and (2) density of saplings (DBH 1–5 cm). In the same way, for species richness, we analyzed: (1) total richness (all individual sampled); (2) overstory richness (DBH > 5 cm); and (3) richness of saplings (DBH 1–5 cm). For all previous ecological indicators, we calculated the mean values per study site. In addition, for all sampled individuals we classified the species origin as: native or nonnative to the study region. Specifically, in the AR site, we evaluated planted tree mortality.

2.4. Soil Sampling

Soil sampling was conducted during the dry season in April 2018. In the middle of each plot, we installed a 50 m transect along the hillslope gradient. At intervals of 15 m, three disturbed soil samples at 0–5 cm depth were collected. Before soil sampling, the litter and a small layer of soil (e.g., organic horizon) of less than 1 cm was removed. We determined the soil particle size distribution (PSD) with sand particle separation, the particle density (*Pd*) and soil organic carbon content (*OC*). The PSD analyses were carried out according to the hydrometer method [45]. Then, soil textures were classified following the USDA standards. The *Pd* was determined using the helium gas pycnometer method [46], and the *OC* analyses were performed following the Walkley–Black method [47]. In the same transects, at 7 m intervals, soil hydraulic measurements were conducted. In the specific case of AR site, to minimize spatial variability and possible induced effects by tillage, we placed the seven sampling points in the inter-plant space of planting lines. At each sampling point, we performed the Beerkan method [48,49]. A total of 126 Beerkan experiments were carried out, using a steel ring with an inner diameter of 16 cm inserted to a depth of about 1 cm into the soil surface. In each infiltration point, a known volume of water (150 mL) was repeatedly poured into the cylinder at a small height above soil surface (e.g., a few cm) and the energy of the water was dissipated with the hand fingers to minimize the soil disturbance. Then, the time needed for complete infiltration was logged. This procedure was repeated until the difference in infiltration time between two or three consecutives trials became negligible. Following a procedure commonly used for Beerkan method, at the beginning of each infiltration run, and near the steel ring, we collected one undisturbed soil core (100 cm^3) at 0–5 cm depth to determine the bulk density (ρ_b) and the initial volumetric soil water content (θ_i). Saturated soil hydraulic conductivity (K_s) values were estimated by the Steady version of the Simplified method based on a Beerkan Infiltration run (SSBI method) [50]. According to previous a investigation carried out by Lozano-Baez et al. [28] on the same area, this method was chosen to avoid uncertainties due to a specific shape of the cumulative infiltration [51,52].

In addition, the undisturbed soil cores were used to determine total soil porosity (*Pt*), soil microporosity (*Mic*) and soil macroporosity (*Mac*). The *Pt* was calculated using ρ_b and mean *Pd* of each plot [53]. The *Mic* was estimated using a tension table with application of 6 kPa suction, and *Mac* was obtained by the difference between *Pt* and *Mic* [54]. Finally, according to Lassabatere et al. [48], at the end of each infiltration test, a disturbed soil sample was collected to determine the saturated gravimetric water content, and ρ_b was used to calculate the saturated volumetric soil water content (θ_s).

2.5. Data Analysis

The hypothesis of normal distribution of both the raw and the log-transformed data was tested by the Kolmogorov–Smirnov test. One-way analysis of variance (ANOVA) was performed with raw or log-transformed data. When ANOVA null hypothesis was rejected, we used multiple comparisons to detect differences between pairs by applying the Tukey's honestly significant difference test. The related *p*-values were computed and compared to the level of significance of 0.05. Alternative non-parametric tests (Kruskal–Wallis ANOVA) were used when even the log-transformed data were non-normally distributed. In this case, when ANOVA null hypothesis was rejected, multiple comparisons between pairs were made with the Bonferroni method (adjusted *p*-values). Variables means were calculated for soil attributes according to the statistical distribution of the data, e.g., geometric mean for log-normal distributions and arithmetic means for normal distributions [55]. According to Lee et al., the appropriate CV expression for a log-normal distribution was calculated for the geometric means, and the usual CV was calculated for the arithmetic means [56]. Pearson's correlation coefficient was calculated to identify correlation among the selected soil attributes: *Pt*, *Mic*, *Mac*, *OC*, θ_i, ρ_b and K_s across all study sites. Furthermore, to compare the soil and vegetation attributes among land use types, Principal Component Analysis (PCA) was performed on standardized variables. All statistical analyses were carried out using the Minitab$^©$ computer program (Minitab Inc., State College, PA, USA).

3. Results

3.1. Vegetation Attributes

A total of 541 saplings and 646 trees distributed in 38 families, 92 genera, and 138 species were sampled. For non-native species, we found 62 saplings and 147 trees, representing 11% and 23% of the total, respectively (Tables S2 and S3). Although the basal area and vegetation height of trees were much higher in RF, these did not differ statistically with both restored forests. The canopy cover showed significant differences in AR with RF and APR. We highlight the higher similarity between RF and APR for density of trees and saplings (Table 1), as a result of the high density of non-native trees *Psidium guajava* and *Tecoma stans*, which represented 24% and 5% of trees in APR, respectively. In contrast, non-native trees in the RF and AR represented 18% and 16%, respectively, of all tree individuals sampled in each site. Additionally, the total richness, the density and richness of trees and saplings were markedly lower in AR, where 14% of planted trees (e.g., initial species) were dead yet there was a higher presence of grasses (e.g., *U. brizantha*) observed in all plots.

Table 1. Mean vegetation attributes ($\pm$ standard error, n = 4) sampled in the forest stands. Different superscript letters denote statistically significant differences between land use types, according to the Tukey's test ($p < 0.05$), except for the basal area and vegetation height of trees where Kruskal–Wallis test ($p < 0.05$) was applied. RF, Reference Forest; APR, Assisted Passive Restoration; AR, Active Restoration; LiP, Low-intensity Pasture; HiP, High-intensity Pasture.

	RF	APR	AR
Basal area (m^2/ha^{-1})	26.4 ± 4.49 [a]	20.8 ± 2.53 [a]	12.5 ± 3.32 [a]
Canopy cover (%)	95.8 ± 2.17 [a]	91.3 ± 1.49 [a]	77.5 ± 3.11 [b]
Vegetation height of trees (m)	10.1 ± 1.16 [a]	7.79 ± 0.57 [a]	7.00 ± 0.11 [a]
Density of trees ($ind \cdot ha^{-1}$)	$1{,}325 \pm 137$ [a]	$1{,}300 \pm 72$ [a]	610 ± 72 [b]
Density of saplings ($ind \cdot ha^{-1}$)	$3{,}950 \pm 172$ [a]	$1{,}963 \pm 959$ [ab]	850 ± 119 [b]
Total richness (tree and non-tree)	82 ± 4 [a]	62 ± 1 [a]	38 ± 2 [b]
Overstory richness	50 ± 2 [a]	41 ± 1 [a]	30 ± 1 [b]
Richness of saplings	62 ± 2 [a]	39 ± 3 [b]	15 ± 1 [b]

Note. Different superscript letters denote statistically significant differences between land use types, according to the Tukey's test ($p < 0.05$), except for the basal area and vegetation height of trees where Kruskal–Wallis test ($p < 0.05$) was applied.

3.2. Soil Physical and Hydraulic Properties

The texture of the upper layers of the soil (0–5 cm) was clay loam in APR and pasture sites, while it was sandy clay loam in RF and AR sites. The clay content at the study sites ranged between 21% and 44%, but only the RF with lower values of clay differed significantly from the other study sites. Moreover, soil samples taken from the HiP showed the highest clay content. The silt ranged between 18% and 37%, with higher silt values in APR that differed significantly from other land-covers. The sand content varied between 31% and 55%, and was significantly lower in APR compared with other study sites (Table 2).

Table 2. Mean values for soil particle size distribution, and textural class according to the USDA classification of the depth 0–5 cm for each land use type. For each variable, different superscript letters denote statistically significant differences between land use types, according to the Tukey's test ($p < 0.05$). RF, Reference Forest; APR, Assisted Passive Restoration; AR, Active Restoration; LiP, Low-intensity Pasture; HiP, High-intensity Pasture.

Land Cover	Clay (%)	Silt (%)	Sand (%)	Sand (%)					Textural Class
				Very Fine	Fine	Medium	Coarse	Very Coarse	
RF	24.8 [b]	25.9 [b]	49.3 [a]	6.23 [a]	14.0 [abc]	12.8 [a]	9.21 [a]	6.99 [a]	Sandy clay loam
APR	30.2 [a]	31.9 [a]	37.9 [c]	6.28 [a]	12.1 [c]	9.11 [c]	5.54 [c]	4.84 [b]	Clay loam
AR	30.0 [ab]	23.9 [b]	46.1 [ab]	5.73 [a]	14.2 [ab]	11.7 [ab]	7.67 [b]	6.68 [a]	Sandy clay loam
LiP	31.7 [a]	22.6 [b]	45.7 [ab]	6.60 [a]	15.1 [a]	12.6 [ab]	6.87 [bc]	4.44 [b]	Clay loam
HiP	33.6 [a]	23.2 [b]	43.1 [bc]	5.82 [a]	12.8 [bc]	10.7 [bc]	7.43 [b]	6.34 [ab]	Clay loam

Note. Number of soil texture samples: RF = 12, APR = 12, AR = 12, LiP = 9 and HiP = 9. Different superscript letters denote statistically significant differences between land use types, according to the Tukey's test ($p < 0.05$).

Comparisons of ρ_b values between study sites revealed higher similarity between RF and APR, while AR presented similar ρ_b values with both pasture sites. The Pd had similar values in all study sites, ranging from 2.61 to 2.71 g cm^{-3}. The OC varied significantly among sites (from 4.6 to 25.6 g kg^{-1}), with higher values in HiP and markedly lower values in AR. The soil Mac ranged from 0.16 to 0.38 cm^3 cm^{-3}, with greater values observed in RF, followed by the APR, AR, LiP and HiP. Similar results were obtained for Pt, which varied from 0.48 to 0.66 cm^3 cm^{-3}. In contrast, the soil Mic (from 0.21 to 0.43 cm^3 cm^{-3}) was much higher in pasture sites and decreased in forest land-covers, with lower values in RF and AR. In addition, the θ_i at the time of the Beerkan infiltration run varied between 0.12 and 0.32 cm^3 cm^{-3}, with significant lower values in the RF. The θ_s varied between 0.29 and 0.75 cm^3 cm^{-3} with significant lower values in restored forests. The K_s at the soil surface in the study land uses ranged from 4 mm h^{-1} to a maximum of 1121 mm h^{-1}. The higher K_s was evidenced in APR, which was only similar with RF and significantly different from other three land uses. The K_s values obtained in the RF were lower than APR. In contrast, the K_s of AR between 15 and 1121 mm h^{-1} was similar to RF and differed significantly for the other three land uses. In addition, across the five land uses, K_s values varied least and differed significantly from each other at the LiP and HiP (Table 3).

Table 3. Mean and associated coefficient of variation (CV, in parenthesis) of soil bulk density (ρ_b in g cm^{-3}), soil particle density (Pd in g cm^{-3}), soil organic carbon content (OC g kg^{-1}), saturated soil hydraulic conductivity (K_s in mm h^{-1}), microporosity (Mic in cm^3 cm^{-3}), macroporosity (Mac in cm^3 cm^{-3}), total soil porosity (Pt in cm^3 cm^{-3}), initial volumetric soil water content (θ_i in cm^3 cm^{-3}) and saturated volumetric soil water content (θ_s in cm^3 cm^{-3}), of the depth 0–5 cm for each land use type. For each variable, different superscript letters denote statistically significant differences between land use types, according to the Tukey's test ($p < 0.05$). RF, Reference Forest; APR, Assisted Passive Restoration; AR, Active Restoration; LiP, Low-intensity Pasture; HiP, High-intensity Pasture.

Land Cover	ρ_b	Pd	OC	K_s	Mic	Mac	Pt	θ_i	θ_s
RF	1.04 [b]	2.66 [ab]	16.2 [a]	215 [ab]	0.29 [ab]	0.32 [a]	0.61 [a]	0.18 [c]	0.48 [ab]
	(7.06)	(1.17)	(24.3)	(90.2)	(14.6)	(9.82)	(4.54)	(12.6)	(22.2)
APR	1.04 [b]	2.68 [a]	16.4 [a]	351 [a]	0.31 [bc]	0.29 [a]	0.60 [a]	0.24 [a]	0.45 [b]
	(6.50)	(1.11)	(21.4)	(58.4)	(9.12)	(9.18)	(2.53)	(14.1)	(19.4)
AR	1.19 [a]	2.68 [a]	10.3 [b]	163 [b]	0.29 [c]	0.25 [b]	0.56 [b]	0.20 [bc]	0.38 [c]
	(7.20)	(0.49)	(35.5)	(135.5)	(12.8)	(10.1)	(4.49)	(13.7)	(14.7)
LiP	1.14 [a]	2.65 [ab]	15.1 [ab]	32.6 [c]	0.33 [ab]	0.22 [c]	0.57 [b]	0.22 [ab]	0.54 [a]
	(7.12)	(0.82)	(12.4)	(155.0)	(10.8)	(11.7)	(5.31)	(10.8)	(15.0)
HiP	1.18 [a]	2.64 [b]	18.6 [a]	10.4 [d]	0.34 [a]	0.20 [c]	0.55 [b]	0.22 [ab]	0.50 [ab]
	(12.0)	(0.67)	(28.4)	(82.9)	(11.9)	(9.90)	(9.59)	(27.0)	(15.5)

Note. For ρ_b, K_s, Mic, Mac, Pt, θ_i and θ_s numbers of soil sample: RF = 28, APR = 28, AR = 28, LiP = 21 and HiP = 21. For Pd and OC number of soil samples: RF = 12, APR = 12, AR = 12, LiP = 9 and HiP = 9. Different superscript letters denote statistically significant differences between land use types, according to the Tukey's test ($p < 0.05$).

Within-site plots, high variability in K_s was observed in the RF plots and within the two restored forest classes. In contrast, smaller variations were evidenced in pasture sites. Figure 4 includes the results of the of the Tukey's test for all sampled plots. The grouping information highlights the significant and non-significant comparisons for all sampled plots. In the first group, the forest plots evidenced not significant differences due to the high K_s variability within these plots (e.g., K_s means from 104 to 407 mm h^{-1}). Then, the second group (RF1, RF3, RF4, AR1, AR3, LiP1 and LiP2) showed significant differences with pasture sites, which were grouped in a third (LiP1, LiP2, LiP3, HiP1 and HiP2) and fourth group (LiP3, HiP1, HiP2 and HiP3). In general, the LiP and HiP plots were similar, and mean K_s values altogether (e.g., from 8 to 47 mm h^{-1}) were very low (Figure 4).

According to the Pearson's correlation coefficient among selected soil attributes across all study sites, significant positive correlations were found for Pt vs. Mac (0.60) and K_s vs. Mac (0.67). In contrast, significant negative correlations were found for ρ_b vs. Pt (-0.99), ρ_b vs. Mac (-0.58) and K_s vs. Mic (-0.49) (Figure S1).

The first and second axis of the PCA for the soil attributes explained 43.0% and 29.3%, respectively, of the variation among all study sites. This analysis revealed a gradient of land-cover types from pastures to forest covers. As expected, the pasture sites were separated from the forest covers due to the higher Mic and ρ_b. Similarly, the higher ρ_b values in AR plots contributed to separating the study site. Then, APR plots were more similar to RF plots, and both forest covers were associated with higher K_s, Mac, Pt, θ_i and OC values (Figure 5A). The PCA correlating the soil and vegetation attributes showed a clear segregation among forest cover sites, explaining 55.2% and 16.5% of the variation in the first and second axis, respectively. This analysis showed that RF plots were mainly related with larger trees, evidencing higher correlation with vegetation attributes such as height of trees, density of saplings, basal area, canopy cover and overstory richness, also it was evidenced intermediate values of Mic. Considering the vegetation attributes in APR plots, the PCA showed positive correlation with total richness of species, density of trees and richness of saplings, also positive correlation and higher values of K_s, θ_i, *Mac* and *Pt* were found in these plots. By contrast, the separation of AR plots was driven by the higher ρ_b values and lower vegetation attributes, since AR site had a more intensive land use history compared to RF and APR sites. In particular, among AR plots, AR3 was the most different plot, composed by few and smaller trees growing in a compacted soil (Figure 5B).

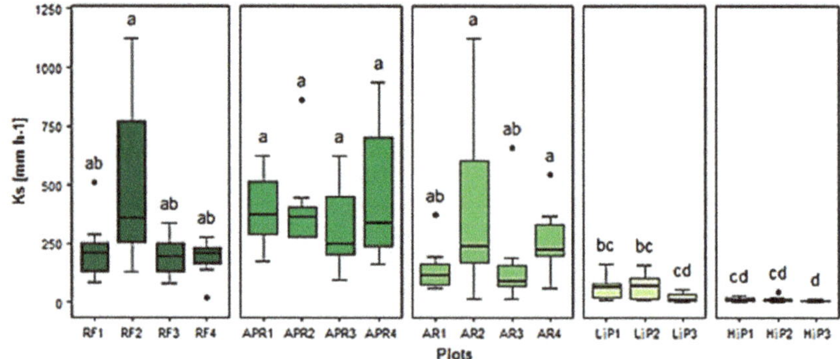

Figure 4. K_s at the surface by individual plots. Different letters above boxplots indicate significant difference based on Tukey's test ($p < 0.05$). RF, Reference Forest; APR, Assisted Passive Restoration; AR, Active Restoration; LiP, Low-intensity Pasture; HiP, High-intensity Pasture. The subscript number refer to plot numbers.

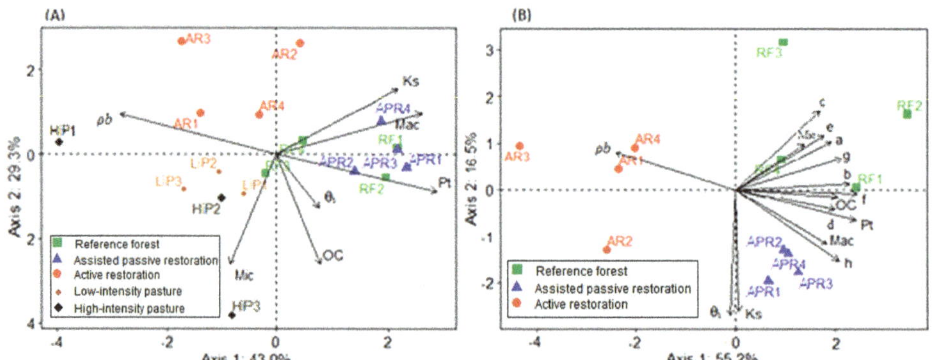

Figure 5. Principal component analysis (PCA) biplot based on soil attributes (**A**); and PCA correlating soil and vegetation attributes (**B**). Symbols represent plot sites for each land-cover type: Reference Forest (RF), Assisted Passive Restoration (APR), Active Restoration (AR), Low-intensity Pasture (LiP) and High-intensity Pasture (HiP). The soil physical and hydraulic are indicated in the vectors as follow: ρ_b, bulk density; θ_i, initial volumetric soil water content; K_s, saturated soil hydraulic conductivity; *OC*, soil organic carbon content; *Mac*, soil macroporosity; *Mic*, soil microporosity; and *Pt*, total soil porosity. The vegetation attributes are indicated in the vectors by the letters: (a) basal area; (b) canopy cover; (c) vegetation height of trees; (d) density of trees; (e) density of saplings; (f) total richness of species; (g) overstory richness; and (h) richness of saplings.

4. Discussion

4.1. Effects of Land-Cover Type and Land Use History on Soil Physical and Hydraulic Properties

Assessing the K_s recovery, soil physical and hydraulic properties of different forest restoration strategies, and investigating their relationships with land use history, vegetation structure and composition, provided the opportunity to identify the extent to which these forest restoration strategies contribute to supplying ecosystem functions as infiltration of rainwater. The variation of K_s among land-cover types was not as we expected, due to the higher K_s evidenced in APR and lower in RF. However, our results supported the first study hypothesis, namely that a more intensive land use history in AR resulted in a lower K_s recovery compared to APR. Importantly, despite both restored forest types being located in the same soil type and landscape position, it was not clear from our measurements whether APR resulted in a faster recovery of K_s compared to AR, due to the high variability in land use history. Similar situations have been reported in several tropical studies [26,29]. In addition, the K_s recovery in APR could be associated with improved soil physical and hydraulic properties, which suggest a higher soil pore connectivity. Hassler et al. [12] found similar K_s at 0–6 cm depth between 100-year-old and 12–15-year-old secondary forests in Panama. Similarly, Leite et al. [24], at Brazilian Caatinga, obtained no significant differences for surface K_s between old-growth forest (more than 55 years) and young secondary forest (7 years). This K_s recovery to pre-pasture levels was also detected at 12.5 cm depth after 15 years of pasture abandonment for an Oxisol in the Brazilian Amazon [11] and by other studies carried out in tropical environments [20,21]. In contrast, after 10 years of natural regeneration in Ecuador, no significant changes of K_s were reported at 12.5 cm depth for an Inceptisol, which was related with invasive species delaying the K_s recovery [33].

Our K_s in the RF plots can be compared with those for Lozano-Baez et al. [28], as both investigations in the same forest biome estimated the K_s with the SSBI method, on the same soil texture (e.g., sandy clay loam) and at the same soil depth used an identical measurement technique (e.g., Beerkan method) and instruments. Our mean K_s in the RF (215 mm h^{-1}) was close to reported value (387 mm h^{-1}) by Lozano-Baez et al. [28] under similar soil conditions. This difference can be explained by the more conserved soil conditions in the study area of Lozano-Baez et al. [28], for example, as their remnant forest was never burned or cultivated. In contrast, our RF was partially cleared and was affected by natural fire disturbances. This observation is in line with several studies, which suggest that in old-growth tropical forests the K_s can be affected by past soil degradation and intensity of forest use [11,15,29,57]. The K_s values obtained in RF plots (e.g., from 23 to 1122 mm h^{-1}) showed the high spatial variability of the infiltration process under forest cover, which can be associated with the heterogeneous soil structure, lower ρ_b and higher *Mac* [57–59]. Another factor to consider is the spatial heterogeneity of our RF, where different landscape conditions such as higher slope and vegetation attributes among sample plots could have influenced the K_s variability. In this sense, we believe that the true reference soil condition could be represented by RF plot RF2, due to observed low soil disturbance in this plot, which is consistent with the higher tree basal area (39.4 m^2 ha^{-1}), vegetation height (average of 12.9 m), species richness (41 trees and non-tree; Table 1 and Table S4), *OC* and soil porosity (Table 3 and Table S5). Unfortunately, it was not possible to find similar forests in the study area, but we could expect that infiltration capacity in other Brazilian Atlantic Forest patches will be directly related with the forest age and forest conditions, which has been shown by several studies [12,24,60]. Therefore, the K_s values in RF could be limited by the number of measurements (n = 28), which should be increased in future studies, considering the gradient and spatial heterogeneity of forest cover.

Our results highlight the importance of land use legacy on K_s recovery after forest restoration. The restoration diagnosis in APR and AR based on the "Diagnostic" protocol [37], allowed evidencing significant differences in the initial environmental situations (e.g., naturally regenerating native plants) and land use history between both sites. In fact, the initial differences in the initial environmental situations in each restored site allowed the restoration practitioners to identify and select at the beginning of the restoration project the most suitable restoration strategy. Figure 3 provides a graphical summary of these differences between APR and AR. Our findings are also in agreement with other studies that reported lower K_s when land use was more intense prior to forest regrowth [11,21,27]. In this sense, our AR site with a more intensive land use history, resulted in significant lower K_s, which could be attributed mainly to greater soil exposure and soil compaction during the land use history. Our mean K_s in the AR site (163 mm h^{-1}) was considerably higher than the reported value (54 mm h^{-1}) by Lozano-Baez et al. [28] on the same soil texture (e.g., sandy clay loam). This difference can be explained by the higher ρ_b values found in the actively restored forest of Lozano-Baez et al. [28].

Despite no statistically significant differences being found between AR and other forest plots, we stress that the specific past land use intensity and management in each plot could also have played an important role in soil degradation. Thus, we found AR plots (AR1 and AR3), where the K_s values were below the mean of other forest plots. The lower K_s in these plots are consistent with their higher ρ_b and *Mic*, lower *Mac* and *OC* (Table S5) and possibly a more intensive past management, suggesting that these plots still retain the "memory" from the previous land use. Similar results have been reported by Bonell et al. [26], for a 10-year-old Acacia plantation in India growing in Ultisols and Oxisols, with lower K_s when compared to less disturbed forests. On the other hand, the higher K_s in restored plots AR2 and AR4 is closer to the RF and APR plots, indicating that in some cases after 11 years the active restoration could reach the infiltration recovery target defined by the reference conditions. This finding agrees with recent literature reviews, which show that K_s recovery after tree planting in the tropics occur across a wide range of soil conditions [61,62] and probably after more than one decade [26,33]. In addition, it is important to underscore that before tree planting at AR site, trampling pressure occurred for 13 years over abandoned agricultural terraces, causing terracing failure. Several previous studies have reported an increase in soil loss, surface runoff, ρ_b and reduction of infiltration rates after terrace abandoning [63,64]. Another important factor that might have influenced the current K_s in AR site is the possible soil compaction during soil preparation, tree planting using bulldozers or tractors is associated with high levels of soil disturbance, and the effects of this soil preparation can persist long after tree planting [65]. Overall, these circumstances suggest that initial soil conditions before forest restoration actions at AR site were more degraded than in APR. Nevertheless, in our study the lack of soil measurements in each moment of the land use history precludes a stronger understanding of the relative impacts of historical land management on K_s recovery, thus future studies should consider the role of previous land use, comparing sites with a truly identical history.

The significantly lower K_s observed in pasture sites compared to forest land-cover types was consistent with several previous studies [20,33,66], supporting the importance of preserving the forest cover and promoting forest restoration actions in the landscape to maintain the infiltration process. This result can be attributed to higher ρ_b and *Mic* in both pasture sites (Table 3). In the present study, we observed a significant higher K_s in the LiP compared to HiP. The differences in K_s between LiP and HiP are mainly related to factors such as cattle-grazing intensity and the duration of pasture cover, which has been similarly reported in several other tropical studies [11,12,66]. Additionally, the similar *OC* between pasture sites and forest covers (RF and APR) is a trend that has been previously noted by other authors [9,22,28,67], suggesting that such similarities are linked to the accumulation of organic matter by the root system of grasses, the animal-derived inputs and application of fertilizers.

4.2. Relationships between Soil, Vegetation and Land Use History

When evaluating vegetation attributes in forest cover sites, our study revealed the significantly lower values in AR and higher values in RF. The higher values of vegetation attributes in the RF could be explained as a consequence of the longer time that this old-growth forest (Figure 3) has remained undisturbed [68,69]. It is interesting to note that basal area and vegetation height of trees in AR could reach statistically similar values to the RF. Similarly, Garcia et al. [70] found in the same biome no significant differences in the basal area between actively restored forests (12, 23 and 55 years old) and the reference condition. Furthermore, it is noteworthy in AR that, while K_s, basal area and vegetation height of trees have reached statistically similar values with the RF, other vegetation attributes, such as canopy cover, tree density and species richness, were significantly lower than those for the RF plots. For instance, the lower canopy cover, density and richness of saplings (e.g., lack of regenerating trees) in AR might compromise the future forest structure [68], which could hamper the recovery of soil hydraulic properties [62]. Although the herbaceous cover was not directly quantified, we observed a higher abundance of grasses (e.g., *U. brizantha*) in AR site, which could be related to the lower canopy cover. The open canopy conditions in AR may have favored the persistence of grasses, hindering the recruitment of new trees species [71]. In contrast, APR site had statistically similar K_s (Figure 4) and vegetation attributes to the RF (Table 1). However, the vegetation attributes in APR were mainly influenced by non-native trees, such as *P. guajava*, an aggressive pioneer species with allelopathic potential [72]. For these reasons, we suggest that both restored forests need management activities to improve soil and vegetation attributes. Canopy cover protects the soil from physical disturbance, and higher species richness and tree density with native species can produce a higher biomass and enhance the K_s [73,74].

One possible explanation for the different outputs between AR and APR are the initial soil conditions at each site. As mentioned above, when the forest restoration actions began in APR, the soil might have had better initial conditions (lower ρ_b, higher K_s, OC and soil porosity), and some degree of natural regeneration, which stimulated the potential recovery of the site and facilitating the recruitment of new trees species [3,37]. In AR site, the more intensive land use history probably led to an area with low resilience, a more compacted soil and poor OC. In particular, the intensity of past land use has been reported as the main factor affecting tropical forest recovery, for example, the vegetation in pastures with a long-lasting land use will regenerate more slowly relative to pastures used less intensively [75,76]. Both restoration approaches (APR and AR) have an important role in the process of restoring degraded ecosystems in tropical landscapes and can be used complementary to enhance the chances of restoration success [37,68]. To understand the soil hydraulic recovery after forest restoration, it is important that future studies consider the role of the duration and intensity of the previous land use, including parameters to assess the land use legacy effects [11,77] and more measurements over time in deeper soil layers, which may reveal further differences among restoration strategies.

The correlation results for *Pt* vs. *Mac* and K_s vs. *Mac* are in agreement with several other studies in the Atlantic Forest [30,58], which reported the positive influence of *Mac* for the pore space and infiltration process. This finding is consistent with the inverse relationships between K_s vs. ρ_b, K_s vs. *Mic* and *Mac* vs. ρ_b, also found by some studies [9,23,28,78]. High OC contributes to the trend of increasing K_s, soil porosity and reducing ρ_b values [9]. However, the reverse occurred in the present study, which can be attributed to the high OC and ρ_b in pastures sites as well as low OC in AR but with a high K_s, suggesting that recovery of soil physical and hydraulic properties is not only dependent on the OC. The result of the PCA indicate the importance of forest cover to promote the infiltration and better soil physical and hydraulic properties. This could be associated with the litter inputs, roots and higher soil faunal activity produced by the trees, which can influence positively the aggregate stability, *Mac* and OC, thereby K_s increase in forest covers [79,80]. Nevertheless, there is a need to further research the plant–soil interactions; for example, little attention has been paid to the effects of individual trees, richness and density of species on K_s and soil hydraulic properties [80].

5. Conclusions

The K_s recovery differed between AR and APR sites. As we expected, the knowledge of the land use history was crucial for understanding the current differences among the study sites for K_s, soil physical and hydraulic properties. This is consistent with our previous work [28] in the same forest restoration program. The K_s and vegetation attributes decreased when land use was more intense prior to forest restoration actions. The influence of land use intensity on soil physical and hydraulic properties could also be evidenced in the comparison between LiP and HiP. The present results further illustrate the positive correlation between K_s and vegetation attributes (tree basal area, vegetation height of trees and overstory richness) in forests undergoing restoration.

Supplementary Materials: The following are available online at http://www.mdpi.com/2073-4441/11/1/86/s1, Figure S1: Correlogram showing the Pearson correlations coefficients between soil attributes across the study sites: Reference Forest, Assisted Passive Restoration, Active Restoration, Low-intensity Pasture and High-intensity Pasture. Table S1: List of species used in the Active Restoration site. Table S2: Species list of the trees with DBH > 5 cm sampled in the studies sites: Reference Forest, Assisted Passive Restoration and Active Restoration. Table S3: Species list of the trees with DBH 1–5 cm sampled in the studies sites: Reference Forest, Assisted Passive Restoration and Active Restoration. Table S4: Vegetation attributes across the study plots. Table S5: Mean for soil attributes in the depth 0–5 cm across the study plots.

Author Contributions: S.E.L.-B. carried out the data collection and wrote the initial draft. M.C. participated in the design of the study and helped to draft and edit the manuscript. S.F.d.B.F. and R.R.R. provided important advice on the concept of structuring of the manuscript and supervised the study. M.C. and S.D.P. analyzed the data and made great contributions to writing the manuscript. All authors reviewed and approved the final manuscript.

Funding: This research was supported by the Fundação de Amparo à Pesquisa do Estado de São Paulo (BIOTA/FAPESP Program: 2013/50718-5 and 1999/09635-0), and Conselho Nacional de Desenvolvimento Científico e Tecnológico (CNPq 561897/2010-7 and Miguel Cooper's, Ricardo Ribeiro Rodrigues's and Silvio Frosini de Barros Ferraz's scientific productivity fellowships).

Acknowledgments: We would like to thank Carolina Bozetti Rodrigues and Luiz Felippe Salemi for their valuable suggestions. We are grateful for Vanessa Erler Sontag who illustrated the land use history, as well as for the taxonomist and researchers who assisted the plant species identification: Vinicius Castro Souza, Fabiano Turini Farah, Thiago Bevilacqua, Gabriel Dalla Colleta and Marcelo Antonio Pinho Ferreira. We thank Daigard Ricardo Ortega, Monica Borda, Miller Ruiz, Raissa Corrêa de Andrade, Renato de Oliveira, Nayana Alves Pereira, Glaucia Santos, Aline Franzosi and all field assistants for their company and help in the field. The authors also would like to thank the landowners who allowed the forest sampling and anonymous reviewers for their constructive comments.

Conflicts of Interest: The authors declare no conflict of interest.

References

1. Chazdon, R.L. Beyond deforestation: Restoring forests and ecosystem services on degraded lands. *Science* **2008**, *320*, 1458–1460. [CrossRef]
2. Keenan, R.J.; Reams, G.A.; Achard, F.; de Freitas, J.V.; Grainger, A.; Lindquist, E. Dynamics of global forest area: Results from the FAO Global Forest Resources Assessment 2015. *For. Ecol. Manag.* **2015**, *352*, 9–20. [CrossRef]
3. Rodrigues, R.R.; Gandolfi, S.; Nave, A.G.; Aronson, J.; Barreto, T.E.; Vidal, C.Y.; Brancalion, P.H.S. Large-scale ecological restoration of high-diversity tropical forests in SE Brazil. *For. Ecol. Manag.* **2011**, *261*, 1605–1613. [CrossRef]
4. Shono, K.; Cadaweng, E.A.; Durst, P.B. Application of assisted natural regeneration to restore degraded tropical forestlands. *Restor. Ecol.* **2007**, *15*, 620–626. [CrossRef]
5. Zahawi, R.A.; Reid, J.L.; Holl, K.D. Hidden costs of passive restoration: Passive restoration. *Restor. Ecol.* **2014**, *22*, 284–287. [CrossRef]
6. Holl, K.D.; Aide, T.M. When and where to actively restore ecosystems? *For. Ecol. Manag.* **2011**, *261*, 1558–1563. [CrossRef]
7. Badalamenti, E.; da Silveira Bueno, R.; Campo, O.; Gallo, M.; La Mela Veca, D.; Pasta, S.; Sala, G.; La Mantia, T. Pine stand density influences the regeneration of *Acacia saligna* Labill. H.L. Wendl. and native woody species in a mediterranean coastal pine plantation. *Forests* **2018**, *9*, 359. [CrossRef]
8. Crouzeilles, R.; Ferreira, M.S.; Chazdon, R.L.; Lindenmayer, D.B.; Sansevero, J.B.B.; Monteiro, L.; Iribarrem, A.; Latawiec, A.E.; Strassburg, B.B.N. Ecological restoration success is higher for natural regeneration than for active restoration in tropical forests. *Sci. Adv.* **2017**, *3*, 1–7. [CrossRef]
9. Gageler, R.; Bonner, M.; Kirchhof, G.; Amos, M.; Robinson, N.; Schmidt, S.; Shoo, L.P. Early response of soil properties and function to riparian rainforest restoration. *PLoS ONE* **2014**, *9*, e104198. [CrossRef]
10. Mendes, M.S.; Latawiec, A.E.; Sansevero, J.B.B.; Crouzeilles, R.; de Moraes, L.F.D.; Castro, A.; Pinto, H.N.A.; Brancalion, P.H.S.; Rodrigues, R.R.; Chazdon, R.L.; et al. Look down—There is a gap—The need to include soil data in Atlantic Forest restoration: Scarcity of soil data in restoration. *Restor. Ecol.* **2018**. [CrossRef]
11. Zimmermann, B.; Elsenbeer, H.; De Moraes, J.M. The influence of land-use changes on soil hydraulic properties: Implications for runoff generation. *For. Ecol. Manag.* **2006**, *222*, 29–38. [CrossRef]

12. Hassler, S.K.; Zimmermann, B.; van Breugel, M.; Hall, J.S.; Elsenbeer, H. Recovery of saturated hydraulic conductivity under secondary succession on former pasture in the humid tropics. *For. Ecol. Manag.* **2011**, *261*, 1634–1642. [CrossRef]

13. Zimmermann, A.; Schinn, D.S.; Francke, T.; Elsenbeer, H.; Zimmermann, B. Uncovering patterns of near-surface saturated hydraulic conductivity in an overland flow-controlled landscape. *Geoderma* **2013**, *195–196*, 1–11. [CrossRef]

14. Alagna, V.; Di Prima, S.; Rodrigo-Comino, J.; Iovino, M.; Pirastru, M.; Keesstra, S.; Novara, A.; Cerdà, A. The impact of the age of vines on soil hydraulic conductivity in vineyards in eastern Spain. *Water* **2017**, *10*, 14. [CrossRef]

15. Di Prima, S.; Marrosu, R.; Lassabatere, L.; Angulo-Jaramillo, R.; Pirastru, M. In situ characterization of preferential flow by combining plot- and point-scale infiltration experiments on a hillslope. *J. Hydrol.* **2018**, *563*, 633–642. [CrossRef]

16. Cullotta, S.; Bagarello, V.; Baiamonte, G.; Gugliuzza, G.; Iovino, M.; La Mela Veca, D.S.; Maetzke, F.; Palmeri, V.; Sferlazza, S. Comparing different methods to determine soil physical quality in a mediterranean forest and pasture land. *Soil Sci. Soc. Am. J.* **2016**, *80*, 1038–1056. [CrossRef]

17. Di Prima, S.; Bagarello, V.; Angulo-Jaramillo, R.; Bautista, I.; Cerdà, A.; del Campo, A.; González-Sanchis, M.; Iovino, M.; Lassabatere, L.; Maetzke, F. Impacts of thinning of a Mediterranean oak forest on soil properties influencing water infiltration. *J. Hydrol. Hydromech.* **2017**, *65*, 276–286. [CrossRef]

18. Elrick, D.E.; Reynolds, W.D. Methods for analyzing constant-head well permeameter data. *Soil Sci. Soc. Am. J.* **1992**, *56*, 320–323. [CrossRef]

19. Deb, S.K.; Shukla, M.K. Variability of hydraulic conductivity due to multiple factors. *Am. J. Environ. Sci.* **2012**, *8*, 489–502. [CrossRef]

20. Godsey, S.; Elsenbeer, H. The soil hydrologic response to forest regrowth: A case study from southwestern Amazonia. *Hydrol. Process.* **2002**, *16*, 1519–1522. [CrossRef]

21. Ziegler, A.D.; Giambelluca, T.W.; Tran, L.T.; Vana, T.T.; Nullet, M.A.; Fox, J.; Vien, T.D.; Pinthong, J.; Maxwell, J.; Evett, S. Hydrological consequences of landscape fragmentation in mountainous northern Vietnam: Evidence of accelerated overland flow generation. *J. Hydrol.* **2004**, *287*, 124–146. [CrossRef]

22. Paul, M.; Catterall, C.P.; Pollard, P.C.; Kanowski, J. Recovery of soil properties and functions in different rainforest restoration pathways. *For. Ecol. Manag.* **2010**, *259*, 2083–2092. [CrossRef]

23. Nyberg, G.; Bargués Tobella, A.; Kinyangi, J.; Ilstedt, U. Soil property changes over a 120-yr chronosequence from forest to agriculture in western Kenya. *Hydrol. Earth Syst. Sci.* **2012**, *16*, 2085–2094. [CrossRef]

24. Leite, P.A.M.; de Souza, E.S.; dos Santos, E.S.; Gomes, R.J.; Cantalice, J.R.; Wilcox, B.P. The influence of forest regrowth on soil hydraulic properties and erosion in a semiarid region of Brazil. *Ecohydrology* **2017**, *11*, 1–12. [CrossRef]

25. Mapa, R.B. Effect of reforestation using Tectona grandis on infiltration and soil water retention. *For. Ecol. Manag.* **1995**, *77*, 119–125. [CrossRef]

26. Bonell, M.; Purandara, B.K.; Venkatesh, B.; Krishnaswamy, J.; Acharya, H.A.K.; Singh, U.V.; Jayakumar, R.; Chappell, N. The impact of forest use and reforestation on soil hydraulic conductivity in the Western Ghats of India: Implications for surface and sub-surface hydrology. *J. Hydrol.* **2010**, *391*, 47–62. [CrossRef]

27. Ghimire, C.P.; Bruijnzeel, L.A.; Bonell, M.; Coles, N.; Lubczynski, M.W.; Gilmour, D.A. The effects of sustained forest use on hillslope soil hydraulic conductivity in the Middle Mountains of Central Nepal: Sustained forest use and soil hydraulic conductivity. *Ecohydrology* **2014**, *7*, 478–495. [CrossRef]

28. Lozano-Baez, S.; Cooper, M.; Ferraz, S.; Ribeiro Rodrigues, R.; Pirastru, M.; Di Prima, S. Previous land use affects the recovery of soil hydraulic properties after forest restoration. *Water* **2018**, *10*, 453. [CrossRef]

29. Zwartendijk, B.W.; van Meerveld, H.J.; Ghimire, C.P.; Bruijnzeel, L.A.; Ravelona, M.; Jones, J.P.G. Rebuilding soil hydrological functioning after swidden agriculture in eastern Madagascar. *Agric. Ecosyst. Environ.* **2017**, *239*, 101–111. [CrossRef]

30. Cooper, M.; Rosa, J.D.; Medeiros, J.C.; de Oliveira, T.C.; Toma, R.S.; Juhász, C.E.P. Hydro-physical characterization of soils under tropical semi-deciduous forest. *Sci. Agricol.* **2012**, *69*, 152–159. [CrossRef]

31. Ziter, C.; Graves, R.A.; Turner, M.G. How do land-use legacies affect ecosystem services in United States cultural landscapes? *Landsc. Ecol.* **2017**, *32*, 2205–2218. [CrossRef]

32. Foster, D.; Swanson, F.; Aber, J.; Burke, I.; Brokaw, N.; Tilman, D.; Knapp, A. The importance of land-use legacies to ecology and conservation. *BioScience* **2003**, *53*, 77–88. [CrossRef]

33. Zimmermann, B.; Elsenbeer, H. Spatial and temporal variability of soil saturated hydraulic conductivity in gradients of disturbance. *J. Hydrol.* **2008**, *361*, 78–95. [CrossRef]

34. Mello, M.H.; Pedro Junior, M.J.; Ortolani, A.A.; Alfonsi, R.R. *Chuva e Temperatura: Cem Anos de Observações em Campinas*; Boletim Tecnico; IAC: Campinas, Brazil, 1994.

35. de Oliveira, L.H.d.S.; Valladares, G.S.; Coelho, R.M.; Criscuolo, C. Soil vulnerability to degradation at Campinas municipality, SP. *Geografia (Londrina)* **2014**, *22*, 65–79.

36. Soil Survey Staff. *Keys to Soil Taxonomy*, 12th ed.; USDA-Natural Resources Conservation Service: Washington, DC, USA, 2014.

37. Rodrigues, R.R.; Lima, R.A.F.; Gandolfi, S.; Nave, A.G. On the restoration of high diversity forests: 30 years of experience in the Brazilian Atlantic Forest. *Biol. Conserv.* **2009**, *142*, 1242–1251. [CrossRef]

38. Molin, P.G.; Gergel, S.E.; Soares-Filho, B.S.; Ferraz, S.F.B. Spatial determinants of Atlantic Forest loss and recovery in Brazil. *Landsc. Ecol.* **2017**, *32*, 857–870. [CrossRef]

39. Preiskorn, G.M.; Pimenta, D.; Amazonas, N.T.; Nave, A.G.; Gandolfi, S.; Rodrigues, R.R.; Belloto, A.; Cunha, M.C.S. Metodologia de restauração para fins de aproveitamento econômico (reservas legais e áreas agrícolas). In *Pacto Pela Restauração da Mata Atlântica—Referencial dos Conceitos e ações de Restauração Florestal*; Rodrigues, R.R., Brancalion, P.H.S., Eds.; LERF/ESALQ: Instituto BioAtlântica: São Paulo, Brazil, 2009; pp. 158–175, ISBN 978-85-60840-02-1.

40. Suganuma, M.S.; Durigan, G. Indicators of restoration success in riparian tropical forests using multiple reference ecosystems: Indicators of riparian forests restoration success. *Restor. Ecol.* **2015**, *23*, 238–251. [CrossRef]

41. Toledo, R.M.; Santos, R.F.; Baeten, L.; Perring, M.P.; Verheyen, K. Soil properties and neighbouring forest cover affect above-ground biomass and functional composition during tropical forest restoration. *Appl. Veg. Sci.* **2018**, *21*, 179–189. [CrossRef]

42. Viani, R.A.G.; Barreto, T.E.; Farah, F.T.; Rodrigues, R.R.; Brancalion, P.H.S. Monitoring young tropical forest restoration sites: How much to measure? *Trop. Conserv. Sci.* **2018**, *11*, 1–9. [CrossRef]

43. Chaves, R.B.; Durigan, G.; Brancalion, P.H.S.; Aronson, J. On the need of legal frameworks for assessing restoration projects success: New perspectives from São Paulo state (Brazil): Legal instruments for assessing restoration. *Restor. Ecol.* **2015**, *23*, 754–759. [CrossRef]

44. Canfield, R. Application of line interception method in sampling range vegetation. *J. For.* **1941**, *39*, 388–394. [CrossRef]

45. Gee, G.; Or, D. Particle-size analysis. In *Methods of Soil Analysis: Physical Methods*; Dane, J.H., Topp, C., Eds.; Soil Science Society of America: Madison, WI, USA, 2002; pp. 255–293, ISBN 978-0-89118-841-4.

46. Flint, A.L.; Flint, L.E. Particle density. In *Methods of Soil Analysis: Physical Methods*; Dane, J., Topp, G.C., Eds.; Soil Science Society of America: Madison, WI, USA, 2002; pp. 229–240.

47. Walkley, A.; Black, I.A. An examination of the degtjareff method for determining soil organic matter, and a proposed modification of the chromic acid titration method. *Soil Sci.* **1934**, *37*, 29–38. [CrossRef]

48. Lassabatère, L.; Angulo-Jaramillo, R.; Soria Ugalde, J.M.; Cuenca, R.; Braud, I.; Haverkamp, R. Beerkan estimation of soil transfer parameters through infiltration experiments—BEST. *Soil Sci. Soc. Am. J.* **2006**, *70*, 521. [CrossRef]

49. Braud, I.; De Condappa, D.; Soria, J.M.; Haverkamp, R.; Angulo-Jaramillo, R.; Galle, S.; Vauclin, M. Use of scaled forms of the infiltration equation for the estimation of unsaturated soil hydraulic properties (the Beerkan method). *Eur. J. Soil Sci.* **2005**, *56*, 361–374. [CrossRef]

50. Bagarello, V.; Di Prima, S.; Iovino, M. Estimating saturated soil hydraulic conductivity by the near steady-state phase of a Beerkan infiltration test. *Geoderma* **2017**, *303*, 70–77. [CrossRef]

51. Alagna, V.; Iovino, M.; Bagarello, V.; Mataix-Solera, J.; Lichner, L. Alternative analysis of transient infiltration experiment to estimate soil water repellency. *Hydrol. Process.* **2018**. [CrossRef]

52. Di Prima, S.; Concialdi, P.; Lassabatere, L.; Angulo-Jaramillo, R.; Pirastru, M.; Cerdà, A.; Keesstra, S. Laboratory testing of Beerkan infiltration experiments for assessing the role of soil sealing on water infiltration. *Catena* **2018**, *167*, 373–384. [CrossRef]

53. Danielson, R.E.; Sutherland, P.L. Porosity. In *Methods of Soil Analysis. Part I. Physical and Mineralogical Methods. Agronomy Monograph No. 9*; Klute, A., Ed.; American Society of Agronomy, Soil Science Society of America: Madison, WI, USA, 1986; pp. 443–461.

54. EMBRAPA. *Manual of Methods of Soil Analysis*, 2nd ed.; Embrapa Soils: Rio de Janeiro, Brazil, 2011.

55. Reynolds, W.D.; Drury, C.F.; Yang, X.M.; Tan, C.S. Optimal soil physical quality inferred through structural regression and parameter interactions. *Geoderma* **2008**, *146*, 466–474. [CrossRef]

56. Lee, D.M.; Elrick, D.E.; Reynolds, W.D.; Clothier, B.E. A comparison of three field methods for measuring saturated hydraulic conductivity. *Can. J. Soil Sci.* **1985**, *65*, 563–573. [CrossRef]

57. Scheffler, R.; Neill, C.; Krusche, A.V.; Elsenbeer, H. Soil hydraulic response to land-use change associated with the recent soybean expansion at the Amazon agricultural frontier. *Agric. Ecosyst. Environ.* **2011**, *144*, 281–289. [CrossRef]

58. Salemi, L.F.; Groppo, J.D.; Trevisan, R.; de Moraes, J.M.; de Barros Ferraz, S.F.; Villani, J.P.; Duarte-Neto, P.J.; Martinelli, L.A. Land-use change in the Atlantic rainforest region: Consequences for the hydrology of small catchments. *J. Hydrol.* **2013**, *499*, 100–109. [CrossRef]

59. Cooper, M.; Medeiros, J.C.; Rosa, J.D.; Soria, J.E.; Toma, R.S. Soil functioning in a toposequence under rainforest in São Paulo, Brazil. *Rev. Bras. Ciência Solo* **2013**, *37*, 392–399. [CrossRef]

60. Ferraz, S.F.B.; Ferraz, K.M.P.M.B.; Cassiano, C.C.; Brancalion, P.H.S.; da Luz, D.T.A.; Azevedo, T.N.; Tambosi, L.R.; Metzger, J.P. How good are tropical forest patches for ecosystem services provisioning? *Landsc. Ecol.* **2014**, *29*, 187–200. [CrossRef]

61. Ilstedt, U.; Malmer, A.; Verbeeten, E.; Murdiyarso, D. The effect of afforestation on water infiltration in the tropics: A systematic review and meta-analysis. *For. Ecol. Manag.* **2007**, *251*, 45–51. [CrossRef]

62. Filoso, S.; Bezerra, M.O.; Weiss, K.C.; Palmer, M.A. Impacts of forest restoration on water yield: A systematic review. *PLoS ONE* **2017**, *12*, 1–26. [CrossRef]

63. Wei, W.; Chen, D.; Wang, L.; Daryanto, S.; Chen, L.; Yu, Y.; Lu, Y.; Sun, G.; Feng, T. Global synthesis of the classifications, distributions, benefits and issues of terracing. *Earth-Sci. Rev.* **2016**, *159*, 388–403. [CrossRef]

64. Atta, H.A.E.; Aref, I. Effect of terracing on rainwater harvesting and growth of Juniperus procera Hochst. ex Endlicher. *Int. J. Environ. Sci. Technol.* **2010**, *7*, 59–66. [CrossRef]

65. Löf, M.; Dey, D.C.; Navarro, R.M.; Jacobs, D.F. Mechanical site preparation for forest restoration. *New For.* **2012**, *43*, 825–848. [CrossRef]

66. Martínez, L.; Zinck, J. Temporal variation of soil compaction and deterioration of soil quality in pasture areas of Colombian Amazonia. *Soil Tillage Res.* **2004**, *75*, 3–18. [CrossRef]

67. Nogueira, L.R.; da Silva, C.F.; Pereira, M.G.; Gaia-Gomes, J.H.; da Silva, E.M.R. Biological Properties and Organic Matter Dynamics of Soil in Pasture and Natural Regeneration Areas in the Atlantic Forest Biome. *Rev. Bras. Ciência Solo* **2016**, *40*. [CrossRef]

68. César, R.G.; Moreno, V.S.; Coletta, G.D.; Chazdon, R.L.; Ferraz, S.F.B.; de Almeida, D.R.A.; Brancalion, P.H.S. Early ecological outcomes of natural regeneration and tree plantations for restoring agricultural landscapes. *Ecol. Appl.* **2018**, *28*, 373–384. [CrossRef]

69. Chazdon, R.L.; Finegan, B.; Capers, R.S.; Salgado-Negret, B.; Casanoves, F.; Boukili, V.; Norden, N. Composition and dynamics of functional groups of trees during tropical forest succession in Northeastern Costa Rica: Functional groups of trees. *Biotropica* **2010**, *42*, 31–40. [CrossRef]

70. Garcia, L.C.; Hobbs, R.J.; Ribeiro, D.B.; Tamashiro, J.Y.; Santos, F.A.M.; Rodrigues, R.R. Restoration over time: Is it possible to restore trees and non-trees in high-diversity forests? *Appl. Veg. Sci.* **2016**, *19*, 655–666. [CrossRef]

71. de Souza, F.M.; Batista, J.L.F. Restoration of seasonal semideciduous forests in Brazil: Influence of age and restoration design on forest structure. *For. Ecol. Manag.* **2004**, *191*, 185–200. [CrossRef]

72. Chapla, T.E.; Campos, J.B. Allelopathic evidence in exotic guava (*Psidium guajava* L.). *Braz. Arch. Biol. Technol.* **2010**, *53*, 1359–1362. [CrossRef]

73. Niemeyer, R.J.; Fremier, A.K.; Heinse, R.; Chávez, W.; DeClerck, F.A.J. Woody vegetation increases saturated hydraulic conductivity in dry tropical Nicaragua. *Vadose Zone J.* **2014**, *13*, 1–11. [CrossRef]

74. Fischer, C.; Tischer, J.; Roscher, C.; Eisenhauer, N.; Ravenek, J.; Gleixner, G.; Attinger, S.; Jensen, B.; de Kroon, H.; Mommer, L.; et al. Plant species diversity affects infiltration capacity in an experimental grassland through changes in soil properties. *Plant Soil* **2015**, *397*, 1–16. [CrossRef]

75. Holl, K.D.; Zahawi, R.A. Factors explaining variability in woody above-ground biomass accumulation in restored tropical forest. *For. Ecol. Manag.* **2014**, *319*, 36–43. [CrossRef]

76. Rocha, G.P.E.; Vieira, D.L.M.; Simon, M.F. Fast natural regeneration in abandoned pastures in southern Amazonia. *For. Ecol. Manag.* **2016**, *370*, 93–101. [CrossRef]

77. Bürgi, M.; Östlund, L.; Mladenoff, D.J. Legacy Effects of Human Land Use: Ecosystems as Time-Lagged Systems. *Ecosystems* **2017**, *20*, 94–103. [CrossRef]
78. Owuor, S.O.; Butterbach-Bahl, K.; Guzha, A.C.; Jacobs, S.; Merbold, L.; Rufino, M.C.; Pelster, D.E.; Díaz-Pinés, E.; Breuer, L. Conversion of natural forest results in a significant degradation of soil hydraulic properties in the highlands of Kenya. *Soil Tillage Res.* **2018**, *176*, 36–44. [CrossRef]
79. Bargués Tobella, A.; Reese, H.; Almaw, A.; Bayala, J.; Malmer, A.; Laudon, H.; Ilstedt, U. The effect of trees on preferential flow and soil infiltrability in an agroforestry parkland in semiarid Burkina Faso. *Water Resour. Res.* **2014**, *50*, 3342–3354. [CrossRef] [PubMed]
80. Ilstedt, U.; Bargués Tobella, A.; Bazié, H.R.; Bayala, J.; Verbeeten, E.; Nyberg, G.; Sanou, J.; Benegas, L.; Murdiyarso, D.; Laudon, H.; et al. Intermediate tree cover can maximize groundwater recharge in the seasonally dry tropics. *Sci. Rep.* **2016**, *6*, 1–12. [CrossRef] [PubMed]

 *water*

Article

Automated Laboratory Infiltrometer to Estimate Saturated Hydraulic Conductivity Using an Arduino Microcontroller Board

Pedro Rodríguez-Juárez, Hugo E. Júnez-Ferreira *, Julián González Trinidad, Manuel Zavala, Susana Burnes-Rudecino and Carlos Bautista-Capetillo

Universidad Autónoma de Zacatecas, Campus Siglo XXI, Carretera Zacatecas-Guadalajara Km. 6, Ejido La Escondida, Zacatecas 98160, Mexico; pedrordz@uaz.edu.mx (P.R.-J.); jgonza@uaz.edu.mx (J.G.T.); mzavala73@uaz.edu.mx (M.Z.); sl_burnesr@uap.uaz.edu.mx (S.B.-R.); baucap@uaz.edu.mx (C.B.-C.)
* Correspondence: hejunez@uaz.edu.mx; Tel.: +52-492-108-2170

Received: 29 November 2018; Accepted: 14 December 2018; Published: 17 December 2018

Abstract: This paper describes the design, calibration and testing processes of a new device named Automated Laboratory Infiltrometer (ALI). It allows to determinate in laboratory, under controlled conditions the saturated hydraulic conductivity (Ks) of altered or unaltered soil samples which is a key parameter to understand the movement of water through a porous medium. The ALI combines the advantages of three different approaches: measures vertical infiltration rates in a soil column, measures the actual volumes of vertically drained water through the soil column, and finally, uses heat as a natural tracer to determinate water flux rates through the porous medium; all those parameters are used to determinate Ks. The ALI was developed using the popular Arduino microcontroller board and commercially available sensors that give the whole system a low cost. Data from the ALI are recorded in a microSD memory so they can be easily read from any spreadsheet software helping to reduce time consuming and avoiding reading errors. The performance of this device was evaluated by comparing the water flow rates determined by the three approaches for which is designed; an excellent correlation among them was observed (worst correlation: $R^2 = 0.9826$ and r-RSME = 0.94%).

Keywords: permeameter; infiltrometer; natural tracer; heat; Darcy flow; Stallman equation; saturated hydraulic conductivity; Arduino

---

1. Introduction

Aquifers have become an important source of water, especially in zones with low precipitation and with scarce availability of surface water. However, due to the excessive withdrawals, some aquifers have been carried out to overexploitation and their levels have been reduced drastically, e.g., in Brazil the piezometric surface has been lowered up to 100 m in some points [1], and worldwide groundwater (GW) estimated depletion is 204 ± 30 km^3 y^{-1} [2]. This condition has several negative effects such as saline water intrusion to aquifers in coastal regions, affectations to rivers, springs and wetlands that depend on groundwater supplies, land subsidence due to pore pressure reductions [3,4] and climate changes [2]. In order to achieve an adequate planning and management of groundwater, it is necessary to improve the knowledge of its functioning; the proper understanding of groundwater flow systems is of paramount importance for hydrogeologists.

The hydraulic conductivity is a parameter used in hydrogeology to define the ease by which groundwater can flow through a porous medium [5], it is fundamental to explain the groundwater movement in the saturated zone, recharge-discharge processes in the vadose zone, transport of contaminants, etc. The hydraulic conductivity values depend on several factors such as fluid density and viscosity, packing and size of the granular material, pore connection, among others. In the vadose zone, it can take several values in time since it additionally depends on the water content and the soil water pressure head, therefore, its characterization in this zone is a difficult task due to both its spatial and temporal variability [6]. For saturated conditions, the saturated hydraulic conductivity (Ks) is applied, and due to minor changes assumed in those zones of an aquifer, values are usually kept constant through time.

Several methods to estimate Ks have been developed and they can be classified in different manners. A first approach classifies them in direct (e.g., infiltrometers, seepage meters, lysimeters, etc.) and indirect methods (use of tracers such as isotopes or heat and numerical modelling). Direct methods provide results in the short term; numerical modeling involves deep analysis of porous media physical properties. Readers can refer to Scanlon et al. [7] or Shanafield and Cook [8] for a method review and to Rosenberry and LaBaugh [9] for a detailed explanation of the most common methods. A second approach classifies them in field (e.g., pump test, slug test, single or double ring infiltrometer, Guelph permeameter) and laboratory methods (e.g., lysimeters, constant or falling head permeameter, column test).

Single ring (SRI) or double ring (DRI) infiltrometers are one of the most used devices for in-situ determination of infiltration rates and saturated hydraulic conductivity. Several studies have used infiltrometers [10–14]; recently, Nestingen et al. [15] made a comparison between three types of infiltrometers (modified Philip-Dunne, double-ring, and minidisk infiltrometers) and they found that the double ring infiltrometer was the most precise. With the recent advances in microelectronics, telecommunications and software, some developments have been done to automate this device [13,16,17], which has been useful to reduce time consumption and increase accuracy. Unfortunately, the infiltrometer just provides information from the surface without considering the possibility of horizontal flows. To compensate for this, some researchers have preferred to carry out infiltration tests in the laboratory under controlled conditions using test columns, so it is possible to measure not only the inflows, but also the outflows as well as others parameters such as water content, temperature, etc. [18,19].

Tracers are used in groundwater studies to determinate Ks and infiltration rates at large scales, surface water—groundwater exchanges, among others. Tracers such as coloured dye, artificial radioactive tracers [20], bromide (Br) or tritiated water (HTO) [21], potassium bromide (KBr) [22], disodium-fluorescein, sulforhodamine B or lithium [23] can be artificially injected into aquifers. However, artificial tracers represent potential environmental risks due to their high persistence in the porous medium and they require expensive equipment to measure it.

On the other hand, it has been shown that the use of heat as a tracer has certain advantages: (1) it is a natural phenomenon, therefore it does not disturb the soil or contaminate it; (2) thermal properties of soils have a narrower range of variation than their analogous hydraulic properties [24], so they can be taken directly from literature introducing small uncertainty to analysis; (3) heat measurement does not require expensive laboratory equipment; and, (4) it is more robust and less expensive than chemical tracers. Several studies have been done using temperature time series to estimate surface water-groundwater interactions [24–31]. Some other studies have been carried out in laboratory using heat as a tracer to estimate infiltration rates in test columns [18,32], see Rau et al. [33] and Halloran et al. [34] for reviews.

In this work, a new device capable to determine Ks of altered or unaltered soil samples is presented. This device named Automated Laboratory Infiltrometer (ALI) was designed to combine the characteristics and advantages of three different approaches. 1. It measures the vertical infiltration rates in a soil column; 2. It measures the real volumes of vertically drained water through the soil column, and finally, 3. It uses heat as a natural tracer to estimate water flux rates through a porous medium; these data can be used to estimate the Ks by using the Darcy equation (drained water fluxes and infiltration rates) or the Stallman equation (temperature time series). The ALI has been developed with an open source Arduino microcontroller board and commercially available sensors, which gives to the whole system a lower cost than in previous works. Details of the construction, calibration and evaluation of the ALI are described in this paper, so it can be easily reproduced at low cost by any researcher. It was tested in laboratory and statistical analyses were done to evaluate its performance.

2. Materials and Methods

2.1. The Automated Laboratory Infiltrometer

The ALI (Figure 1) is made in a 1.0 × 0.15 m PVC tube which can be filled with any soil sample; it is wrapped with one layer of flexible polyurethane foam 1 cm thick (thermal conductivity = 0.03 Wm^{-1} K^{-1}) to insulate it from the environment. Its lower end is connected to a plastic funnel and a clear hose to drain out the water; to keep the lower hydraulic head constant, the hose is connected to a small tank from the bottom and once it reaches the desired level it is drained trough the upper outlet to the container of the digital scale.

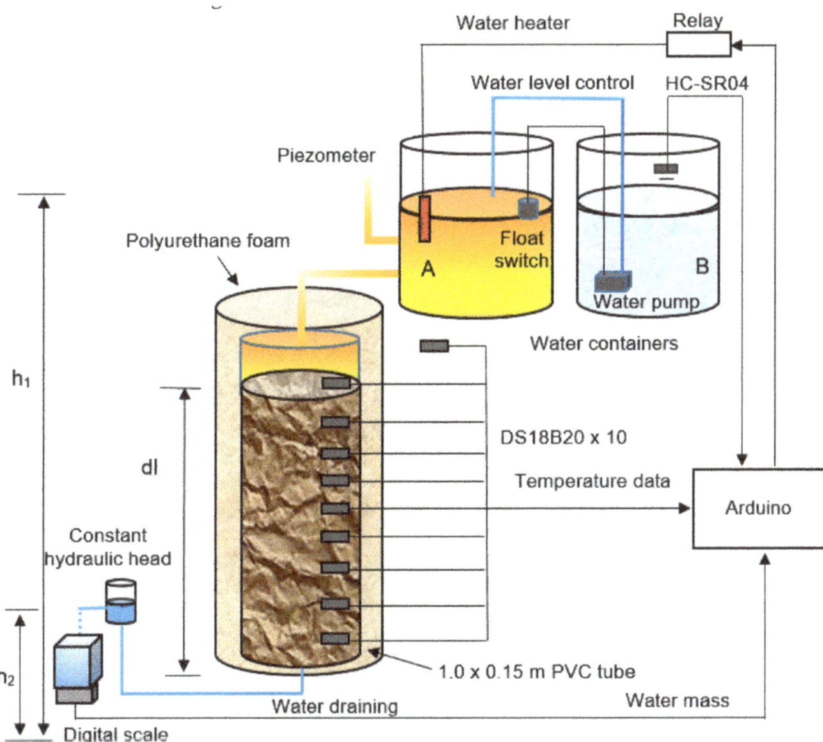

Figure 1. The Automated Laboratory Infiltrometer.

In order to estimate water flux rates through the use of temperature time series, quasi-sinusoidal heating cycles with a 6-h period are induced in the water container A using one 400 W submersible electrical heater. The heater is turned on by the Arduino for 3 h heating up the water and then switched off for 3 h, allowing the water to cool down. To measure temperature, the ALI has installed every 0.10 m along the tube, nine waterproof 1-wire digital temperature sensors DS18B20 Dallas semiconductors brand with accuracy of ±0.5 °C and 9–12 bits of resolution [35] previously calibrated (standard deviation (σ) = 0.0181 °C, coefficient of variation (cv) = 0.0626%). Additionally, one more sensor is installed outside the tube to measure the ambient temperature.

One Cytron ultrasonic ranging module HC-SR04 with resolution of 3.0×10^{-3} m [36] is installed on the top of the container B to measure infiltration volumes. A generic 10 kg load cell (accuracy of $\pm 0.05\%$ [37]) connected to an amplifier module HX711 is used to measure drained water mass.

The hydraulic head on the infiltration surface is kept quasi-constant by using a generic liquid level float switch and a submersible 6 volts water pump.

Data from all sensors are recorded in a microSD memory card and displayed on a 20×4 LCD. Power circuitry is handled by using a single channel relay.

A boundary condition of Dirichlet type (of constant hydraulic head) and quasi-sinusoidal periodic heating cycles were imposed at the top of the column test (water input). Water circulates downwards with free drainage at the bottom (water output) until it reaches the steady state, at this moment, a Dirichlet type boundary condition is imposed at the bottom because the water level is controlled at the container that receives the output water. The ALI can be changed to hydraulic head falling conditions just by eliminating container B and placing the HC-SR04 on container A.

The ALI has a maintenance cycle that depends of the soil under study; during this maintenance, container B is refilled, the digital scale container is emptied, and the system reestablished. This cycle can be augmented by increasing the capacity of the container B and the digital scale. Small changes on the Arduino program have to be done to increase the digital scale capacity. Container A is refilled automatically, so it does not require any maintenance during the experiment. Details of the main functions for the ALI are described below.

2.2. Data Acquisition System

The data acquisition system reads and records data from all the sensors into the microSD memory card every 5 min; these data can be easily read in any spreadsheet program such as Microsoft Excel. It controls the electrical heaters and displays the information on a 20×4 LCD. It is developed using the very popular and low-cost Arduino Mega 2560 microcontroller board based on the ATmega 2560 microcontroller and programmed using the open source Arduino software.

2.3. Drained Water Flux Rates Measurement

A digital scale is made by using a 10 kg load cell (a transducer capable to translate pressure into a proportional electrical signal) mounted on a wood base. Once the system is turned ON, the digital scale starts an auto-calibration subtracting the tare weight. The drained water mass is measured every 5 min and the previous value is subtracted from the new value and accumulated per hours, then it is converted to volume by $V_w = m_w/\rho_w$, where V_w is the volume (m^3), m_w is the mass (kg) and ρ_w density (kg m^{-3}) of the water. These data are used to calculate the drained water rate (Q) per time unit (m^3 h^{-1}). Finally, the water flux rates q (m h^{-1}) are determined by Equation (1), where A is the cross-sectional area of the column (m^2),

$$q = \frac{Q}{A} \tag{1}$$

Calibration was done by applying a constant load for 12 h. Figure 2 is a control chart showing stable behavior, with $\sigma = 7.0 \times 10^{-4}$ kg and cv = 0.11%; the upper limit (UL) is the sum of 2σ and the mean; the lower limit (LL) is the result of the mean minus 2σ. A pre-processing step was done by taking 20 samples at time, then the data were filtered extracting the median, which is more robust than the mean in presence of outliers or skewed data.

2.4. Water Flux Rates Determination by Measuring Infiltration Rates

As shown in Figure 1, there are two water containers. The container A is used to heat up the water supplied to the ALI and keep the upper hydraulic head (h_1) quasi-constant. This level is maintained by the use of a float level switch with a range of 5.0×10^{-3} m; once the water reaches its lower level (h_1-5.0×10^{-3} m), a water pump submerged in the container B is activated until the level reaches the upper level (h_1). The infiltration rate is measured in container B by the use of an ultrasonic ranging module HC-SR04. Basically, its transmitter emits a 40 KHz ultrasonic wave when it is triggered and a timer is started. An ultrasonic pulse travels outward until it encounters an object (in this case water), the water reflects back the pulse, the ultrasonic receiver detects the reflected wave and stops the timer; this is the traveling time, which divided by two and multiplied by the sound speed, gives the distance between the sensor and the water. The increase of this distance multiplied by the cross-sectional area of the container B gives the infiltrated volume, which is accumulated per hours and considered the infiltration rate per time unit (m h^{-1}). The water flux rates are determined by Equation (1).

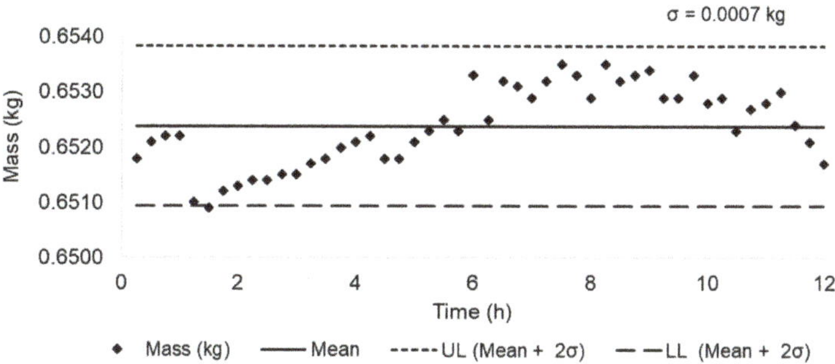

Figure 2. Digital scale calibration along 12 h by applying a constant load.

The sound speeds depend on the air temperature and humidity, so, calibration was done by measuring a constant level of water for 12 h to observe the impact of the temperature and humidity daily changes. Figure 3 is a control chart showing a stable behavior through time, with σ equal to 3.0×10^{-4} m and cv equal to 0.23%; the upper limit (UL) is the sum of 2σ and the mean; the lower limit is the result of the mean minus 2σ. A pre-processing step was done by taking 20 samples at a time, then the data were filtered extracting the median.

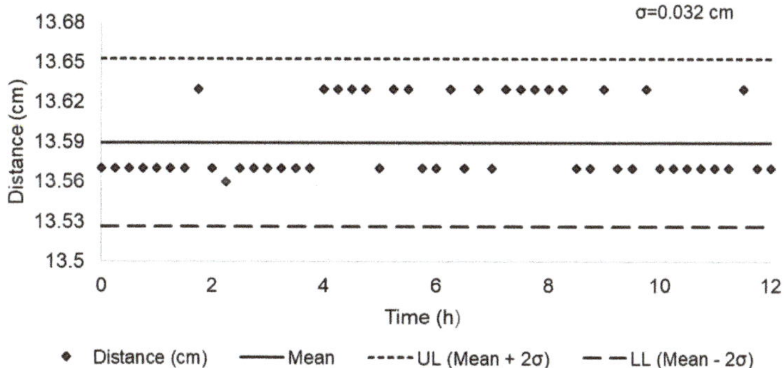

Figure 3. Ultrasonic ranging module calibration along 12 h.

2.5. Low Pass Filter

Data from infiltration rates, drained water flux rates and digital scale are filtered by using a moving average filter (MA). This is a low pass Finite Impulse Response (FIR) filter, used to reduce random noise, which is common on sensors and electronics devices. It is frequently used on digital signal processing and is a premier filter for time domain encoded signals [38]. The moving average filter (2) operates by averaging a number of points (M) from the input signal $x()$, to produce each point in the output signal $y()$,

$$y(i) = \frac{1}{M} \sum_{j=0}^{M-1} x(i+j) \qquad (2)$$

2.6. Water Flux Rates Determination by Using Temperature Time Series

2.6.1. The Heat and Fluid Transport Equation

The conduction-convection equation describes one-dimensional, anisothermal and vertical flow of an incompressible fluid through an isotropic, homogeneous porous medium,

$$K_t \frac{\delta^2 T}{\delta z^2} - qC_w \frac{\delta T}{\delta z} = C_s \frac{\delta T}{\delta t}$$

(3)

where K_t is the thermal conductivity of the saturated porous medium (W m^{-1} °C^{-1}), T is the temperature (°C), q is the infiltration flux (m s^{-1}), C_w and C_s are the volumetric heat capacity of the water and soil porous media respectively (J m^{-3} °C^{-1}), z is the distance along the vertical axis (m) and t is the elapsed time (s). There are some analytical solutions proposed for (3) in the scientific literature; the reader can refer to Irvine et al. [30] for a detailed review. The proposal by Hatch et al. [39] is one of the most commonly used to calculate water flux rates through porous media by using either the amplitude (4) or phase differences (5) of a periodic temperature signal between two points at different depths. This model assumes a quasi-sinusoidal temperature oscillation at the upper boundary and one directional water flow path.

$$q = \frac{C_s}{C_w} \left(\frac{2K_e}{\Delta z} \ln A_r + \sqrt{\frac{\alpha + v^2}{2}} \right)$$

(4)

$$|q| = \sqrt{\alpha - 2 \left(\frac{4\pi \Delta t K_e}{P \Delta z} \right)}$$

(5)

where

$$\alpha = \sqrt{v^4 + \left(\frac{8\pi K_e}{P \Delta z} \right)^2}$$

(6)

$$K_e = \frac{\lambda_0}{C_s} + \beta |v_f|$$

(7)

$$v_f = \frac{q}{n_e}$$

(8)

where K_e is the effective thermal diffusivity (m^2 s^{-1}), A_r and Δt are the amplitude and phase relations (dimensionless), v is the water front velocity (m s^{-1}), P is the period of the sinusoidal signal (s), λ_0 is the baseline thermal conductivity of the saturated sediment (J s^{-1} m^{-1} °C^{-1}), β is the thermal dispersivity (m), v_f is the linear particle velocity (m s^{-1}) and n_e is the effective porosity (dimensionless). Both A_r and Δt are less influenced by errors in thermal conductivity at higher flow rates, as expected when advection becomes more important for heat transport [40]. As demonstrated by Stallman [41], analysis of diurnal temperature fluctuations may yield accurate detection of velocity to a minimum of 0.3 cm/day, so it is useful for most of the soil textures.

2.6.2. Dynamic Harmonic Regression (DHR)

Dynamic harmonic regression, developed by Young et al. [42] is a non-stationary extension of Discrete Fourier Transform (DFT), it works in both time and frequency domains, so it is useful to detect amplitude and phase variations in time. DHR is particularly useful for adaptive seasonal adjustment, signal extraction and interpolation over gaps, as well as forecasting or backcasting [42]; it is a simplification of the unobserved component model (9),

$$y_t = T_t + C_t + e_t$$

(9)

where y_t is the observed time series, T_t is the trend or the zero frequency component, C_t is the cyclical component and e_t is an irregular component assumed to be Gaussian white noise with constant variance. The cyclical component (10) is the sum of the fundamental signal and its associated harmonics

$$C_t = \sum_{i=1}^{N} [a_{i,t} \cos(\omega_i t) + b_{i,t} \sin(\omega_i t)], \tag{10}$$

where $a_{i,t}$ and $b_{i,t}$ are the stochastic time-varying parameters (TPV's) and $\omega_1, \omega_2, \ldots, \omega_N$ are the fundamental frequency (ω_1) and its harmonics ($\omega_i = i\omega_1$) up to the Nyquist frequency (ω_N).

2.6.3. Temperature Time Series Processing

Raw data from temperature sensors are analyzed in both frequency and time domain. The analytical solution from Hatch et al. [39] uses the amplitude ratio (4) or the phase shift (5) of temperature time series to derivate seepage fluxes. So, as a first step, data are filtered by using an anti-aliasing low-pass finite impulse response (FIR) filter designed with a Kaiser window to remove high frequency noise (inherent to any electronic device) and resampled to 48 samples per fundamental cycle (day) which is recommended in order to eliminate spurious filtration artifacts. Dynamic harmonic regression (DHR) is used to isolate fundamental signals and for the extraction of phase and amplitude information. Water flux rates are determined using the analytical solutions developed by Hatch et al. [39]. Resampling, DHR, filtering process as well a water flux rate determination is done through the use of VFLUX, which is a MATLAB toolbox developed by Gordon et al. [43].

Equations (4)–(8) need certain parameters for water flux rate determination. Those parameters are soil physical and thermal properties and they can be determined either by laboratory tests or from literature as proposed by Munz et al. [44]. In this paper, thermal dispersivity and baseline thermal conductivity were taken from literature, and then they were varied until the best results of the relative root mean square error (r-RSME) were reached. Parameters of soil such as effective porosity (n_e), volumetric heat capacity of soil (C_s) and water (C_w), thermal dispersivity (β) and baseline thermal conductivity (λ) were determined experimentally in laboratory as described by Rodríguez et al. [32]. Thermal properties values were found similar to the ones reported in the literature [18,39,44], which supports the idea of a narrower variation, which is one of the advantages of the heat as a tracer. Table 1 shows the physical and thermal parameters for a sand texture soil used to estimate water flux rates with Hatch equations.

Table 1. Physical and thermal parameters of the soil and water used for water flux rates calculation by using analytical solution for heat and transport Equation (3).

Parameters	Determined	Units
Effective porosity (n_e)	0.28	dimensionless
Volumetric heat capacity of soil (C_s)	0.5	Cal cm^{-3} °C^{-1}
Volumetric heat capacity of water (C_w)	1.0	Cal cm^{-3} °C^{-1}
Thermal dispersivity (β)	0.001	M
Baseline thermal conductivity (λ_0)	0.0045	Cal s^{-1} cm^{-1} °C^{-1}

2.6.4. Soil Sample Preparation

A soil sample with sand texture according to the U.S. Department of Agriculture (USDA) soil texture classification, is previously passed through a #20 sieve (nominal sieve opening = 0.841 mm) in order to remove rocks and coarse sand before being used on the column test.

2.7. Saturated Hydraulic Conductivity

Saturated hydraulic conductivity (K_s) is estimated by using the Darcy Equation (11) which is commonly used to calculate the Darcy velocity (one-dimensional and equivalent to the specific discharge) of water for a saturated homogeneous and isotropic porous medium,

$$q = -K_s \frac{dh}{dl} \tag{11}$$

where q is the the Darcy velocity or the water flux rate (m s^{-1}), K_s (m s^{-1}) is the saturated hydraulic conductivity and dh/dl is the hydraulic gradient (dimensionless); the negative sign indicates that the flow of water is in the direction of decreasing head.

3. Results

To evaluate the ALI performance, the PVC tube was filled with the altered sand texture soil sample described in the previous section. To ensure the proper soil compaction and a proper particle accommodation, water circulation through the column was maintained for 12 h with a free drainage at the bottom.

3.1. Hydraulic Boundary Conditions and Data Processing

Dirichlet boundary conditions were established as a constant hydraulic head ($2.05 \pm 5 \times 10^{-3}$ m) at the top boundary and free drainage and a constant hydraulic head at the bottom boundary (0.45 m). Infiltration rates and drained rates were filtered by using a low pass filter; water flux rates were determined and fitted using a MATLAB polynomial fitting.

3.2. Temperature Boundary Conditions and Data Processing

Six hours quasi-sinusoidal heating cycles with an amplitude of 10 °C were induced in container A. Six-hour cycles (cyclical component) were chosen to easily distinguish from the daily temperature cycle (trend).

Temperature data were sampled every five minutes for all the sensors. Figure 4 shows raw temperature signal from eight sensors indicating the depth at which they were installed along the vertical axis. The artificially induced 6-h cycles are easily observable as well as the long-term trend; the signal looks noisy, amplitude ratio is not easy to distinguish due to the proximity of the sensors but phase delays (Δt) are more visible, especially between the points 0.1 cm and 90 cm, which is more distinguishable. Raw temperature data were resampled to 48 samples per fundamental cycle (24 h); this was done to reduce noise. Figure 5 shows temperature signals after being resampled; it can be seen that the noise has been reduced but some still remains, especially on the crests and troughs. Figure 6 shows the temperature signals after filtering and trend removal; now the signal appears without noise, relative temperature changes are clearly observable as well as the phase delays (Δt). Because amplitude ratio between signals is small, water flux rates calculations were done using only (5) and fitted using a MATLAB polynomial fitting.

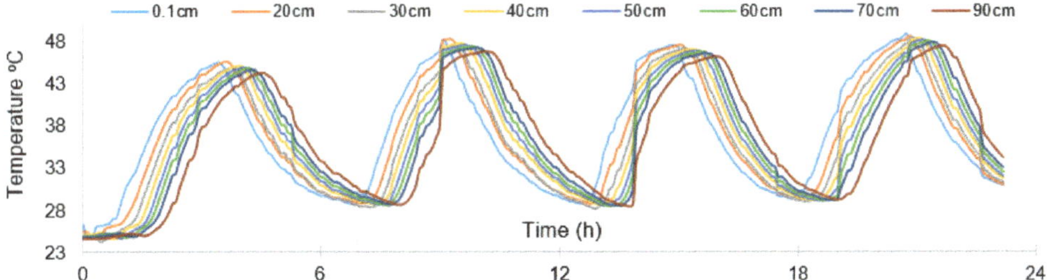

Figure 4. Raw temperature time series from eight sensors installed at different depths and sampled every 5 min.

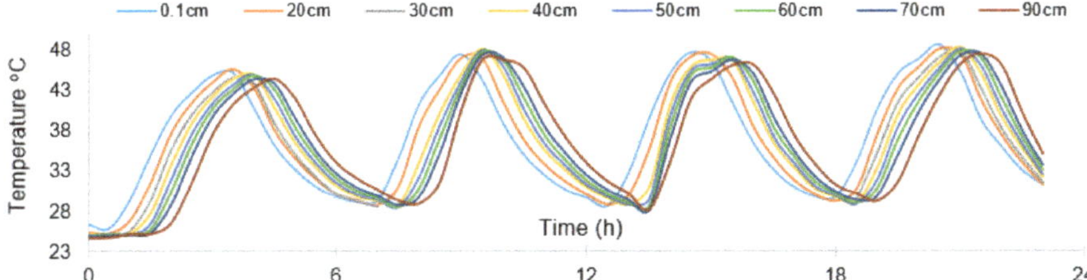

Figure 5. Resampled temperature time series from eight sensors, sampling rate was reduced to 48 samples per fundamental cycle. There is still some noise on the crests and troughs. Phase delays can be observed, amplitude ratios are not so observable due to the proximity between the temperature sensors.

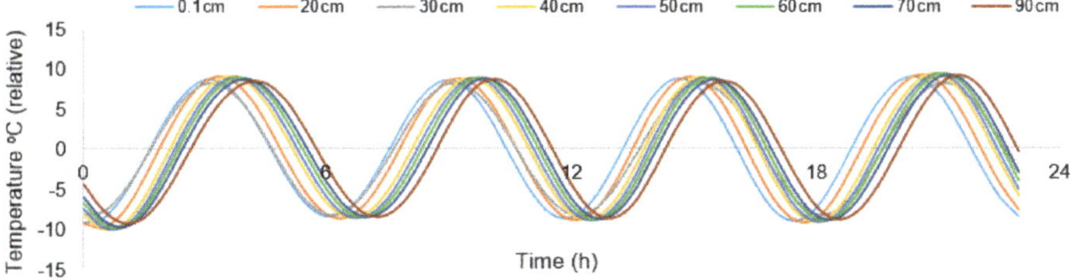

Figure 6. Resampled and filtered temperature time series. A low-pass FIR filter was used to remove high-frequency noise and Dynamic Harmonic Regression (DHR) to remove the long-term trend. Only the changes in amplitude due to artificially induced heating cycles remain.

3.3. Water Flux Rates Comparison

Figure 7 shows the water flux rates (q) determined by the three different approaches under saturated conditions: Infiltrated (determined by measuring the infiltration rates), Drained (determined by measuring the drained water) and Hatch (calculated by VFLUX). As it can be seen, there is a clear trend for all the curves; they start with higher q values and finally they tend to stabilize (after 15 h) once the soil gets almost fully saturated (trapped air bubbles prevent full saturation). Stabilization time depends on different factors such as physical properties and initial boundary conditions of the porous media. Initially, Infiltrated water flux rates are bigger than Drained because it takes more time for the water reaching the lower end and the steady state flow condition. On the other hand, the analytical solution from Hatch et al. [39] was developed for saturated conditions, so Hatch water flux rates must be considered once the steady state is reached. As Figure 7 shows, the three water flux rates look very similar after stabilization.

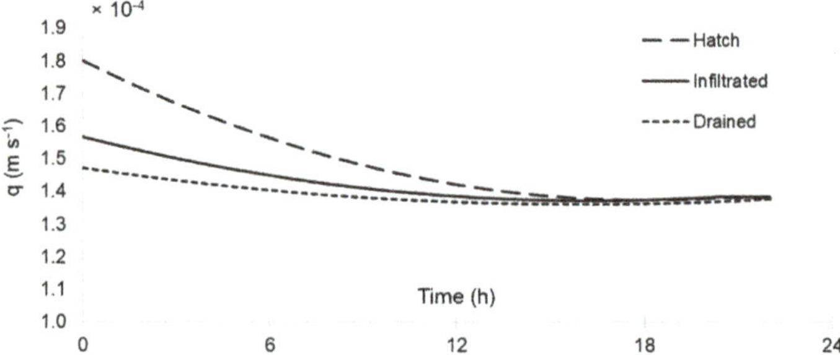

Figure 7. Comparison of water flux rates determined by three approaches. There is a trend to start with higher q values and then the curves tend to stabilize. A high similarity among the curves can be observed once the steady state has been reached.

Correlations between the results of the three different approaches were determined after the steady state was reached. Figure 8 shows the correlation between the results obtained by Hatch and Drained. Figure 9 shows the correlation between the results obtained by Hatch and Infiltrated; and Figure 10 shows the correlation between Drained and Infiltrated. Values of the coefficient of determination (R^2) and relative root mean square error (r-RSME) are also shown.

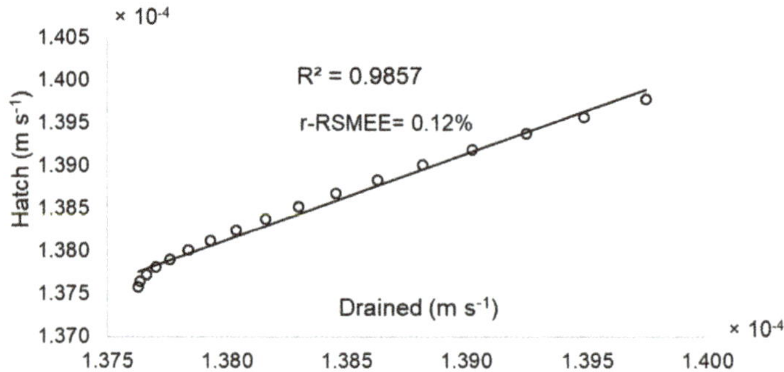

Figure 8. Correlation between Hatch water flux rates and Drained water flux rates.

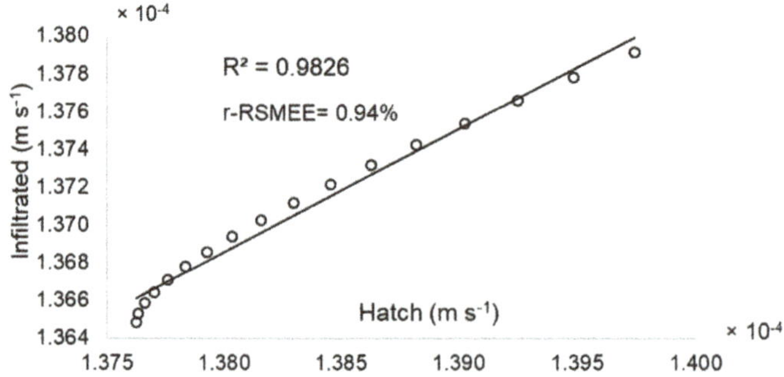

Figure 9. Correlation between Infiltrated water flux rates and Hatch water flux rates.

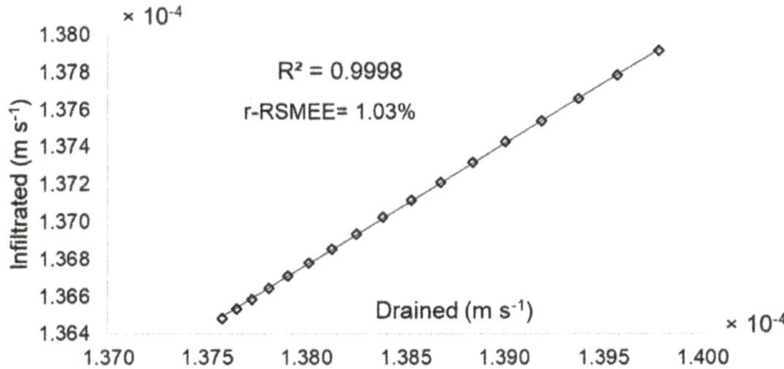

Figure 10. Correlation between Infiltrated water flux rates and Drained water flux rates.

As described by Loutfi et al. [45], R^2 expresses the correlation between the measured values and the estimated ones; the best approximation corresponds to the highest R^2 (closer to 1), RMSE shows the difference between the measured values and the predicted ones; it indicates the scattering of data around a straight line inclined 45°. The approximation is better if RMSE is minimal (tends to 0). Normalized root means square error (nRSME) or relative root mean square error (r-RSME) is calculated by dividing RMSE with the average value of measured data. According to Despotovic et al. [46], model accuracy is considered excellent when r-RMSE is less than 10%, good if 10% < r-RMSE < 20%, fair if 20% < r-RMSE < 30%, and poor if r-RMSE > 30%. Therefore, an excellent correlation among the methods was found, which indicates that the developed device (ALI) is able to estimate efficiently water flux rates by three different methods, giving more confidence in its results.

3.4. Determination of Saturated Hydraulic Conductivity

Saturated hydraulic conductivity Ks was determined by using (11), considering h_1 = 2.05 m, h_2 = 0.45 m, dl = 1.0 m and the water flux rates in steady state conditions from each of the three approaches. On average, the saturated hydraulic conductivity was 8.61×10^{-5} m s^{-1}, which is within the range of the values reported on the saturated hydraulic conductivity in relation to the soil texture table from the U.S. Department of Agriculture (USDA) [47] which gives a range of 4.2×10^{-5} to 1.41×10^{-4} m s^{-1} for soils with sand textures. Table 2 shows the average Ks from the three approaches.

Table 2. Average saturated hydraulic conductivity determined for the three approaches and range of values reported by the U.S. Department of Agriculture (USDA) for sand texture soils.

Approach	Ks	Units
Heat as a tracer [a]	8.6318×10^{-5}	m s^{-1}
Infiltrated [b]	8.6384×10^{-5}	m s^{-1}
Drained [c]	8.5680×10^{-5}	m s^{-1}
USDA [d]	4.2×10^{-5} to 1.41×10^{-4}	m s^{-1}

[a] Ks determined using heat as tracer and the analytical solution from Hatch et al. [39]. [b] Ks determined by measuring infiltration rates from the top. [c] Ks determined by measuring drained flux rates from the bottom. [d] Ranges of values of Ks for soils with sand texture reported by the USDA [47].

4. Conclusions

The saturated hydraulic conductivity is a key parameter to understand the groundwater movement, natural and artificial recharge-discharge of aquifers as well as others hydrological processes such as contaminant transport, among others. There are several methods and devices used to determine Ks for a porous medium; in this paper, a new device and the methodology needed to calculate Ks by three different approaches simultaneously is presented.

A new Automated Laboratory Infiltrometer (ALI) is developed by using a very popular Arduino Mega 2560 Microcontroller board and commercially available sensors; giving the total system a cost of around 200.00 USD, which is very low compared to a single commercial grade sensor. Furthermore, the ALI provides temperature time series, infiltration rates and drained water volumes which allow researchers to calculate water flux rates in altered or unaltered soil samples by three different approaches: one indirect by using heat as a tracer (Stallman equation) and two direct methods, i.e., vertical infiltration rates and drained water volumes (Darcy).

Drained water volumes provided by the ALI are used to verify that the steady state flow is reached; it is approximately the time where the curves for the three approaches used to calculate the water flux rates coincide, therefore, the most trustable value for Ks is determined from this point onwards. Furthermore, temperature time series could be used to determine hydraulic conductivity in unsaturated conditions without using water content sensors or tensiometers which are more expensive in comparison with temperature sensors.

Thermal properties of soils determined by the ALI in laboratory can be used as an alternative way to determine Ks in the field and to measure water interaction (upwards and downwards flows) in the vadose zone in a continuous manner. Long-term temperature data are easier and cheaper to take than the commonly used tracers that require more specialized equipment and knowledge.

The ALI was tested and excellent correlations between the three approaches used to determinate Ks were obtained (R^2 = 0.9826 and r-RSME = 0.94% between Infiltrated and Hatch; R^2 = 0.9857 and r-RSME = 0.12% among Hatch and Drained; and R^2 = 0.9826 and r-RSME = 0.94% between Infiltrated and Drained). The saturated hydraulic conductivity determined for the three methods falls inside the ranges established by the USDA for the analyzed soil, which proved its efficiency.

Analytical solutions from Hatch et al., the Darcy equation as well as the infiltrometer, are very useful to measure water flux rates and hydraulic conductivity in saturated conditions. However, in order to obtain a better picture of the whole infiltration process (unsaturated and saturated conditions) some improvements can be done to the ALI, such as the incorporation of water content sensors and electronic tensiometers. Those additional devices will provide more parameters, allowing researchers to use, for instance, the Richards equations for estimating hydraulic conductivity values for unsaturated conditions.

Author Contributions: Conceptualization, Methodology and Writing-Original Draft Preparation, P.R.-J.; Supervision, Writing—Review & Editing, H.E.J.-F.; Investigation and Resources, M.Z. and J.G.T.; Software, S.B.-R.; Project Administration, C.B.-C.

Funding: This research received no external funding.

Acknowledgments: First author would like to acknowledge the Mexican National Council for Science and Technology (CONACYT) for the scholarship awarded for doctoral studies in engineering sciences.

Conflicts of Interest: The authors declare no conflict of interest.

References

1. De Luna, R.M.R.; Garnés, S.J.D.A.; Cabral, J.J.D.S.P.; dos Santos, S.M. Groundwater overexploitation and soil subsidence monitoring on Recife plain (Brazil). *Nat. Hazards* **2017**, *86*, 1363–1376. [CrossRef]
2. Taylor, R.G. Ground water and climate change. *Nat. Clim. Chang.* **2012**, *3*, 322–329. [CrossRef]
3. Custodio, E. Aquifer overexploitation: What does it mean? *Hydrogeol. J.* **2002**, *10*, 254–277. [CrossRef]
4. Fetter, C.; Boving, T.; Kreamer, D. *Contaminant Hydrogeology*; Waveland Press Inc.: Long Grove, IL, USA, 2017; ISBN 978-14786-3279-5.
5. Tang, Y.; Zhou, J.; Yang, P.; Yan, J.; Zhou, N. *Groundwater Engineering*; Tongji University Press/Springer: Shanghai, China, 2016; ISBN 978-981-10-0669-2.
6. Mertens, J.; Madsen, H.; Feyen, L.; Jacques, D.; Feyen, J. Including prior information in the estimation of effective soil parameters in unsaturated zone modelling. *J. Hydrol.* **2004**, *294*, 251–269. [CrossRef]
7. Scanlon, B.R.; Healy, R.W.; Cook, P.G. Choosing appropriate techniques for quantifying groundwater recharge. *Hydrogeol. J.* **2002**, *10*, 18–39. [CrossRef]

8.   Shanafield, M.; Cook, P.G. Transmission losses, infiltration and groundwater recharge through ephemeral and intermittent streambeds: A review of applied methods. *J. Hydrol.* **2014**, *511*, 518–529. [CrossRef]

9.   Rosenberry, D.O.; LaBa ugh, J.W. *Field Techniques for Estimating Water Fluxes between Surface Water and Ground Water*; USGS: Reston, VA, USA, 2008. Available online: https://pubs.usgs.gov/tm/04d02/ (accessed on 7 December 2018).

10.   Xu, X.; Lewis, C.; Liu, W.; Albertson, J.D.; Kiely, G. Analysis of single-ring infiltrometer data for soil hydraulic properties estimation: Comparison of BEST and Wu methods. *Agric. Water Manag.* **2012**, *107*, 34–41. [CrossRef]

11.   Lai, J.; Luo, Y.; Ren, L. Numerical evaluation of depth effects of double-ring infiltrometers on soil saturated hydraulic conductivity measurements. *Soil Sci. Soc. Am. J.* **2011**, *76*, 867–875. [CrossRef]

12.   Kadam, A.S. Determination of infiltration rate for rite selection of artificial water recharge: An experimental study. *Int. J. Sci. Res.* **2016**, *5*, 699–705. [CrossRef]

13.   Fatehnia, M.; Paran, S.; Kish, S.; Tawfiq, K. Automating double ring infiltrometer with an Arduino microcontroller. *Geoderma* **2016**, *262*, 133–139. [CrossRef]

14.   Vand, A.S.; Sihag, P.; Singh, B.; Zand, M. Comparative evaluation of infiltration models. *KSCE J. Civ. Eng.* **2018**, *22*, 4173–4184. [CrossRef]

15.   Nestingen, R.; Asleson, B.C.; Gulliver, J.S.; Hozalski, R.M.; Nieber, J.L. Laboratory Comparison of Field Infiltrometers. *J. Sustain. Water Built Environ.* **2018**, *4*. [CrossRef]

16.   Arriaga, F.J.; Kornecki, T.S.K.; Balkcom, K.S.B.; Raper, R.L. A method for automating data collection from a double-ring infiltrometer under falling head conditions. *Soil Use Manag.* **2010**, *26*, 61–67. [CrossRef]

17.   Di Prima, S.; Lassabatere, L.; Bagarello, V.; Iovino, M.; Angulo-Jaramillo, R. Testing a new automated single ring infiltrometer for Beerkan infiltration experiments. *Geoderma* **2015**, *262*, 20–34. [CrossRef]

18.   Lautz, L.K. Observing temporal patterns of vertical flux through streambed sediments using time-series analysis of temperature records. *J. Hydrol.* **2012**, *464–465*, 199–215. [CrossRef]

19.   Salas-García, J.; Garfias, J.; Martel, R.; Bibiano-Cruz, L. A low-cost automated test column to estimate soil hydraulic characteristics in unsaturated porous media. *Geofluids* **2017**, *2017*, 6942736. [CrossRef]

20.   IAEA. *Use of Artificial Tracers in Hydrology IAEA-TECDOC-601*; International Atomic Energy Agency: Vienna, Austria, 1991; ISSN 1011-4289.

21.   Yeh, Y.; Lee, C.; Chen, S. A tracer method to determinate hydraulic conductivity and effective porosity of saturated clays under low gradients. *Groundwater* **2000**, *38*, 522–529. [CrossRef]

22.   Hwang, H.; Jeen, S.; Suleiman, A.A.; Lee, K. Comparison of saturated hydraulic conductivity estimated by three different methods. *Water* **2017**, *9*, 942. [CrossRef]

23.   Mosthaf, K.; Brauns, B.; Fjordbøge, A.S.; Rohde, M.M.; Kerrn-Jespersen, H.; Bjerg, P.L.; Binning, P.J.; Broholm, M.M. Conceptualization of flow and transport in a limestone aquifer by multiple dedicated hydraulic and tracer tests. *J. Hydrol.* **2018**, *561*, 532–546. [CrossRef]

24.   Stonestrom, D.A.; Constantz, J. *Heat as a Tool for Studying the Movement of Ground Water near Streams*; USGS Circular 1260; USGS: Reston, VA, USA, 2003. Available online: https://pubs.usgs.gov/circ/2003/circ1260/pdf/Circ1260.pdf (accessed on 12 July 2018).

25.   Ronan, A.D.; Prudic, D.E.; Thodal, C.E.; Constantz, J. Field study and simulation of diurnal temperature effects on infiltration and variably saturated flow beneath an ephemeral stream. *Water Resour. Res.* **1998**, *34*, 2137–2153. [CrossRef]

26.   Thomas, C.L.; Steward, A.E.; Constantz, J.E. *Determination of Infiltration and Percolation Rates along a Reach of the Santa Fe River near La Bajada New Mexico*; U.S. Geological Survey, Water-Resources Investigations Report 00-4141; USGS: Reston, VA, USA, 2000. [CrossRef]

27.   Arriaga, M.A.; Leap, D.I. Using solver to determine vertical groundwater velocities by temperature variations. *Hydrogeol. J.* **2006**, *14*, 253–263. [CrossRef]

28.   Birkel, C.; Soulsby, C.; Irvine, D.I.; Malcolm, I.; Lautz, L.K.; Tetzlaff, D. Heat-based hyporheic flux calculations in heterogeneous salmon spawning gravels. *Aquat. Sci.* **2016**, *78*, 203–213. [CrossRef]

29.   Kurylyk, B.L.; Irvine, D.J.; Carey, S.K.; Briggs, M.A. Heat as a groundwater tracer in shallow and deep heterogeneous media: Analytical solution, spreadsheet tool, and field applications. *Hydrol. Process.* **2017**, *31*, 2648–2661. [CrossRef] [PubMed]

30. Irvine, D.J.; Briggs, M.A.; Lautz, L.K.; Gordon, R.P.; McKenzie, J.M.; Cartwright, I. Using Diurnal Temperature Signals to Infer Vertical Groundwater-Surface Water Exchange. *Groundwater* **2016**, *55*, 10–26. [CrossRef] [PubMed]

31. Rodríguez-Rodríguez, M.; Fernández-Ayuso, A.; Hayashi, M.; Moral-Martos, F. Using water temperature, electrical conductivity, and pH to Characterize surface–groundwater relations in a shallow ponds system (Doñana National Park, SW Spain). *Water* **2018**, *10*, 1406. [CrossRef]

32. Rodríguez, P.; Júnez-Ferreira, H.E.; González, J.; de la Rosa, J.I.; Galván, C.; Burnes, S. Vadose zone hydraulic conductivity monitoring by using an Arduino data acquisition system. In Proceedings of the 2018 International Conference on Electronics, Communications and Computers, Cholula, Mexico, 21–23 February 2018; pp. 80–85. [CrossRef]

33. Rau, G.C.; Andersen, M.S.; McCallum, A.M.; Roshan, H.; Acworth, R.I. Heat as a tracer to quantify water flow in near-surface sediments. *Earth Sci. Rev.* **2014**, *10*, 41–58. [CrossRef]

34. Halloran, L.J.; Rau, G.C.; Andersen, M.S. Heat as a tracer to quantify processes and properties in the vadose zone: A review. *Earth Sci. Rev.* **2016**, *159*, 358–373. [CrossRef]

35. Programmable Resolution 1-Wire Digital Thermometer. Available online: https://datasheets.maximintegrated.com/en/ds/DS18B20.pdf (accessed on 12 July 2018).

36. Product Users Manual HC-SR04 Ultrasonic Sensor. Available online: https://docs.google.com/document/d/1Y-yZnNhMYy7rwhAgyL_pfa39RsB-x2qR4vP8saG73rE/edit (accessed on 12 July 2018).

37. Micro Load Cell Datasheet. Available online: https://www.robotshop.com/media/files/pdf/datasheet-3133.pdf (accessed on 12 July 2018).

38. Smith, S.W. The Scientist and Engineer's Guide to Digital Signal Processing. *DSP Guide* **1997**, 423–450. Available online: http://www.dspguide.com/pdfbook.htm (accessed on 12 July 2018). [CrossRef]

39. Hatch, C.E.; Fisher, A.T.; Revenaugh, J.S.; Constantz, J.; Ruehl, C. Quantifying surface water-groundwater interactions using time series analysis of streambed thermal records: Method development. *Water Resour. Res.* **2006**, *42*. [CrossRef]

40. Shanafield, M.; Hatch, C.; Pohll, G. Uncertainty in thermal time series analysis estimates of streambed water flux. *Water Resour. Res.* **2011**, *47*. [CrossRef]

41. Stallman, R.W. Steady One-Dimensional Fluid Flow in a Semi-Infinite Porous Medium with Sinusoidal Surface Temperature. *J. Geophys. Res.* **1965**, *70*, 2821–2827. [CrossRef]

42. Young, P.; Pedregal, D.; Tych, W. Dynamic harmonic regression. *J. Forecast.* **1999**, *18*, 369–394. [CrossRef]

43. Gordon, R.P.; Lautz, L.K.; Briggs, M.A.; McKenzie, J.M. Automated calculation of vertical pore-water flux from field temperature time series using the VFLUX method and computer program. *J. Hydrol.* **2012**, *420–421*, 142–158. [CrossRef]

44. Munz, M.; Oswald, S.E.; Schmidt, C. Sand box experiments to evaluate the influence of subsurface temperature probe design on temperature based water flux calculation. *Hydrol. Earth Syst. Sci.* **2011**, *15*, 3495–3510. [CrossRef]

45. Loutfi, H.; Bernatchou, A.; Raoui, Y.; Tadili, R. Learning Processes to Predict the Hourly Global, Direct, and Diffuse Solar Irradiance from Daily Global Radiation with Artificial Neural Networks. *Int. J. Photoenergy* **2017**, *2017*, 4025283. [CrossRef]

46. Despotovic, M.; Nedic, V.; Despotovic, D.; Cvetanovic, S. Evaluation of empirical models for predicting monthly mean horizontal diffuse solar radiation. *Renew. Sustain. Energy Rev.* **2016**, *56*, 246–260. [CrossRef]

47. Saturated Hydraulic Conductivity in Relation to Soil Texture. Available online: https://www.nrcs.usda.gov/wps/portal/nrcs/detail/soils/survey/office/ssr10/tr/?cid=nrcs144p2_074846 (accessed on 12 July 2018).

 *water*

Article

How do Soil Moisture and Vegetation Covers Influence Soil Temperature in Drylands of Mediterranean Regions?

Javier Lozano-Parra [1],*, Manuel Pulido [2], Carlos Lozano-Fondón [3] and Susanne Schnabel [2]

[1] Instituto de Geografía, Pontificia Universidad Católica de Chile, Avda. Vicuña Mackenna 4860, Santiago de Chile 7820436, Chile
[2] Research Institute for Sustainable Land Development, University of Extremadura, Avda. Universidad s/n. 10071 Cáceres, Spain; mapulidof@unex.es (M.P.); schnabel@unex.es (S.S.)
[3] Department of Chemistry, Life Sciences and Environmental Sustainability, University of Parma, Viale delle Scienze 11/A, 43124 Parma, Italy; lzncls@unife.it
* Correspondence: jlozano@outlook.es

Received: 27 October 2018; Accepted: 25 November 2018; Published: 28 November 2018

Abstract: Interactions between land and atmosphere directly influence hydrometeorological processes and, therefore, the local climate. However, because of heterogeneity of vegetation covers these feedbacks can change over small areas, becoming more complex. This study aims to define how the interactions between soil moisture and vegetation covers influence soil temperatures in very water-limited environments. In order to do that, soil water content and soil temperature were continuously monitored with a frequency of 30 min over two and half hydrological years, using capacitance and temperature sensors that were located in open grasslands and below tree canopies. The study was carried out on three study areas located in drylands of Mediterranean climate. Results highlighted the importance of soil moisture and vegetation cover in modifying soil temperatures. During daytime and with low soil moisture conditions, daily maximum soil temperatures were, on average, 7.1 °C lower below tree canopies than in the air, whereas they were 4.2 °C higher in grasslands than in the air. As soil wetness decreased, soil temperature increased, although this effect was significantly weaker below tree canopies than in grasslands. Both high soil water content and the effect of shading were reflected in a decrease of maximum soil temperatures and of their daily amplitudes. Statistical analysis emphasized the influence of soil temperature on soil water reduction, regardless of vegetation cover. If soil moisture deficits become more frequent due to climate change, variations in soil temperature could increase, affecting hydrometeorological processes and local climate.

Keywords: soil temperature; soil water; vegetation cover; hydrometeorology; ecohydrology

---

1. Introduction

Soil temperature is a key factor in determining energy and mass exchange with the atmosphere. It strongly affects the water balance and ecohydrological processes, such as evapotranspiration ratios and water uptake by plants. Conversely, soil water influences the energy balance and hydrometeorological processes by determining the partition of available solar energy between latent and sensible heat, both manifested as evaporated water and as heat of air and land, respectively [1]. Soil wetness and vegetation will determine the gradient of these energy fluxes. For example, soil moisture deficits could lead to more frequent and severe hot summer temperatures [2]. Similarly, variations in vegetation cover or changes in vegetation growth and renewal could modify feedbacks between the surface and the atmosphere and, therefore, the local climate [3]. Therefore, an appropriate understanding of the relationships between hydrometeorological processes and vegetation covers would enable better decision-making and could contribute to provide a more resilient environment.

The role of hydrometeorological processes could be particularly important in drylands of Mediterranean regions, where the upcoming climate is expected to be warmer and drier [4]. Some environments included in these lands are the silvopastoral systems of California, Australia, Chile or the Iberian Peninsula. They occupy important extensions of a human-made landscape which is the consequence of long-term management, and result in an open woodland where scattered evergreen trees play a critical role, both economically and environmentally [5]. Trees usually present a canopy cover up to 40% [6], meaning that a large land surface is below their canopies. Because of this vegetation distribution, hydrometeorological processes could present heterogeneous behavior since they may change over small areas, i.e., between open grasslands and tree canopies [7]. For example, spaces below tree canopies can show different microclimate and plant phenology than those from open grasslands [8]. The ability to assess hydrometeorological processes across scales constitutes a challenge due to the heterogeneity of the land surface and the complexity of the processes involved [9].

Vegetation covers influence the distribution of several variables above and below ground. A consensus is that tree canopies modify water and energy balances below them by intercepting rainfall and solar radiation. Thus, temperatures have usually been considered to be lower below tree canopies than in grasslands in summer, and the opposite in winter [8,10–12]. Conversely, the influence of vegetation cover on soil water content has been a more controversial issue since it has been found to be lower under tree crowns than in grasslands in drylands and humid lands, although opposite results have also been observed [13–16]. However, interactions between soil moisture and soil temperature have been a less frequent topic, and existing studies have usually been carried out under controlled laboratory conditions or without using continuous readings of the target variables. For example, some studies have found inverse relationships between soil water content and daily surface soil temperatures [17–20]. Despite this, few field studies have compared patterns of soil and air temperatures using continuous and high-frequency measurements, and considering the combined effect of soil moisture and vegetation covers [12].

The interactions between vegetation covers and hydrometeorological processes have been receiving increased interest, nevertheless most of the studies come mainly from the climate modelling community and have usually been focused at continental or global scale [7,21,22]. Likewise, techniques based on spacecraft sensors constitute a very powerful approach to analyze these processes, although the current coarse resolution could increase the uncertainty of measurements when the spatial variability of environmental processes is large. For example, it has been observed that soil wetness could reduce the daily temperature range (DTR) in semiarid regions by increasing the nighttime soil temperature, so soil moisture would reduce the impact of daily evaporative cooling by reducing nocturnal cooling [23]. However, it has also been hypothesized that the reduction of soil moisture and vegetation cover would decrease the diurnal temperature range during episodes of drought and human mismanagement by increasing nighttime surface air temperature [24]. These conflicting results provide evidence that the combined effect of soil moisture and vegetation covers on soil temperatures is not completely understood, and emphasize the necessity to develop studies using more accurate measurements in order to capture the spatial variability occurring between vegetation covers, i.e., in grasslands and under tree crowns.

The focus of this work was, therefore, to define the combined effect of soil moisture and vegetation covers on soil temperatures in dryland ecosystems, in order to understand the complexity of involved feedbacks and overcome the scarcity of field observations for characterizing the relevant processes in these water-sensitive areas. Thus, the following questions were addressed: (I) How does the interaction between soil moisture and different vegetation covers affect the relationships between soil temperature and air temperature? (II) What are the main factors affecting soil drying under different vegetation covers? An understanding of the relationships between these variables could help to explain the feedback between soil water and temperature.

2. Study Areas

The study was conducted in three study areas of the SW of the Iberian Peninsula: Cuartos, Naranjero, and Parapuños (Figure 1). They are representative of a human-made agrosilvopastoral system called dehesa and constitute semi-natural ecosystems composed of scattered evergreen trees and grasslands regularly grazed by livestock. These ecosystems have important sustainability issues, particularly related to tree decline and renewal [25].

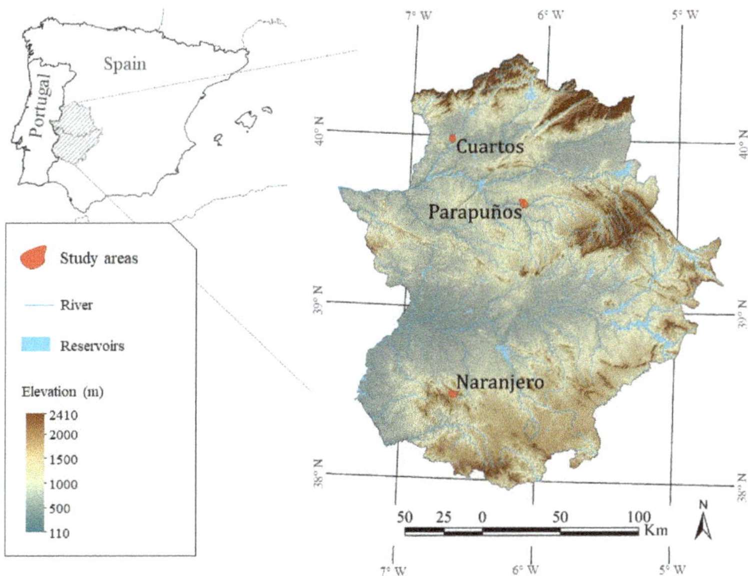

Figure 1. Study areas located in Extremadura (Spain).

The study areas have a gently undulated landscape, with an average elevation between 350 and 650 m. Annual mean precipitations were 594, 518, and 631 mm in Cuartos, Parapuños, and Naranjero, respectively, for a 35 year period (1977–2012) [13]. Although rainfall can vary strongly, both seasonally and inter-annually, rainy periods usually span from mid-autumn to mid-spring, whereas the period from late spring to early autumn is characterized by its almost total absence of rain. The mean annual air temperature is 16 °C, with an average daily minimum and maximum close to 3 °C in January and 32 °C in July, respectively. Climate can be considered as Mediterranean with oceanic and continental influences, comprising semiarid to dry sub-humid conditions.

There are three different vegetation layers (tree, shrub, and grassland) that present different densities and combinations. Predominant trees are mainly holm oaks (*Quercus ilex* L.), which present a density of 19, 34, and 68 trees/ha and a canopy cover of 13%, 16%, and 38% in Cuartos, Parapuños, and Naranjero, respectively [13]. The observed leaf area index in these ecosystems was 3.25 [26]. Ranchers frequently take out the shrub layer to promote herbaceous growth as resource for livestock. The grassland layer is composed of natural pastures of annual and perennial herbaceous plants, abounding annual grasses (such as *Vulpia bromoides* L. Gray, *Bromus* sp. or *Aira caryophyllea* L.) and annual legumes (*Ornithopus compressus* L., *Lathyrus angulatus* L., and several species of *Trifolium*). The growth period follows an approximately sigmoidal curve, starting in autumn and decaying in late spring [27], while summer constitutes a non-vegetative period.

Soils are shallow ($\approx$40 cm), poor in nutrients and have variable organic matter content ($\pm$3%), except in the uppermost soil layer below tree covers. They show a high bulk density ($\approx$1.5 g/cm^3)

although topsoil can present higher porosity ($\approx$45%) because of concentration of roots of herbaceous plants [28]. They are classified as Cambisols, Luvisols, and Leptosols.

3. Material and Methods

3.1. Soil Water and Soil Temperature Measurements

Soil moisture (m^3/m^3) and soil temperature ($^{\circ}$C) were measured by means of capacitive sensors and temperature sensors, respectively (Decagon Devices, Inc., Washington, DC, USA, models EC-5 and EC-T). They were continuously registered with a frequency of 30 min. Soil temperature was always measured at 5 cm depth, whereas soil moisture was monitored at 5, 10, and 15 cm, and a fourth measurement was taken at 5 cm above the bedrock depending on soil thickness. The sensors were gathered in Soil Moisture Stations (SMS) in two contrasting situations characterized by their vegetation covers: open spaces (Grassland) and below tree canopies (Tree). In order to avoid soil ponding, SMS in grasslands were positioned between hilltops and valley bottoms. Besides, below tree canopies, SMS were also installed with a southwest orientation and midway between the tree trunk and the limit of the crown. The terrain slopes in both situations were always lower than 4.4° [13]. Sensors were calibrated in the laboratory in order to improve their accuracy from $\pm$4% to $\pm$2% [29]. For this, a soil wetting processes was applied using soil samples from each study area and following the method proposed by Cobos and Chambers [29] for standard mineral soils. Soil temperature and moisture were measured over more than two complete hydrological years, from 1 April 2010 to 31 August 2012 (a hydrological year spanning from 1 September to 31 August). The SMS were distributed among the three study areas as shown in Table 1.

Table 1. Soil Moisture Stations (SMS) with the location and the symbols used to identify them.

Study Area	Vegetation Cover	SMS	Symbol
Cuartos (C)	Grassland (G)	1	CG1
	Tree (T)	1	CT1
Parapuños (P)	Grassland (G)	1, 2, 3	PG1, PG2, PG3
	Tree (T)	1	PT1
Naranjero (N)	Grassland (G)	1	NG1
	Tree (T)	1	NT1

3.2. Determining Soil Properties and Meteorological Variables

The experimental catchment of Parapuños is equipped with a meteorological station which has continuously registered precipitation, air temperature, relative humidity, wind velocity, as well as global and net radiation with a frequency of 5 min from the year 2000. Precipitation and air temperature were also continuously monitored in Naranjero and Cuartos every 5 min during the study period (by means of HOBO Data Logging Rain Gauge, models RG3). Relative humidity, wind velocity and global radiation were monitored in both study areas by meteorological stations close to them (<10 km). They belong to a regional network of meteorological stations (Redarex) which continuously register with a resolution of 30 min.

Soil properties were determined for each sensor depth in all soil moisture stations (Table 2). Total porosity was calculated using bulk density values, which were determined using three replicates of undisturbed samples of $\approx$100 cm^3. Soil organic matter was determined by standard methods [30] and grain size distribution following the USDA classification [31]. Soil properties values between 5 and 15 cm depth were calculated as an average between the three sampled depths corresponding to the soil surface (Table 2).

Table 2. Soil properties in each soil moisture station (SMS).

SMS	Soil	Depth (cm)	Clay (%)	Silt (%)	Sand (%)	OM (%)	Porosity (%)	BD (g/cm3)
CG1	Epileptic Cambisol	5–15	9.3	49.6	41.2	2.3	43.2	1.50
		30	14.1	50.0	35.9	1.3	39.0	1.62
CT1	Epileptic Cambisol	5–15	10.3	53.4	36.2	1.3	41.6	1.55
		30	13.9	54.9	31.2	0.2	42.9	1.51
PG1	Chromic Luvisol	5–15	10.1	42.1	47.9	0.8	38.5	1.63
		45	64.0	24.2	11.9	0.0	41.5	1.55
PG2	Chromic Luvisol	5–15	6.6	46.5	46.8	1.5	38.3	1.64
		40	n.a	n.a	n.a	n.a	50.5	1.31
PG3	Epileptic Cambisol	5–15	9.5	56.0	34.5	2.2	46.5	1.42
		30	13.1	59.2	27.7	1.0	39.1	1.61
PT1 *	Distric Leptosol	5–15	7.1	51.3	40.5	3.8	45.2	1.50
NG1	Distric Cambisol	5–15	6.2	51.6	42.3	3.7	42.0	1.54
		35	14.3	50.6	35.1	0.9	32.1	1.80
NT1	Distric Leptosol	5–15	4.8	47.5	47.8	3.2	52.4	1.26
		20	8.4	47.4	44.2	1.5	45.2	1.45

OM = Organic Matter; BD = Bulk Density; n.a = not available. * SMS with the fourth soil moisture sensor installed at 15 cm due to the shallowness of the soil profile. Soil type was described by FAO method [32].

3.3. Determining the Influence of Vegetation Covers and Soil Moisture on Soil Temperature

To determine the influence of soil water content on soil temperature, four different soil moisture states were defined: wet, medium wet (MW), medium dry (MD), and dry. For this, daily averages of soil moisture between the two upper sensors of each SMS were calculated and then, the average plus-minus one standard deviation method was used for each SMS, such as already applied in previous studies [13].

The combined effect of vegetation covers and soil moisture on temperatures was analyzed by daily variations of the temperature wave at 30 min intervals in soils and air. Because variations of mean temperatures can be due to changes in either maximum or minimum temperature, or relative changes in both, these latter could provide more information than the mean alone [33]. For this reason, several properties of the daily temperature wave were individually defined for each SMS and for the air: daily peaks of maximum and minimum temperatures, time of peaks, daily temperature range (i.e., amplitude between daily peaks), and time lag to reach the temperature peaks between air and soil (Figure 2).

Finally, to determine if significant differences in soil temperatures were found between vegetation covers and soil moisture states, the t-test statistical test was applied for each soil moisture state grouped by vegetation cover. Before, a normal distribution test was applied by the Shapiro–Wilk statistical test. Analyses were carried out by Statistica 8.0 software.

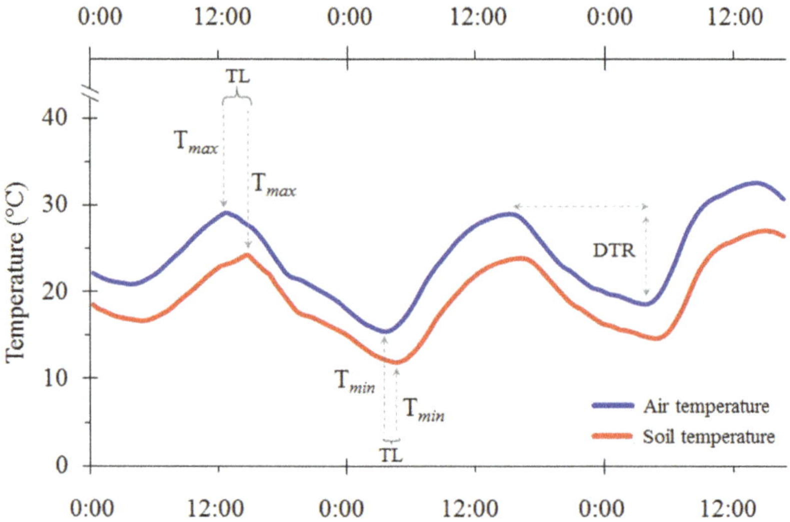

Figure 2. Methodology used to determine the daily cycle of the soil temperature wave. T_{max} and T_{min} = maximum and minimum temperature, respectively; TL = Time lag between T_{max} in air and T_{max} in soil, and T_{min} in air and T_{min} in soil, respectively; DTR = daily temperature range, i.e., T_{max} minus the next T_{min}.

3.4. Determining the Importance of Factors Involved in Soil Water Decrease

3.4.1. Database Processing

To determine the importance of factors implied in the decreases of soil water in different vegetation covers, daily variations of soil moisture were calculated for each SMS by:

$$\Delta\theta = \theta_n - \theta_{n-1} \tag{1}$$

where $\Delta\theta$ is the daily variation of soil moisture, θ_n and θ_{n-1} are soil moisture at day n and soil moisture the day before, respectively. Soil moisture on a daily scale was calculated using the daily mean of the two upper sensors of each SMS. Finally, soil water increases were removed from the database.

Additionally, as soil moisture can also decrease by vertical drainage or evapotranspiration processes, two procedures were considered. On the one hand, soil water retention curves were determined for each SMS by a drying process based on sand and kaolin boxes and the Richards plate. For this, 70 undisturbed soil samples that were taken at the depth at which the sensors were placed were used. When some points of the curves were not able to be calculated because of the difficulty in taking out soil samples, they were estimated by hierarchical pedotransfer functions. These latter used soil texture, organic matter, bulk density, and potential suction data, such as has already been described in previous studies [34]. Then, both measured and estimated points were adjusted by the van Genuchten model [35], commonly used to determine unsaturated soil hydraulic properties. Finally, only soil moisture values for suctions from −33 kPa to the lowest potentials were selected. This is justified because water at suctions from 0 kPa to −33 kPa is supposed to be freely drained by gravity as larger pores cannot hold it [36]. On the other hand, the variables used to estimate the reference evapotranspiration by the FAO–Penman–Monteith method [37] were separately used on a daily scale and for every site.

3.4.2. Modelling Factors that Influence Soil Water Decrease

To determine the importance of factors implied in soil water decrease under different vegetation covers, the Multivariate Adaptive Regression Splines (MARS) modeling technique [38] was used to generate two models, one for grasslands and another for below tree canopies. Thus, soil moisture was used as a target variable, whereas five variables related to the energy balance and hydrometeorological factors were used as independent continuous predictors. These latter were: air temperature, solar radiation, relative humidity, wind velocity, and soil temperature.

In order to test the models, the original databases of soil water decrease, one for grasslands (n = 2798) and the other for below tree canopies (n = 1202), were separated into two aleatory independent databases, respectively. For running the models the first databases were used (75% of data) and for validating them the second ones (25% of data). Irrespective of the size of these latter, independent databases, they showed similar statistical properties to those used for training the models (Table 3). The importance of the independent variables was evaluated as has already been described in previous studies [39]. The resulting equations obtained by this data mining method were not applied in other environments, but they were used with the explicit aim of determining the importance of the factors involved in soil water decreases in both vegetation covers of these ecosystems.

Table 3. Statistical summary of training and test databases for soil water decrease (m^3/m^3).

Cover	Database	n	Mean	Median	SD	Max	Min
Grassland	Training	2239	0.178	0.156	0.082	0.350	0.060
	Test	559	0.167	0.146	0.075	0.349	0.064
Tree	Training	962	0.167	0.155	0.076	0.351	0.032
	Test	240	0.165	0.149	0.084	0.358	0.033

n = sample size; SD = Standard deviation; Max and Min = maximum and minimum, respectively.

4. Results

4.1. Effects of Vegetation Covers and Soil Moisture on Soil Temperatures

The daily cycle of soil temperatures was mainly explained by heating and cooling during the day and night, respectively (Figure 3). However, factors such as soil moisture and vegetation cover strongly affected the daily temperature cycle. Soil temperature decreased as soil wetness increased, although this effect was weaker and significantly different in soils below tree canopies than those in grasslands, regardless of soil moisture conditions becoming wetter or drier (Table 4). Additionally, the daily amplitude of soil temperatures increased from wetter to drier soil moisture conditions, and this range was usually more reduced below tree canopies than in grasslands (Figure 4). Thus, both low soil water content and the absence of shading were reflected in an increase of the daily amplitude (Figure 4).

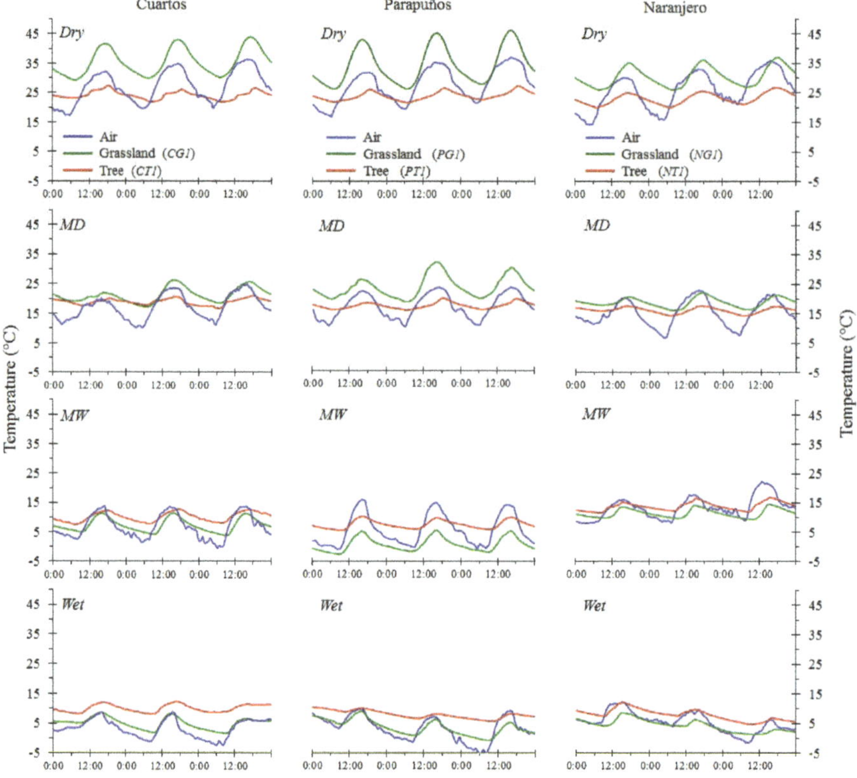

Figure 3. Air temperature and soil temperature at 30-min intervals in grasslands and below tree canopies, and under different soil moisture states. MD and MW = medium dry and medium wet soil conditions, respectively.

Table 4. Soil temperature (°C) in different vegetation covers and under different soil water states.

Cover	Dry		MD		MW		Wet	
	Mean	SD	Mean	SD	Mean	SD	Mean	SD
Grasslands	29.4 *	4.5 *	21.7 *	3.6 *	13.7 *	2.4 *	9.5 *	1.7 *
Trees	22.5 *	1.6 *	17.6 *	1.3 *	14.7 *	1.0 *	12.3 *	0.9 *

MD and MW = medium dry and medium wet soil moisture conditions, respectively. SD = daily standard deviation of soil temperature. Statistical differences between vegetation cover have been tested by *t*-test statistical test (* = $p < 0.00$).

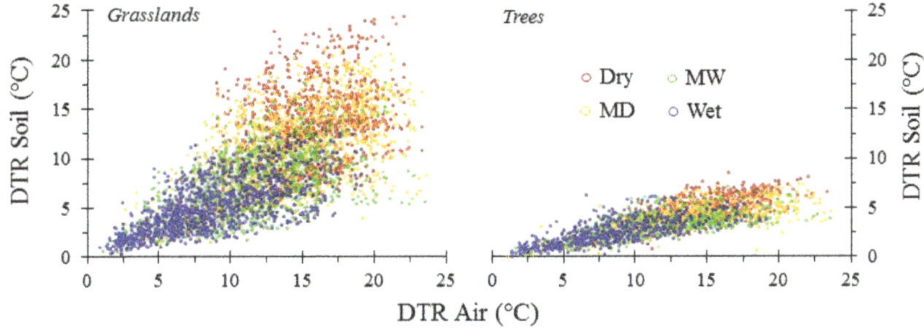

Figure 4. Daily air temperature range (DTR) and its relationship with soil DTR under different vegetation covers and different soil moisture states. MD, MW = Medium dry and medium wet soil moisture conditions, respectively.

The influence of vegetation covers and soil moisture on soil temperatures was also reflected in the daily peaks of soil temperature and on the time lag between soils and air to reach these peaks (Figure 5). During daytime, daily peaks of maximum soil temperature under dry soil moisture conditions were, on average, 7.1 °C lower below tree canopies than in the air, whereas they were 4.2 °C higher in soils of grasslands than in the air (Figure 5 (1.A)). Nevertheless, soil wetness decreased these differences in both vegetation covers. Daily peaks of maximum soil temperatures were reached, on average, 90 min later below tree canopies than in the air, whereas these peaks were reached approximately at the same time in soils of grasslands than in the air (Figure 5 (2.A)). On the other hand, during night-time and under dry soil moisture conditions, daily peaks of minimum soil temperature were always higher in grasslands (6.3 °C) and below tree canopies (3.2 °C) than in the air (Figure 5 (1.B)). However, in this case soil wetness only decreased these differences in grasslands. Finally, daily peaks of minimum soil temperature were reached, on average, more than 100 min later in both vegetation covers than in the air, and the time lag between soils and air increased with soil wetness (Figure 5 (2.B)).

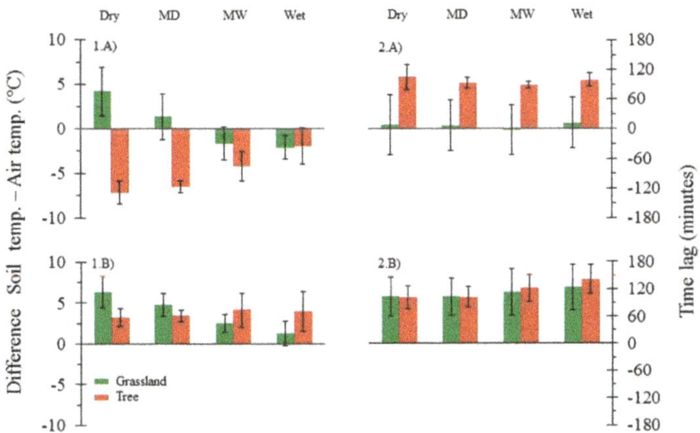

Figure 5. Differences between daily peaks of maximum temperatures of soil and air (**1.A**) and the minimum ones (**1.B**), under different vegetation covers and different soil moisture states. Differences between time lag of soil and air to reach daily peaks of maximum temperature (**2.A**) and the minimum ones (**2.B**). MD, MW = medium dry and medium wet soil moisture conditions, respectively. Bar errors depicts one standard deviation.

4.2. Factors Influencing the Decrease of Soil Water

The quality fit presented by the two models obtained to explain the variables influencing soil water decreases was good. Coefficients of determinations were 0.73 and 0.71 in both cases (Figure 6). Models showed a good performance with low error for grasslands and tree, respectively (0.001 and 0.003), confirming the ability of this modelling technique to discriminate the most important variables to explain soil water decreases in soils of two vegetation covers.

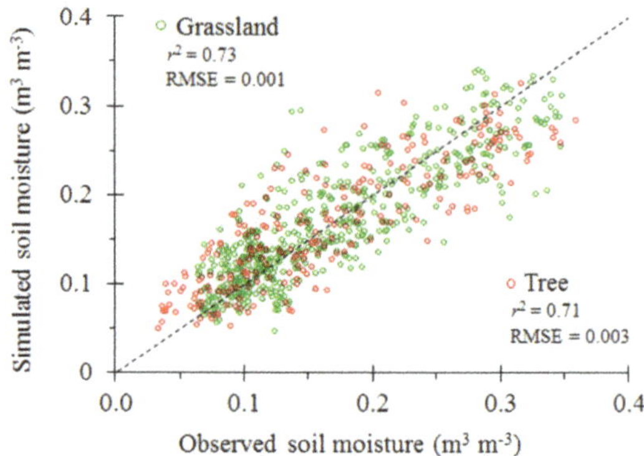

Figure 6. Observed and predicted soil moisture decrease in both grasslands and below tree canopies. RMSE = Root mean square error.

MARS selected all variables to build the two models and explain soil water decrease in both grasslands and below tree crowns (Figure 7). The most important individual variable in both cases was soil temperature, with an importance of 64.5% over 100% for trees, and 55.2% for grasslands. Relative humidity was also important in both cases (36.5% in grasslands and 20.8% in tree), whereas solar radiation and wind velocity showed lower importance in the two models.

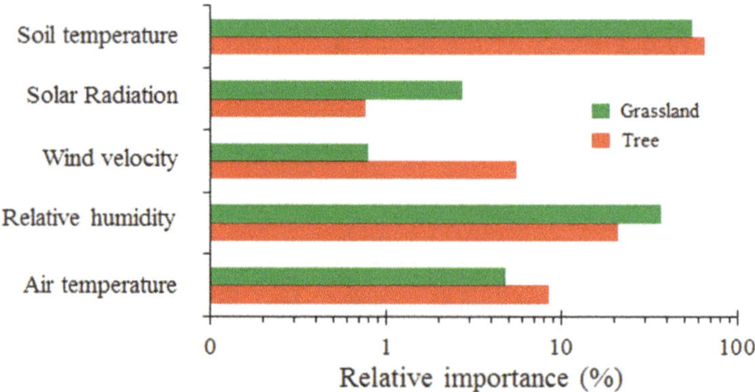

Figure 7. Independent factors chosen by the two models and their relative weight to explain soil moisture decrease in soils of both grasslands and below tree crowns.

5. Discussion

The daily cycle of air and soil temperatures (both in grasslands and below tree canopies) was mainly due to heating and cooling during the day and night, respectively. However, factors such as soil moisture and vegetation cover produced significant variations on the daily temperature cycle (Figure 5).

Soil water showed a very strong effect on soil temperatures, given that wet soils were prone to be colder than dry soils (Table 4). Likewise, daily amplitude of soil temperatures was also smaller under wet soil conditions than under drier ones (Figure 4), regardless of vegetation covers. Connections between soil moisture and soil temperature have also been observed in several studies [18–20], which asserted that during soil drying the surface temperature shows an increase that corresponds to a decrease in soil moisture, which coincides with our results. Indeed, soil temperatures played a key role on soil water decrease in both vegetation covers (Figure 7). When solar energy reaches the surface, it can be returned to the atmosphere or be partitioned into latent and sensible heat fluxes, both manifested as evaporated water and as heat of air and land, respectively. Consequently, an increase of soil temperature could be reflected in an increase of potential evapotranspiration and the subsequent decrease of soil water content (Table 4).

The moderation of soil temperatures by soil wetness has been explained by water properties, such as its high specific heat capacity and its high thermal conductivity. Thus, when soil moisture increases, more heat energy input for raising the soil temperature is required, since water has a higher specific heat capacity than soil materials [40]. Similarly, soils close to saturation can transport heat downward through water more quickly and, therefore, without significantly warming the upper soil layers [41]. Additionally, evapotranspiration processes from wet soils use energy that might otherwise be used to heat the soil surface [2]. With progressive drying, the air increasingly occupies the porous system and the soil warms up faster than a wet soil (Table 4), since less energy is required to change the soil temperature when it is dry [17,41].

The interactions between soil temperature and soil moisture could become more critical in dryland regions as compared to wetter ones, since low soil moisture limits the total energy used for cooling by latent heat fluxes, so more energy is available for increasing the soil and air temperature by sensible heating [2,42]. This could explain the observed influence of relative humidity on soil water decrease (Figure 7). Because of the soil-atmosphere water exchange is mainly controlled by atmospheric demand for water vapor, the influence of relative humidity on soil water decrease reduces as atmospheric water increases. Similar observations have mainly been found in modelling studies. For example, in areas with high soil moisture, latent heat fluxes by evapotranspiration dominate over sensible heat fluxes, enhancing the formation of clouds and the tendency to atmospheric cooling [23]. By contrast, if a long soil moisture deficit occurs, the atmospheric pressure deficit will increase and the sensible heat fluxes will be dominant, resulting in a warmer atmosphere that could inhibit the formation of clouds by convection and create a positive feedback loop [3,43]. This fact could be reflected in an intensification or lengthening of extreme climate events, as has been described in some studies [2,44], which have observed that feedbacks between land-atmosphere played a key role in the development of heatwaves in some regions of Europe. The importance of the results found here become more relevant because it has been observed that in the studied ecosystems dry periods can be independent of seasonality, i.e., they can arise in any climatic season [13], so if soil moisture deficits become more frequent, local climate could be affected.

Despite this, the type of vegetation cover also affected the energy reaching soils, incrementing the complexity of the aforementioned processes. For example, the protective effect of vegetation cover against direct solar radiation was not only reflected in lower soil temperature amplitudes below tree canopies than beyond them, but also because soils below canopies reached the daily peak of maximum temperature later than those in grasslands (Figure 5). The effect of tree canopies on temperatures below them has also been highlighted in several studies. For example, some of them have observed that the mean temperature ranges were smaller below tree canopies than in the air beyond them, whereas other studies have found similar results but only for mid and low latitudes [12,45]. Nevertheless, most of the studies addressing this topic emphasize the critical role of trees to modify water and energy balances and their influence on ecological processes. For example, many of them have suggested that phenological patterns could respond to microclimate modifications below tree canopies, which could partially explain the spatio-temporal variations on biomass yield and tree-grassland differences on ecohydrological processes [27,46,47]. Therefore, our results could become more significant because the studied environments present a great land surface below their tree crowns, since they can present a canopy cover up to 40%.

The feedbacks between vegetation covers, soil water content, and soil temperatures can Longer soil moisture deficits could increase the daily peaks of maximum soil temperatures and their amplitudes, although this effect slightly differs between vegetation covers. This could result in a warmer atmosphere, which could affect local climate conditions. If dry episodes become recurrent due to a likely climate change, ecosystem functioning, as well as their economic resources could be compromised.

6. Conclusions

Interactions between soil water and vegetation covers directly affected hydrometeorological processes, given that wet soils were prone to be colder than dry soils. Under this latter soil state, maximum soil temperatures were both lower below tree canopies and higher in grasslands than in the air. Nevertheless, these differences decreased in both vegetation covers as soil wetness increased. Conversely, minimum soil temperatures were higher in both vegetation covers than in the air, regardless of soil moisture conditions. Thus, both low soil water content and absence of shading were reflected in an increase of maximum soil temperatures and of their daily variability.

Modelling analysis proved that soil temperatures played a key role on soil moisture decrease of both vegetation covers, since an increase of soil temperature was reflected in a decrease of soil water content. The influence of relative humidity on soil water decrease was reduced as atmospheric water increased. This could presumably occur because soil-atmosphere water exchange is mainly controlled by atmospheric demand for water vapor.

Relationships between soil moisture and vegetation covers can modify hydrometeorological processes, which could affect ecosystem function. If soil moisture deficits become longer and more frequent due to climate change, daily temperature variations of soil temperature could increase, which could impact on local climate. These results suggest that more research should be carried out to quantify the feedbacks between land and atmosphere, in order to contribute to the standard of long-term weather forecasts, including a likely increase of droughts.

Author Contributions: J.L.-P. conceived the idea and wrote the original draft. M.P., C.L.-F., and S.S. reviewed and edited the draft in the same way.

Funding: This research was funded by the Fondo Nacional de Desarrollo Científico y Tecnológico [CONICYT/FONDECYT 11161097] and by the Programa de Cooperación Internacional para la Formación de Redes Internacionales de Investigación [REDI170640], both granted by the Government of Chile. Likewise, it was also funded by the former Spanish Ministry of Education and Science [CGL2008-01215].

Acknowledgments: Authors would like to thank the anonymous referees for their suggestions.

Conflicts of Interest: The authors declare no conflict of interest. The funders had no role in the design of the study; in the collection, analyses, or interpretation of data; in the writing of the manuscript, or in the decision to publish the results.

References

1. Collier, C.G. *Hydrometeorology*; Wiley Blackwell: Hoboken, NJ, USA; University of Leeds: Leeds, UK, 2016.
2. Alexander, L. Climate science: Extreme heat rooted in dry soils. *Nat. Geosci.* **2011**, *4*, 12–13. [CrossRef]
3. Zemp, D.C.; Schleussner, C.-F.; Barbosa, H.; Hirota, M.; Montade, V.; Sampaio, G.; Staal, A.; Wang-Erlandsson, L.; Rammig, A. Self-amplified amazon forest loss due to vegetation-atmosphere feedbacks. *Nat. Commun.* **2017**, *8*, 1–10. [CrossRef] [PubMed]
4. IPCC. *Climate Change 2014: Synthesis Report. Contribution of Working Groups I, II and III to the Fifth assessment Report of the Intergovernmental Panel on Climate Change*; IPCC: Geneva, Switzerland, 2014; p. 151.

5. Campos, P.; Huntsinger, L.; Oviedo, J.L.; Starrs, P.F.; Diaz, M.; Standiford, R.B.; Montero, G. *Mediterranean Oak Woodland Working Landscapes. Dehesas of Spain and Ranchlands of California*; Springer Netherlands: Dordrecht, The Netherlands, 2013; p. 508.

6. Joffre, R.; Rambal, S.; Ratte, J.P. The dehesa system of southern spain and portugal as a natural ecosystem mimic. *Agrofor. Syst.* **1999**, *45*, 57–79. [CrossRef]

7. Seneviratne, S.; Corti, T.; Davin, E.; Hirschi, M.; Jaeger, E.; Lehner, I.; Orlowsky, B.; Teuling, A. Investigating soil moisture–climate interactions in a changing climate: A review. *Earth-Sci. Rev.* **2010**, *99*, 125–161. [CrossRef]

8. Callaway, R.M. *Positive Interactions and Interdependence in Plant Communities*; Springer: Dordrecht, The Netherlands, 2007; p. 418.

9. Asbjornsen, H.; Goldsmith, G.R.; Alvarado-Barrientos, M.S.; Rebel, K.; Osch, F.P.V.; Rietkerk, M.; Chen, J.; Gotsch, S.; Tobón, C.; Geissert, D.R.; et al. Ecohydrological advances and applications in plant–water relations research: A review. *J. Plant Ecol.* **2011**, *4*, 3–22. [CrossRef]

10. Eviner, V.; Chapin, S. Functional matrix: A conceptual framework for predicting multiple plant effects on ecosystem processes. *Annu. Rev. Ecol. Evol. Syst.* **2003**, *34*, 455–485. [CrossRef]

11. Jose, S.; Gillespie, A.R.; Pallardy, S.G. Interspecific interactions in temperate agroforestry. *Agrofor. Syst.* **2004**, *61–62*, 237–255. [CrossRef]

12. Song, Y.; Zhou, D.; Zhang, H.; Li, G.; Jin, Y.; Li, Q. Effects of vegetation height and density on soil temperature variations. *Chin. Sci. Bull.* **2013**, *58*, 907–912. [CrossRef]

13. Lozano-Parra, J.; Schnabel, S.; Ceballos-Barbancho, A. The role of vegetation covers on soil wetting processes at rainfall event scale in scattered tree woodland of mediterranean climate. *J. Hydrol.* **2015**, *529*, 951–961. [CrossRef]

14. Joffre, R.; Rambal, S. How tree cover influences the water balance of mediterranean rangelands. *Ecology* **1993**, *74*, 570–582. [CrossRef]

15. Calder, I.; Rosier, P.; Prasanna, K.; Parameswarappa, S. Eucalyptus water use greater than rainfall input-a possible explanation from southern india. *Hydrol. Earth Syst. Sci.* **1997**, *1*, 249–256. [CrossRef]

16. García-Estringana, P.; Latron, J.; Llorens, P.; Gallart, F. Spatial and temporal dynamics of soil moisture in a Mediterranean mountain area (Vallcebre, NE Spain). *Ecohydrology* **2013**, *6*, 741–753. [CrossRef]

17. Idso, S.; Schmugge, T.; Jackson, R.; Reginato, R. The utility of surface temperature measurements for the remote sensing of surface soil water status. *J. Geophys. Res.* **1975**, *80*, 3044–3049. [CrossRef]

18. Reginato, R.; Idso, S.; Vedder, J.; Jackson, R.; Blanchard, M.; Goettelman, R. Soil water content and evaporation determined by thermal parameters obtained from ground-based and remote measurements. *J. Geophys. Res.* **1976**, *81*, 1617–1620. [CrossRef]

19. Al-Kayssi, A.; Al-Karaghouli, A.; Hasson, A.; Beker, S. Influence of soil moisture content on soil temperature and heat storage under greenhouse conditions. *J. Agric. Eng. Res.* **1990**, *45*, 241–252. [CrossRef]

20. Lakshmi, V.; Jackson, T.; Zehrfuhs, D. Soil moisture–temperature relationships: Results from two field experiments. *Hydrol. Process.* **2003**, *17*, 3041–3057. [CrossRef]

21. Collatz, J.; Bounoua, L.; Los, S.; Randall, D.; Fung, I.; Sellers, P. A mechanism for the influence of vegetation on the response of the diurnal temperature range to changing climate. *Geophys. Res. Lett.* **2000**, *27*, 3381–3384. [CrossRef]

22. Lim, Y.-K.; Cai, M.; Kalnay, E.; Zhou, L. Impact of vegetation types on surface temperature change. *J. Appl. Meteorol. Clim.* **2008**, *47*, 411–424. [CrossRef]

23. Cheruy, F.; Dufresne, J.; Aït Mesbah, S.; Grandpeix, J.; Wang, F. Role of soil thermal inertia in surface temperature and soil moisture-temperature feedback. *J. Adv. Model. Earth Syst.* **2017**, *9*, 2906–9019. [CrossRef]

24. Zhou, L.; Dickinson, R.; Tian, Y.; Vose, R.; Dai, Y. Impact of vegetation removal and soil aridation on diurnal temperature range in a semiarid region: Application to the sahel. *PNAS* **2007**, *104*, 17937–17942. [CrossRef] [PubMed]

25. Herguido, E.; Lavado-Contador, J.F.; Pulido, M.; Schnabel, S. Spatial patterns of lost and remaining trees in the Iberian wooded rangelands. *Appl. Geogr.* **2017**, *87*, 170–183. [CrossRef]

26. Hoff, C.; Rambal, S. An examination of the interaction between climate, soil and leaf area index in a quercus ilex ecosystem. *Ann. For. Sci.* **2003**, *60*, 153–161. [CrossRef]

27. Lozano-Parra, J.; Schnabel, S.; Pulido, M.; Gómez-Gutiérrez, Á.; Lavado-Contador, F. Effects of soil moisture and vegetation cover on biomass growth in water-limited environments. *Land Degrad. Dev.* **2018**. [CrossRef]

28. Pulido-Fernández, M.; Schnabel, S.; Lavado-Contador, J.F.; Miralles-Mellado, I.; Ortega-Pérez, R. Soil organic matter of Iberian open woodland rangelands as influenced by vegetation cover and land management. *Catena* **2013**, *109*, 13–24. [CrossRef]

29. Cobos, D.R.; Chambers, C. Calibrating ECH2O Soil Moisture Sensors. Decagon Device Application Note. 2010. Available online: www.decagon.com (accessed on 15 October 2011).

30. Walkley, A.; Black, L.A. An examination of degtjareff method for determining soil organic matter and a proposed modification of the chromic acid titration method. *Soil Sci.* **1934**, *37*, 29–38. [CrossRef]

31. USDA. Soil survey laboratory methods manual. In *Soil Survey Investigations*; Report No. 42, Version 4.0; USDA-NCRS: Lincoln, OR, USA, 2004.

32. FAO. *World Reference Base for Soil Resources 2014. International Soil Classification System for Naming Soils and Creating Legends for Soil Maps*; FAO: Rome, Italy, 2014; Volume 106, p. 191.

33. Braganza, K.; Karoly, D.; Arblaster, J. Diurnal temperature range as an index of global climate change during the twentieth century. *Geophys. Res. Lett.* **2004**, *31*, 1–4. [CrossRef]

34. Schaap, M.G.; Leij, F.J.; Van Genuchten, M.T. Rosetta: A computer program for estimating soil hydraulic parameters with hierarchical pedotransfer functions. *J. Hydrol.* **2001**, *251*, 163–176. [CrossRef]

35. van Genuchten, M.T.; Leij, F.J.; Yates, S.R. *The Retc Code for Quantifying the Hydraulic Functions of Unsaturated Soils*; Version 6.02; EPA Report 600/2-91/065; U.S. Department of Agriculture, Agricultural Research Service: Riverside, CA, USA, 1991; p. 93.

36. Cassel, D.; Nielsen, D. Field capacity and available water capacity. In *Methods of Soil Analysis, Part 1. Physical and Mineralogical Method*; Klute, A., Ed.; American Society of Agronomy-Soil Science Society of America: Madison, WI, USA, 1986; Volume 9, pp. 901–924.

37. Allen, R.G.; Pereira, L.S.; Raes, D.; Smith, M. *Crop Evapotranspiration—Guidelines for Computing Crop Water Requirements- Fao Irrigation and Drainage Paper 56*; FAO—Food and Agriculture Organization of the United Nations: Rome, Italy, 1998; Volume 56, p. 322.

38. Friedman, J.H. Multivariate adaptive regression splines. *Ann. Stat.* **1991**, *19*, 1–67. [CrossRef]

39. Lozano-Parra, J.; Van Schaik, L.; Schnabel, S.; Gómez-Gutiérrez, Á. Soil moisture dynamics at high temporal resolution in a mediterranean watershed with scattered tree cover. *Hydrol. Process.* **2016**, *30*, 1155–1170. [CrossRef]

40. Abu-Hamdeh, N. Thermal properties of soils as affected by density and water content. *Biosyst. Eng.* **2003**, *86*, 97–102. [CrossRef]

41. Nassar, I.; Globus, A.; Horton, R. Simultaneous soil heat and water transfer. *Soil Sci.* **1992**, *154*, 465–472. [CrossRef]

42. Gu, L.; Meyers, T.; Pallardy, S.; Hanson, P.; Yang, B.; Heuer, M.; Hosman, K.; Riggs, J.; Sluss, D.; Wullschleger, S. Direct and indirect effects of atmospheric conditions and soil moisture on surface energy partitioning revealed by a prolonged drought at a temperate forest site. *J. Geophys. Res.* **2006**, *111*, 1–13. [CrossRef]

43. Scheffer, M.; Carpenter, S.; Foley, J.A.; Folke, C.; Walker, B. Catastrophic shifts in ecosystems. *Nature* **2001**, *413*, 591–596. [CrossRef] [PubMed]

44. Hirschi, M.; Seneviratne, S.I.; Alexandrov, V.; Boberg, F.; Boroneant, C.; Christensen, O.B.; Formayer, H.; Orlowsky, B.; Stepanek, P. Observational evidence for soil-moisture impact on hot extremes in southeastern Europe. *Nat. Geosci.* **2011**, *4*, 17–21. [CrossRef]

45. Longobardi, P.; Montenegro, A.; Beltrami, H.; Eby, M. Deforestation induced climate change: Effects of spatial scale. *PLoS ONE* **2016**, *11*, e0153357. [CrossRef] [PubMed]

46. Lozano-Parra, J.; Maneta, M.; Schnabel, S. Climate and topographic controls on simulated pasture production in a semiarid Mediterranean watershed with scattered tree cover. *Hydrol. Earth Syst. Sci.* **2014**, *18*, 1439–1456. [CrossRef]

47. Breshears, D.; Nyhan, J.; Heil, C.; Wilcox, B. Effects of woody plants on microclimate in a semiarid woodland: Soil temperature and evaporation in canopy and intercanopy patches. *Int. J. Plant Sci.* **1998**, *159*, 1010–1017. [CrossRef]

MDPI

St. Alban-Anlage 66

4052 Basel

Switzerland

Tel. +41 61 683 77 34

Fax +41 61 302 89 18

www.mdpi.com

Water Editorial Office

E-mail: water@mdpi.com

www.mdpi.com/journal/water